RECENT ADVANCES IN ANIMAL NUTRITION — 1999

Cover design

Shire Mare "Starlight", the Champion Mare at the London Shire Horse Shows, 1890, 1891 and 1892. From Youatt's "The Complete Grazier" by William Fream (1893), Crosby Lockwood and Son, London

Recent Advances in Animal Nutrition

1999

P.C. Garnsworthy, PhD

J. Wiseman, PhD

University of Nottingham

Nottingham University Press
Manor Farm, Main Street, Thrumpton
Nottingham, NG11 0AX, United Kingdom

NOTTINGHAM

First published 1999

British Library Cataloguing in Publication Data
Recent Advances in Animal Nutrition — 1999:
University of Nottingham Feed Manufacturers
Conference (33rd, 1999, Nottingham)
I. Garnsworthy, Philip C. II. Wiseman, J.

ISBN 1-897676-06-9

Typeset by Nottingham University Press, Nottingham
Printed and bound by Redwood Books, Trowbridge, Wiltshire

PREFACE

The 33rd University of Nottingham Feed Manufacturers Conference was held at the University's Sutton Bonington Campus and this book contains the papers from the meeting. The conference programme was devised by the Conference Committee to address issues that are topical for the animal industry as it moves towards the Third Millennium. Feed manufacturers in Europe are facing many challenges and it is vital that technical knowledge produced by researchers is disseminated and put into an appropriate context that will meet the needs of producers and consumers alike.

In dairy cattle, two major problems are fertility and lameness. Nutrition has been implicated in both of these disorders, so it is appropriate that two papers reviewed the evidence for nutritional effects whilst also remembering the importance of alternative influences. In both cases, the authors concluded that whilst nutrition does play a role, other factors, such as high genetic merit and management, also have a bearing on the health and welfare of the modern dairy cow.

For many years reports on human health have maligned dairy products in the human diet because of the high proportion of saturated fatty acids they contain. We now have much better knowledge of the manipulation of milk fat to meet market demands. This includes reducing the proportion of trans fatty acids, currently considered to be detrimental to human health, and possibly increasing the proportion of conjugated linoelic acid, which is the most powerful naturally-occurring anticarcinogen known. Whilst reductions in butterfat can be achieved by low-roughage diets, the value of physical structure in the diet of dairy cows must be acknowledged for the welfare of the cow. Historically, this was accommodated in ration formulation by ensuring a minimum fibre content. Recent work has moved beyond this concept to consider the benefits of physical structure in relation to chewing activity and rumination by dairy cows. Sufficient data are now available to permit the use of critical roughage as a pseudo-nutrient in ration specifications.

Dietary fibre was also the main theme of papers on pig nutrition and three papers discussed the role of fibre in diets for piglets, growing pigs and sows. Fibre can have major effects on nutrient utilisation, gut microflora and feeding behaviour. The withdrawal of sub-therapeutic levels of antibiotics for pigs has focussed attention on the microflora of the intestine, since a healthy microflora can be beneficial in reducing the incidence of diarrhoea in growing pigs. Fibre also affects the rate of passage of material through the intestine, which in turn affects the digestion and absorption of nutrients. In older animals, where it is desirable to restrict nutrient intake, high-fibre diets allow animals to eat a greater quantity of food for a particular energy intake. This keeps sows occupied and reduces stereotypic behaviour patterns.

The interactions between nutrition and environment have major effects on food intake, growth and body composition in young piglets. Simulation models can help to predict these interactions and help towards achieving optimal performance in growing pigs.

During the course of 1998 new legislation affecting most feed manufacturers was introduced. The implications of this, and forthcoming legislation, are discussed.

Most feed supplements for horses are based on cereal grains, yet recent research has suggested that high-starch supplements may lead to health problems. An optimal supplement for grazing horses has been developed that features fibre and fat. The value of such a supplement is discussed in relation to the general nutrient requirements of horses.

Phosphorus is an important component of a bird's skeleton and is also the currency for energy exchange within the body. However, phosphorus normally has a low availability in the organic form, resulting in high levels of secretion to the environment. The addition of phytase enzymes to poultry diets releases phosphorus from organic compounds and thereby reduces the need for supplementary phosphorus and reduces environmental pollution. The nutritional decisions involved in feeding broiler-breeders are aimed at balancing the requirements of reproduction, egg numbers and egg formation. It is important to control body weight in these birds, but this can be difficult if the flock is not uniform. Therefore a range of management techniques must be employed to keep the birds as similar as possible.

We would like to thank all the speakers for their presentations and written papers. We would also like to thank members of the Programme Committee for suggesting topics for papers, and Trouw Nutrition for their financial support of the conference. The sessions at the meeting were chaired by P. Lake, S. Papasolomontos, J. Twigge and P.Toplis.

P.C. Garnsworthy
J. Wiseman

CONTENTS

1

LEGISLATION AND ITS EFFECT ON THE FEED COMPOUNDING INDUSTRY

M. MARSDEN[1] AND J. NELSON[2]
[1] *J Bibby Agriculture, PO Box 250, Oundle Road, Woodston, Peterborough*
[2] *UKASTA Ltd, 3 Whitehall Court, London, SW1A 2EQ*

Introduction

During the course of 1998 new legislation covering the activities of manufacturers of feed additives and all types of compound feedingstuffs was brought into operation. This paper discusses issues arising from this legislation which are still outstanding. It then outlines legislation shortly to be brought into operation in the United Kingdom or still in draft form awaiting adoption in Brussels. A summary is also given of activities, with which UKASTA and its member companies have been involved, which are, respectively, the Responsible Use of Medicines in Agriculture (RUMA) Alliance, the UKASTA Feed Assurance Scheme and the BSE Inquiry.

Definitions

Both the Medicated Feedingstuffs and the Feedingstuffs (Zootechnical Products) Regulations introduce new definitions which will have to be used in place of those terms with which the industry is much more familiar. UKASTA has met with Veterinary Medicines Directorate (VMD) officials to discuss the distinction between a zootechnical additive/authorised medicated pre-mix and a zootechnical premixture/authorised intermediate product. The definitions put to officials are as follows:-

- A zootechnical additive is the product form sold by the pharmaceutical company (for example Maxus G200 whose active ingredient is Avilamycin); similarly, an authorised medicated pre-mix will be the product form sold by the pharmaceutical company (for example Aurofac 100 whose active ingredient is chiortetracycline hydrochloride);

- A zootechnical premixture, containing a zootechnical additive, as well as vitamins and/or trace elements is the product sold by a supplement manufacturer; similarly an authorised intermediate product containing an authorised medicated pre-mix can be the product sold by a supplement manufacturer, though this can include products produced by the feed compounder for example protein concentrates.

The definitions are to be discussed shortly with VMD officials.

Feedingstuffs (Zootechnical Products) Regulations 1998

These regulations implemented, in full, the EC Additives Directive (70/524/EEC) and also took the use of the zootechnical feed additives, that is antibiotics and other growth promoters and coccidiostats and other medicinal substances, from the control of the medicines legislation. These regulations also implemented, in part, the Establishments Directive (95/69/EC) which required the introduction throughout the EC of a system of approval for establishments handling zootechnical additives and putting zootechnical additives or premixtures into circulation. This system is similar to the scheme which was introduced in the UK in the late 1980's whereby manufacturers of medicated feedingstuffs had to be registered with the Royal Pharmaceutical Society of Great Britain (RPSGB) or the Department of Agriculture for Northern Ireland (DANI).

There are three particular aspects, associated with the Feedingstuffs (Zootechnical Products) Regulations 1998 which have either caused particular problems or still need resolving. The first one concerns top dressing. The aim in introducing the new controls was not to outlaw this practice but rather bring it within control so far as EC legislation permitted. Under the EC Additives Directive, a zootechnical additive or premixture can only be used by incorporation into a compound feedingstuff. The term "incorporation" is not however defined and is thus subject to interpretation ultimately by the Courts.

Since the introduction of the new legislation, UKASTA has sought clarification from VMD on the question of top dressing. Recent advice was that if a farmer purchased a supplementary feedingstuff containing zootechnical additives then mixed it with another feedingstuff to feed to his animals, he or she would also need the approval of the RPSGB or DANI. UKASTA has, however, made the point to VMD that mixing might not take place when a number of different types of zootechnical supplementary feedingstuffs, ranging from a free access farm mineral (feeding rate 100g/h/day) to a beef nut (feeding rate 4kg/h/day) are fed alongside/on top of feed materials and/or forage.

In the light of the wide range in concentration of zootechnical additives in the different types of feedingstuffs which could be fed alongside/on top of other feedingstuffs, UKASTA has suggested that the industry would know whether or not a farmer needed to be approved or not if it used the cut-off point of a mineral feedingstuff when it contained a zootechnical additive. A mineral feedingstuff is defined in the Compounds Directive, and hence the Feedingstuffs Regulations 1995, as meaning:-

> "A complementary feedingstuff which is composed mainly of minerals and which contains at least 40% by weight of ash".

It has been suggested to the VMD that anything more dilute than a mineral feedingstuff containing a zootechnical additive, when used on farm in conjunction with other feed materials and/or forage would not require a farmer to seek approval with the RPSGB. The reaction of the VMD is now awaited on this proposition.

Another issue still unresolved concerns the level of zootechnical additives in supplementary feedingstuffs (reference Regulation 57(1) and (2)). UKASTA has not accepted the interpretation of the VMD that the Additives Directive lays down compulsory maximum permitted levels of zootechnical additives in supplementary feedingstuffs. The industry considers that there is already sufficient control on the incorporation and use of zootechnical additives and supplementary feedingstuffs in those provisions of the Additives Directive which are compulsory and which have been in operation since the mid-1980's.

If safety is the concern of Ministry lawyers and VMD officials, UKASTA has pointed out that the imposition of Regulation 57(2) could result in a less satisfactory situation from the point of view of administration of zootechnical products. Whilst the industry is awaiting resolution of the issue, UKASTA has asked VMD to confirm that the advice agreed in the summer of 1998 is still extant. Namely, the feed industry should not make any changes to formulations of supplementary feedingstuffs even though Regulation 57(2) is now technically in force. A response from the VMD is awaited.

Another new ruling introduced by the Feedingstuffs (Zootechnical Products) Regulations 1998 was that a zootechnical additive could only be used in accordance with its entry in the Annex to the Additives Directive. As zootechnical additives are no longer "medicinal products" they cannot be prescribed for use outside the relevant feed additive authorisation. Consequently, it has been necessary for certain products to become authorised medicated pre-mixes, as defined under the Medicated Feedingstuffs Regulations. These include:-

- Copper sulphate - when used for the correction of defiency in dairy cows and goats.

- Deccox (Decoquinate) when used for the prevention and treatment of coccidiosis in lambs and calves, and for the prevention of coccidiosis in lambs by medication of ewe feed. (This new Marketing Authorisation does not effect the use of

Deccox Prescription Premix which is authorised for the treatment of toxoplasmosis by medication of ewe feed).

- Avatec (lasalocid sodium) 15% CC (game birds).

The incorporation of the above three products is subject to an MFS prescription in accordance with the Medicated Feedingstuffs Regulations 1998. With particular reference to copper sulphate, the UKASTA Scientific Committee is producing guidance on when an MFS prescription is required on farm for the correction of copper defiency in dairy cows and goats.

Medicated Feedingstuffs Regulations 1998

These regulations control the addition of veterinary medicines to feedingstuffs. The regulations reimplement the Medicated Feedingstuffs Directive that was adopted in 1990 and which has been in operation through the Medicines (Medicated Animal Feedingstuffs) Regulations since that time. As with the earlier legislation, manufacturers of medicated feedingstuffs need to be registered with the RPSGB/DANI.

The Medicated Feedingstuffs Regulations introduced the MFS Prescription in place of the Veterinary Written Direction (VWD). The MFS Prescription has to be kept for a period of three years beginning with the date specified in the prescription as the "to be used before" date. Under the legislation, the MFS Prescription has to be in a form containing the headings shown in Annex A to the Medicated Feedingstuffs Directive and personally signed and dated by a veterinary surgeon. The MFS Prescription is valid for three months but may only cover a 31 days supply of feed. A veterinary surgeon can issue an MFS Prescription authorising the use of more than one authorised medicated pre-mix.

The BVA has produced a MFS Prescription the preparation of which was discussed with, amongst other organisations, UKASTA. This form includes a section on "special precautions". In this reference may be made to the presence of a zootechnical feed additive with which the authorised medicated pre-mix incorporated into the medicated feedingstuff is incompatible. UKASTA has suggested that feed manufacturers might wish to maintain the practice, introduced when VWDs were used and PML additives could be listed on a VWD, of advising veterinary surgeons of the zootechnical additive they incorporated into their zootechnical feedingstuffs.

The introduction of the Medicated Feedingstuffs Regulations 1998 and the Feedingstuffs (Zootechnical Products) Regulations 1998 has meant that feed labels have had to be revised. In order to assist members, UKASTA produced a blueprint for a feedingstuff label. This was divided into three sections covering not only the provisions laid down in the Feedingstuffs Regulations and the Medicated Feedingstuffs

Regulations but also accommodating a space for declarations to be made if the feedingstuff contained zootechnical additives. Feed manufacturers should be moving towards finalising the wording to be used in those sections of a label covering medicated feedingstuffs and zootechnical feedingstuffs. VMD advised in the summer that labels for medicated and/or zootechnical feedingstuffs needed to be in line with the new legislation by the end of March 1999.

A fee is payable to the RPSGB/DANI for manufacturers seeking authorisation under the Medicated Feedingstuffs Regulations 1998 and the Feedingstuffs (Zootechnical Products) Regulations 1998. Only one fee is payable. Merchants authorised to handle and sell medicated feedingstuffs also have to pay a fee. A fee is not required, at the moment, for any feed manufacturer seeking approval or registration with the Local Trading Standards Department but this may change in relation to approvals once the proposed EU Directive on fees comes into operation.

Additives

ANTIBIOTIC GROWTH PROMOTERS

At the Agriculture Council meeting in Brussels in December 1998, Ministers voted in favour of a proposal to ban the use of zinc bacitracin, spiramycin, virginiamycin and tylosin phosphate. The legislation was to be brought into effect from 1 January 1999 pending the results of a monitoring programme which were to be examined by the end of December 2000. VMD has advised that the legislation allows a period of six months until 30 June 1999 during which the growth promoters, and premixtures and feedingstuffs containing them must be removed from the market and must not be fed to animals.

Earlier texts of the draft proposal had a three month transition period until 31 March 1999 before the products had to be removed from the market. However, following representations by, amongst others, UKASTA the UK Government successfully obtained a six month transition period. This amendment had been supported by a number of other Member States. In preparing briefing for the Minister, VMD had asked interested organisations for an estimate of the costs to livestock production of the ban. UKASTA advised that the cost of this partial ban would be that of the use of alternative antibiotic growth promoters. Thus it was estimated that the legislation would cost the broiler industry approximately £1 million and the pig industry approximately £500,000. In providing the VMD with these estimates the point had been made that the industry has had no experience in managing the withdrawal of these products. A further point which UKASTA has stressed with the VMD was the inconsistency of the Commission's approach if it was to permit the importation of animal products derived from animals which had been fed the banned antibiotic growth promoters.

CARBADOX AND OLAQUINDOX

At a meeting in December, the Standing Committee on Animal Nutrition adopted a proposal to ban the use of Carbadox and Oiaquindox. This was on the grounds of protecting workers health. The legislation becomes effective from 1 January 1999 with an eight month transition period until 31 August 1999 during which these growth promoters and premixtures and feedingstuffs containing them must be removed from the market and must not be fed to animals.

DINITOLMIDE, IPRONIDAZOI AND ARPRINOCIDE

The Commission Services confirmed to FEFAC, the European feed federation, that the first ban on coccidiostats under the 5th Amending Directive to the Additives Directive was decided unanimously by the Standing Committee on Animal Nutrition at its December 1998 meeting, This ban is a direct consequence of a provision in the amended Additives Directive. As no new product dossiers had been presented for the three products by the deadline of 1 October 1998, there would be an automatic withdrawal of these coccidiostats from the market after a one year transition period ending on 30 September 1999.

Future legislation

IMPLEMENTATION OF ADOPTED DIRECTIVES

In early 1999, MAFF will commence consultation on at least four new Regulations. These are:-

i) Feedingstuffs (Establishments and Intermediaries) Regulations 1999. These will revoke the Feedingstuffs (Establishments and Intermediaries) Regulations 1998 which were brought into operation on 6 May 1998. These will be a consolidating set of regulations which implement Commission Directive 98151/ EC on certain measures on the approval and registration of third country establishments. It is understood that these will require the creation of a list of:-

- those countries from which the EU industry can source additives, premixtures and feedingstuffs containing authorised additives;
- those establishments in third countries which meet the criteria in the Establishments Directive.

It is not clear when such a list could, however, be drawn up. The deadline is 2001. As an interim measure, by 1 May 1999, each additive and premixture from a third country will need to have a representative within the EU who will be responsible for that product in the Community. Such a representative could be an international trading house and/or a company selling in a number of different Member States. The aim of MAFF is to ensure that the list is sufficiently comprehensive so that supplies of tried and tested products can continue to be available within the EU.

VMD is also to commence consultation on the draft Feedingstuffs (Zootechnical Products) Regulations 1999.

ii) The Feedingstuffs Regulations 1999

The Feedingstuffs Regulations were last consolidated in 1995, since when there have been three amendments. MAFF also has to implement a number of EU Directives and thus the opportunity is being taken for further consolidation of the regulations.

The consolidated Feedingstuffs Regulations will implement Council Directive 96/25/EC on the circulation of feed materials. This is a single market measure which removes national derogations under Council Directive 77/101 on straight feedingstuffs. The main purpose of the Feed Materials Directive is to:-

- ensure sufficient transparency throughout the feed chain;
- harmonise the Straights Directive which has been in operation throughout the European Union for 20 years;
- introduce a new definition of "feed materials" which covers both straight feedingstuffs and raw materials intended for use in the oral feeding of animals;
- require specific compulsory declarations including analytical constituents relative to the feed material being supplied.

The Directive also recognises that the seller does not need to provide declarations on an analytical composition of a feed material if the purchaser indicates that he does not wish to receive this information. The Directive does not require analytical constituents to be declared on labels associated with a feed material such as unprocessed arable crops i.e. wheat and barley.

The consolidated Regulations will also implement Council Directive 97/8/EC amending the Undesirable Substances Directive (74/63/EEC). The amendments to the Undesirable Substances Directive introduce additional requirements on the declaration of undesirable substances covering a wider range of substances and feed materials. At the moment, the aflatoxin Bl level must, for example, be declared in a consignment of groundnut if it is between 0.05 and 0.2mg/kg. Under the new legislation, it will also, for example, be necessary for the level of lead to be declared if it exceeds 10mg/kg in any feed material except:-

- grass meal, lucerne meal or clover when the level must be declared if it exceeds 40mg/kg;
- phosphates if it exceeds the level of 30mg/kg;
- yeast if it exceeds 5mg/kg.

It will also be necessary for feed manufacturers to introduce control programmes to ensure compliance with the maximum permitted levels of undesirable substances in finished feed. This new legislation on undesirable substances has also implications for whether or not a feed manufacturer, whether a commercial compounder, integrated producer or home-mixer should seek approval under the Feedingstuffs (Establishments and intermediaries) Regulations 1998.

Amendments to feed additive authorisations are usually made several times a year. In the past this had been achieved by a Commission Directive, amending the annexes to the Additives Directive. For non-zootechnical additives this has previously required MAFF to issue subsequent amendments to the Feedingstuffs Regulations. Now, however, amendments to feed additive authorisations are effected by EC regulation and are thus directly applicable. A recent example was the authorisation of two phaffia yeasts. MAFF now advise that, subject to legal advice, officials hope that the EC regulations, although directly applicable, can also be reflected in consolidated versions of the Feedingstuffs Regulations. This is a move which the feed industry would very much welcome.

The background to the use of Commission Regulations, rather than Commission Directives, for confirming the authorisation of additives is as follows. The 5th Amendment to the Additives Directive replaced Annex 1 and Annex 2 with a new system of authorisation. An additive on Annex 1 had satisfied the authorities that it met the criteria of, respectively, safety, quality and efficacy. An Annex 2 entry had partial authorisation whereby a product has met the safety criteria but not, for example, been proven to be fully efficacious. On the introduction of the 5th Amending Directive to the Additives Directive this system was replaced by a full authorisation and a provisional authorisation procedure. Consequently, a product which has provisional authorisation is allowed to be marketed in the

European Union as it will have satisfied the safety criteria. The company seeking authorisation will have five years within which to provide the further information. As the two criteria to be satisfied are quality and efficacy it is possible that any product, with provisional authorisation, on the market for a five year period will also be subjected to commercial assessment during the five year period.

iii) Feedingstuffs Sampling and Analysis Regulations 1999

Revised regulations are to be produced. These will introduce new methods of analysis on, respectively, amino acids and crude oils and fats. The new methods of analysis will be implemented by reference to the relevant Official Journal in which the text is published. This is in order to ensure full implementation. The regulations will also delete some current determinations.

The regulations will also implement the recently adopted Commission Directive 98/88 establishing guidelines for the microscopic identification and estimation of constituents of animal origin for the official control of feedingstuffs. The guidelines are to be used where detection of constituents of animal origin (defined as products from processing bodies and body parts of mammals, poultry and fish) in feedingstuffs is carried out by means of microscopic examination. MAFF advise that consideration has not yet been given to the collaborative work on samples of feedingstuffs which were circulated to laboratories within the European Union earlier in the year. The trial involved the use of microscopic techniques to try and identify animal feed materials, such as MBM, in a sample of feedingstuffs.

The results will be used to assess the extent to which a full declaration of ingredients could be policed.

iv) Feedingstuffs (Enforcement) Regulations 1999

These will implement the main requirements of the Inspections Directive 95/53/ EC. These regulations will be addressed to the enforcement authorities and, as such, are not expected to have any direct impact on the feed industry.

EMOPS

With specific reference to enzyme and micro-organism products (EMOPs), voting on various tranches listing individual products is to be completed by June 1999. Although this legislation is being set out in Commission Regulations, MAFF does intend to introduce a complete list of the authorised products. MAFF has also confirmed to UKASTA that, as individual enzymes and micro-organisms are not linked to the person responsible to marketing, it is possible, in theory, for a company to produce and sell this technical type of additive without having submitted a dossier. This situation differs to the zootechnical additives which are brand specific.

Other Directives to be implemented

A Council Decision on a Community system of fees in the animal feedingstuffs sector was adopted in December. This Directive establishes the rules for determining how, respectively, fees are to be calculated for approved establishments and fees are to be set by Member States when acting as a rapporteur on individual dossiers. The Directive is required to be brought into operation in June 2000. MAFF advise that it is unlikely that implementing legislation will be made much before that date.

Approval and Registration of Establishments and Intermediaries

MAFF is currently consulting on a proposed numbering system to be implemented by local authorities (for approvals and registrations under the Feedingstuffs (Establishments and Intermediaries) Regulations 1998) and the RPSGB (for approvals under the Feedingstuffs (Zootechnical Products) Regulations 1998). DANI is responsible for both sets of regulations in Northern Ireland.

The proposal is for a dual numbering system. The RPSGB will continue with its current system of numbering, while local authorities produce a new system. The main implication of a dual numbering system is the requirement to label numbers on products. Although the label requirements are not likely to come into effect until 2001, MAFF advise that farmers are approaching local authorities for a number because of interest, in turn, by major retailers.

In its consultation paper, MAFF advised that it did not envisage that there would be a need to label both approval and registration numbers. Consequently, it is envisaged that zootechnical additives, premixtures and compound feedingstuffs that only contain zootechnical additives will be labelled with the RPSGB approval number. These types of products which only contain non-zootechnical additives will have to bear the local authority registration number. Where these products contained both zootechnical and non-zootechnical additives it is proposed that the approval (RPSGB) number will have to be labelled.

UKASTA has advised MAFF that a two numbering system would create problems for the industry and add unnecessary and unacceptable cost. The preference would be for a requirement for a manufacturer registered with the RPSGB to declare the

RPSGB code which would apply to an individual feed mill. This code number would be put on the feed label whether or not the product contained a zootechnical additive. The mill would be on the LACOTS Register but that number would not be used for labelling purposes. Obviously, where manufacturers are registered only with LACOTS then that number should be given on the label.

UKASTA also has concerns over the actual numbering system to be used. MAFF has advised that the Greek "alpha" will precede all numbers. This requirement will, it is considered, cause companies great difficulties in the preparation of their labels. Thus, UKASTA has asked MAFF to reconsider is proposal and agree to introducing something more in line with the current systems in operation in the UK. MAFF's formal reaction to these comments has yet to be received.

Whilst on enforcement agencies, a point which UKASTA has put, over the years, to officials is the need for one enforcement body. MAFF's response is that active consideration is not being given to this issue at the moment since much will depend on the stand to be taken in connection with the future Food Standards Agency. In the absence of such developments, the enforcement of food and feed legislation will be considered in a comparable way; the work on food and certain aspects of the feedingstuffs legislation being carried out by the Trading Standards Department.

The industry has also to bear in mind that the report of the BSE Inquiry, expected to be available by the middle of 1999 could have something to say on this issue. UKASTA has also advised MAFF that its long term aim is formal recognition to be given by Government to the fact that independent audits carried out under the UKASTA Feed Assurance Scheme will largely duplicate the activities of the current enforcement agencies.

Council proposals

On genetically modified materials, MAFF advise that there is a question as to whether the Specific Feed Materials Directive will be circulated in advance of the review of Council Directive 90/220 on the deliberate release into the environment of GM0s. A point to be clarified in the context of Directive 90/220 is the extent to which consideration should be given in authorisations for an individual GMO to the feeding aspects of either the GMO itself or material derived from it.

MAFF has also advised UKASTA that the report of the work undertaken at Leeds University is expected to be released shortly. Under this project, research was carried out on the fate of DNA in a number of samples of feed materials which had undergone a variety of processes. The results of the work are, it is understood, indicative rather than definitive, but is expected to be helpful in future assessment of the feed materials concerned. MAFF is also considering having further work carried out. UKASTA is to be sent a copy of the report of this work, for information, once it is available.

Standing Committee for Animal Nutrition Discussions: Swedish Derogations:

The Swedish Government was seeking a reduction of the maximum permitted level (MPL) for aflatoxin Bl in feedingstuffs. This was based on a concern that the current limit could result in aflatoxin Ml levels being in excess of the 50 nanogram per kilogram level in milk. The UK Government took the view that the level of Bl in complementary feedingstuffs for cows could only be reduced if there was a thorough risk assessment as nothing else would satisfy the WTO.

A proposal to reduce the limits of variation for phosphorus in fish feedingstuffs had been circulated. However, in the light of evidence put forward by a number of Member States, including the UK, the proposal had not been accepted by the full Standing Committee.

CITRUS PULP PELLETS

Legislation is now in place which sets a limit of 500pg I-TEQ/kg (upper bound detection limit) of dioxin in citrus pulp pellets. This was a temporary measure introduced until the cause of the dioxin contamination in Brazilian citrus pulp pellets had been identified. The current view is that contaminated lime has caused the problem. This, in turn, raises a difficulty as lime is an integral part of the process of drying citrus pulp. It should also be remembered that dioxin can be present in a range of other feed materials and so MAFF advise that there is some difficulty in having legislation which refers to one specific feed material. Also, forage produced in certain parts of the country can contain elevated levels of dioxin.

The possibility of the Commission sending a mission to Brazil in January is under consideration. This will be part of the review of the legislation introduced on dioxin which is promised by the end of 1999. Consequently, the legislation setting a limit of dioxin in citrus pulp pellets will remain in place. There is also agreement in Brussels that more information is needed on dioxin in a wider range of feed materials.

CANTHAXANTHIN

Member States and the European Commission have now started to consider possible changes to the maximum permitted levels for canthaxanthin in feeds that would be more in line with those levels used in farming. It is reported that SCAN might consider maximum permitted levels for other feed colourants, such as astaxanthin, at a later date. MAFF asked UKASTA to consider:-

i) the effect that reductions to the limits for layers and salmonids might have (assuming revised maximum levels of 2mg/kg for layers, 15mg/kg for trout and 75mg/kg for salmon);

ii) what actual reductions in maximum permitted levels would be acceptable.

The initial response of UKASTA was that 2mg/kg was too low for both broiler and layer feedingstuffs. It was considered that the level for broiler and layer feedingstuffs should be set at 5mg/kg at the very least. With regard to fish feedingstuffs, it was considered that there was no justification to reduce them below the level of 80mg/kg for both trout and salmon. Furthermore, canthaxanthin was used at these levels throughout the EU not just within the UK and Eire.

USE OF FILTRATION AND BLEACHING EARTHS IN ANIMAL FEED

A letter to the Commission requesting information on the incorporation of filtration and bleaching earths further to the refinement of edible oils for use in animal feedingstuffs was circulated. In order to formulate a UK line, comments were sought.

MAFF was advised by UKASTA that both the extracted oil and the clay used for filtration were used as feed materials. The only problem would be if there were heavy metals/catalysts present. The most appropriate way to control the presence of such contaminants would be through the Undesirable Substances Directive not through legislation on individual feed materials.

RUMA Alliance

In November 1997, a Conference organised jointly by the BVA, NOAH and UKASTA was held on the Responsible Use of Medicines. The following month, the NFU called at the BAFSAM AGM for co-operation within the food chain on the use of medicines in agriculture. In May 1998 the first meeting of the RUMA (Responsible Use of Medicines in Agriculture) Alliance was held in the UKASTA offices. The membership of the Alliance is BAFSAM, BVA, BPMF, BRC, BVA, MLC, NFU, NOAH and UKASTA. A representative of the consumers is also to be invited to join the Alliance. As well as having the main Committee, all Species Sub-Groups covering, respectively, pigs, poultry, sheep and beef, and dairy report to the Alliance.

The objectives of the Alliance are as follows:-

- production of guidelines for the responsible use of medicines on farm. To be incorporated into QA Scheme;

- develop "Codes of Best Practice" designed to reduce the dependence on the use of antibiotics for disease prevention and control;

- consider the implications of the selective withdrawal of antibiotic dietary enhancers for livestock production and availability of effective, safe, licensed alternatives;

- if necessary, develop a plan to manage the consequences of the measured and selected withdrawal of antibiotic digestive enhancers from livestock production,

- share information on the potential food scares and develop common messages to avoid the scares being used as marketing tools. This is to avoid stories which unfairly damage the image of British farming.

As a consequence of recent developments within the industry, UKASTA is to consider alternatives to antibiotic digestive enhancers. A senior official in DGVI of the Commission has already expressed concern at the use of products containing certain organic and inorganic acids which are authorised as preservatives and that have been used recently for a growth promotional effect in animals. The VMD has appointed an official who looks specifically at the use of alternative products to the antibiotic growth promoters. Thus, the view within UKASTA is that in order to safeguard the reputation of the UK feed industry a Working Group should be established in order to consider what alternative products are currently on the market.

UKASTA Feed Assurance Scheme

Version 2 of the Code of Practice for the Manufacture of Safe Compound Animal Feedingstuffs, the Guidelines implementing the Code of Practice and the Audit Checklist were circulated to all Feed members of UKASTA together with a list of independent auditors in September 1998. The first version of the Code of Practice and Guidelines had been issued in October 1997. Changes introduced into the revised text included amendment of terminology to bring it in line with the new legislation.

With regard to the administrative arrangements for the Scheme, a Register of "Assured Feed Mills" was being created. At least in the early stages this was to be split into the following sections:-

- Fully approved
- Provisionally approved
- Scheduled for audit

A series of meetings with organisations downstream of the feed industry are being held during the latter part of 1998 and early part of 1999. This included meetings with major retailers. The aim is to seek endorsement and support of the UKASTA Feed Assurance Scheme by companies in these sectors.

Codes, some of which have yet to be finalised, that are associated with the UKASTA Code of Practice for the Manufacture of Safe Compound Animal Feedingstuffs include:-

- UKASTA Code of Practice for Road Haulage (of combinable crops, animal feed materials and as-grown seeds);
- UKASTA Code of Safe Practice for the Storage, Packaging and Supply of Animal Feed Materials which are supplied to farms;
- ABTA Code of Practice for the Supply of Moist Co-Products to Farm;
- National Dairy Farm Assurance Scheme;
- NFU Code of Practice for on-farm mixers;
- NFU Code of Practice for the Safe Storage, Handling and Feeding of Feed Materials on Farm;
- FEFAC General Code of Practice;
- BVA Code of Practice for Medicines.

BSE update

MAFF recently commenced consultation on two pieces of legislation. The BSE (No. 2) (Amendment) Order 1999 and The BSE (Feedingstuffs and Surveillance) Regulations 1999. These latter regulations concern the identification of ruminant protein in feed intended for ruminants manufactured in mixed species mills using the ELISA technique. It is considered that the new legislation should not have any impact on feed manufacturers.

BSE INQUIRY

Evidence was taken by representatives of the feed industry for Phase I Stage 3 of the Inquiry in April. This covered the period up to the end of 1986. Particular attention had been drawn to the difference of opinion in the 1980's between UKASTA and the NFU on the declaration of ingredients although this was not a subject of relevance to BSE. For the rendering industry, attention had been drawn to the lack of progress on the introduction of controls on protein processing.

Again, UKASTA and industry representatives gave evidence in Phase I Stage 5. This covered the period following the identification of BSE up until March 1996. Both the individual companies and UKASTA had submitted written evidence in advance of the oral hearing held in October 1998. Amongst other things, UKASTA had submitted a chronological list of the action taken by the Association on behalf of its members in connection with BSE. The issues raised at the hearing included the immediate effectiveness of the 1988 ruminant feed ban, the SBO controls, cross-contamination and the industry's response to the emerging knowledge of BSE. Phase 11, concerning clarifications, conflicts of potential criticism commenced at the beginning of 1999.

References

Bovine Spongiform Encephalopathy (No. 2) (Amendment) Order 1999 (draft).

Bovine Spongiform Encephalopathy (Feedingstuffs and Surveillance) Regulations 1999 (draft).

Commission Directive 98151/EC laying down certain measures for implementing Council Directive 95169/EC (1998). (OJ L208, p43).

Commission Directive 98188/EC establishing guidelines for the microscopic identification and estimation of constituents of animal origin for the official control of feedingstuffs (1998). (OJ L318, p45).

Council Directive 74/63/EC on undesirable substances and products in animal nutrition (1974). (OJ No. L38, p.31).

Council Directive 771101/EEC on the marketing of straight feeding stuffs (1977). (OJ No. L32, pl).

Council Directive 79/373/EEC on the marketing of compound feeding stuffs (1979). (OJ No. L86, p.30).

Council Directive 901167/EEC laying down the conditions governing the preparation, placing on the market and use of medicated feedingstuffs in the Community (1 990). (OJ L92, p42).

Council Directive 90/220/EEC on the deliberate release into the environment of genetically modified organisms (1990). (OJ L117, p15).

Council Directive 95169/EC laying down the conditions and arrangements for approving and registering certain establishments and intermediaries operating in the animal feed sector (1995). (OJ No. L332, p.15).

Council Directive 95153/EC) fixing the principles governing the organisation of official inspections in the field of animal nutrition (1995). (OJ No. L265, p.17).

Council Directive 701524/EEC concerning additives in feeding stuffs as amended by Council Directive 96151/EC (1996). (OJ No. L270, p.11).

Council Directive 96/25/EC on the circulation of feed materials (1996). (OJ L125, p35).

Council Directive 9718/EC amending Council Directive 74163/EEC on undesirable substances and products in animal nutrition (1997). (OJ L48, p22).

Feeding Stuffs Regulations (1995). (S.1. 1995/No. 1412). London, HMSO.

Feedingstuffs (Zootechnical Products) Regulations (1998). (S.1. 1998/No. 1047). London, SO.

Feedingstuffs (Establishments and Intermediaries) Regulations (1998). (S.1. 1998/N0. 1049). London, SO.

Feeding Stuffs Regulations 1999 (draft).

Feedingstuffs (Establishments and Intermediaries) Regulations 1999 (draft).

Medicated Feedingstuffs Regulations (1998). (S.1. 1998/No. 1046). London, SO.

Proposal for a Council Decision concerning a Community system of fees in the animal feed sector. VI/8071/97 Rev. 3.

The Novel Food and Novel Food Ingredients Regulations 1997 (S.1. 1997/No. 1335). London, HMSO.

UKASTA 1998 Code of Practice for the Manufacture of Safe Compound Animal Feedingstuffs. London. United Kingdom Agricultural Supply Trade Association.

Guidelines for the Implementation of the UKASTA Code of Practice for the Manufacture of Safe Compound Animal Feedingstuffs. London. United Kingdom Agricultural Supply Trade Association.

2

NUTRIENT REQUIREMENTS FOR GRAZING HORSES: DEVELOPMENT OF AN OPTIMAL PASTURE SUPPLEMENT

RHONDA M. HOFFMAN[1] AND DAVID S. KRONFELD[2]
[1]*Department of Animal Science, The University of Connecticut, Storrs, CT, USA*
[2]*Department of Animal and Poultry Sciences, Virginia Polytechnic Institute and State University, Blacksburg, VA, USA*

Introduction

Most commercial horse feeds are supplements for use with forages – pastures or hays. Most supplements are based on cereal grains containing abundant starch, which is alien to the nutritional heritage of the horse. Common experience has been supported by epidemiological and experimental studies that associate grain concentrates with several digestive and metabolic disorders – colic, laminitis, gastric ulcers and, perhaps, developmental orthopedic disease and some forms of exertional rhabdomyolysis (Sprouse, Garner and Green, 1987; Clarke, Roberts and Argenzio, 1990; Valberg, 1995). The key etiologic factors may be overloads of hydrolyzable and rapidly fermentable carbohydrates (Figure 2.1). To circumvent these problems and promote performance, a series of studies at the Middleburg Agricultural Research and Extension Center have been developing a pasture supplement that features fibre and fat in lieu of starch and sugar (Hoffman, Kronfeld, Lawrence, Cooper, Dascanio and Harris, 1996). The biological studies have been complemented by statistical investigation of optimal ranges of nutrients and the application of sensitivity analysis to ration evaluation and supplement formulation (Kronfeld, Ferrante and Grandjean, 1994; Kronfeld, 1998b). This paper will summarize major findings and address certain dilemmas, starting with interpretation of nutrient requirements.

Requirements, Allowances and Optimal Ranges

Current nutrient requirements of horses (NRC, 1989) are conceptually like those of the dog and cat – best regarded as mean minimums intended to avoid undesired effects (Figure 2.2). In contrast, the nutrient requirements of production animals are mean

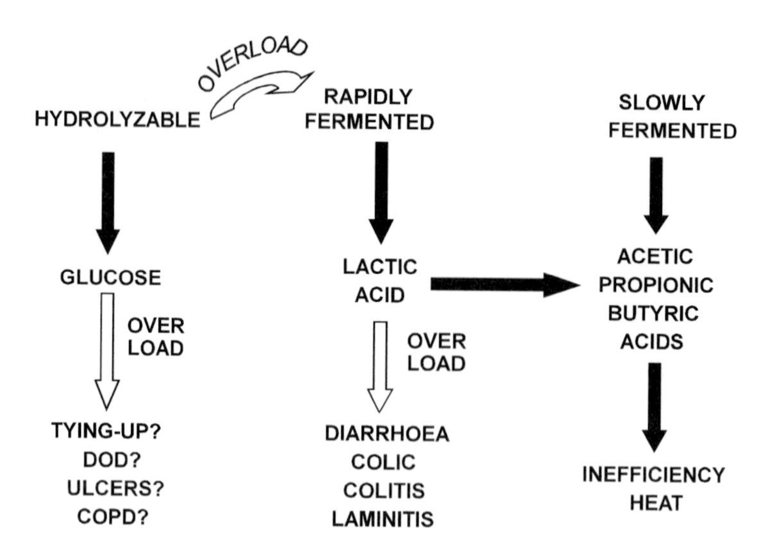

Figure 2.1 Overloads of sugars (molasses and rapidly growing pastures), hydrolyzable starches (grains), and rapidly fermentable carbohydrates (soluble fibres) are commonly thought to precipitate certain digestive and metabolic disorders in the horse. Slow fermentation of insoluble fibres (mainly cellulose) sustains healthy function of the large bowel. (Modified from Kronfeld, 1998a).

values for desired effects. Recommended dietary allowances (RDAs) for humans are aimed at the mean plus two standard deviations, statistically sufficient for 0.98 of the population (Food and Nutrition Board, 1991). Noting that the nutrient requirements of dogs and cats, as proposed by the National Research Council, had little practical value for formulating pet foods, the Association of American Feed Control Officials (an alliance of government and industry, to which American regulatory agencies usually defer) developed nutrient profiles for dogs and cats. These canine and feline allowances emphasized practical experience and, unlike the human RDAs, lacked any systematic statistical basis.

The nutrient requirements of production animals and the allowances for humans, dogs and cats aim to reach the lower ends of optimal ranges of nutrient intakes (Figure 2.2). No such allowances have been offered by national committees for the horse. Following the plan of the human RDAs, equine allowances for essential nutrients should be at least 1.3-times the minimum requirements (NRC, 1989). Experimental and epidemiological studies suggest, however, that at least double the current minimums are needed for vitamin A (Donoghue, Kronfeld, Berkowitz and Copp, 1981; Greiwe-Crandell, Kronfeld, Gay, Sklan, Tiegs and Harris, 1997), zinc (Bridges and Moffitt, 1990), and copper (Knight, Weisbrade, Schmall and Gabel, 1988; Hurtig, Green, Dobson and Burton, 1990). Given the numerous interactions among micronutrients, perhaps the minimums for all, rather than only three, should be doubled to yield useful allowances.

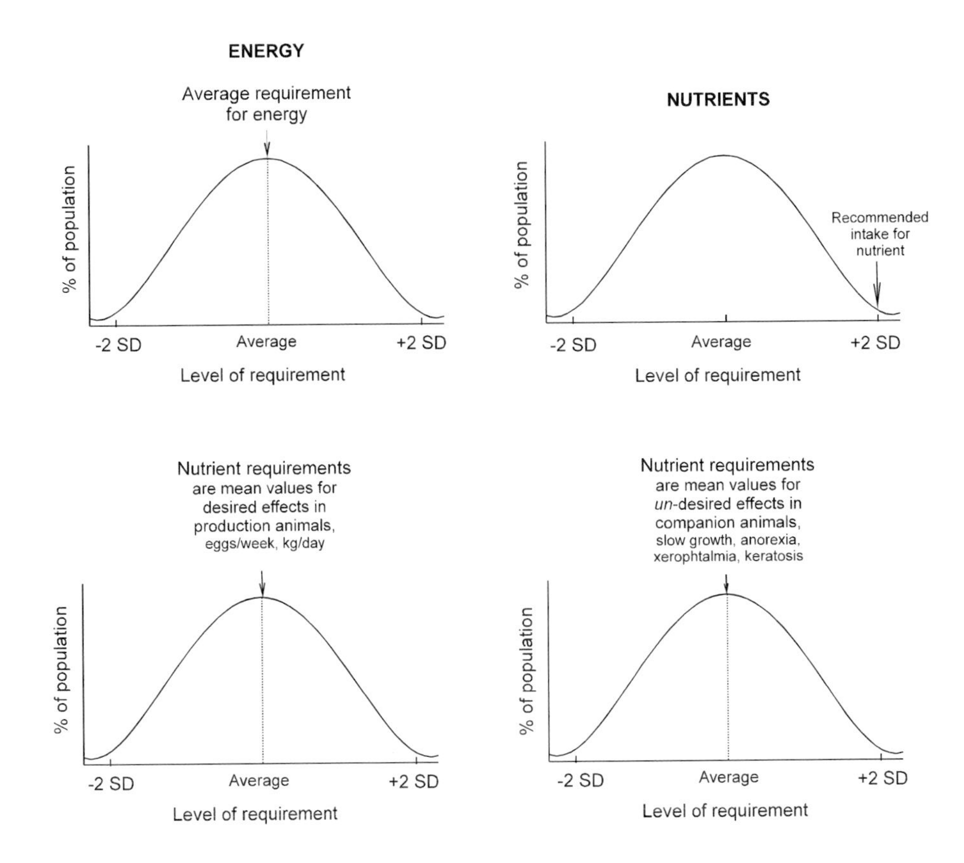

Figure 2.2 The human recommended dietary allowances (RDAs) are means for specified populations (upper left), but the RDAs for nutrients are two standard deviations above mean requirements, so are adequate for 98% of the population (upper right, modified from Food and Nutrition Board, 1989). The nutrient requirements for livestock are mean values for specified production levels (lower left). In contrast, the nutrient requirements of companion animals are mean minimums to avoid undesired effects (lower right), hence are adequate for only half of the population.

Current digestible energy (DE) requirements for maintenance are mean estimates calculated from body weight (NRC 1989). The data were from four horses confined in metabolism stalls and multiplied by 1.43 for activity. Variation was negligible and failed to reflect the large variation in dry matter (DM) intake found in practice (Table 5-4 in NRC, 1989). If these ranges represent 90% confidence intervals, then the coefficient of variation ranges from 8% for maintenance to 12% for lactation and intense work, and 16% for growth. These coefficients of variation for DM and presumably DE intakes could be applied to nutrients for which corresponding coefficients are unknown, following the precedent of the human RDAs. Also, these coefficients of variation need to be kept in mind when estimating pasture consumption of grazing animals for the purposes of ration evaluation or supplement formulation (Kronfeld,

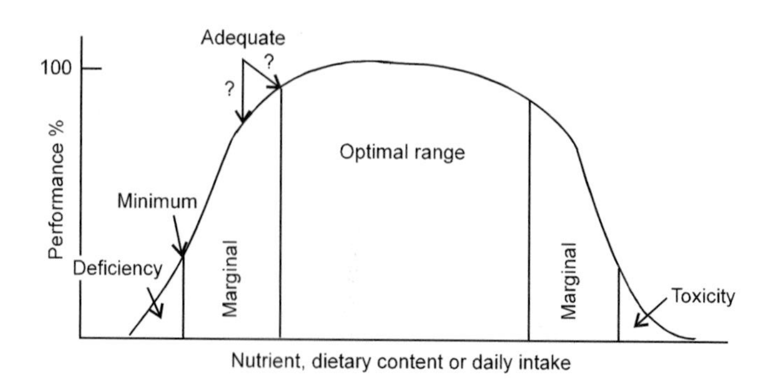

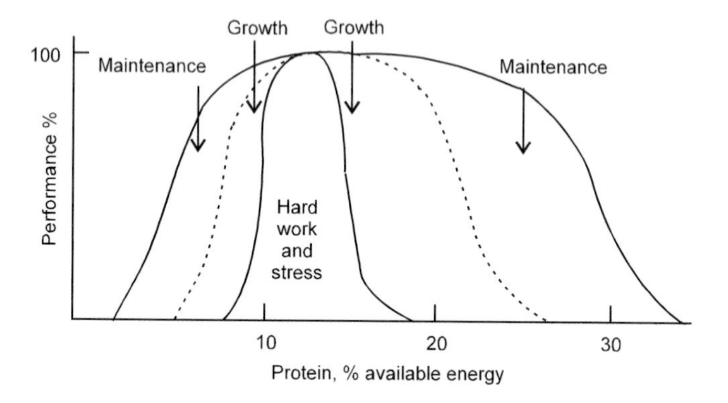

Figure 2.3 The beneficial influence of dietary content or nutrient intake rises to a plateau, the optimal range, then declines. The plateau is broad for undemanding situations, but narrows as demands increase, for example, for growth, lactation, hard work and stress. (Modified from Kronfeld at al., 1994).

1998a). A palatable and presumably completely consumed supplement throws all of the variation into the estimate of pasture intake (Figure 2.4).

Reconciling American with European recommendations for horses involves multiple interconversions and numerous assumptions. The USA uses DE in MCal (NRC, 1989), the Germans use DE in MJ (Meyer, 1992), and the French use net energy (NE) in MCal (Martin-Rosset, 1990). The USA uses crude protein (CP) per kg DM, the Germans digestible protein (DP) per MJ DE, the French DP per MCal NE.

The NE systems are preferable because of large differences in efficiency of utilization of nutrients that follow one of five paths: glucose and other simple sugars; lactate and pyruvate (metabolized largely via glucose); acetate and butyrate; long-chain fatty acids; and amino acids (Vemorel and Martin-Rosset, 1993; Kronfeld, 1996). In formulating feeds, however, expressing nutrients on a DM basis is usually better than any energy

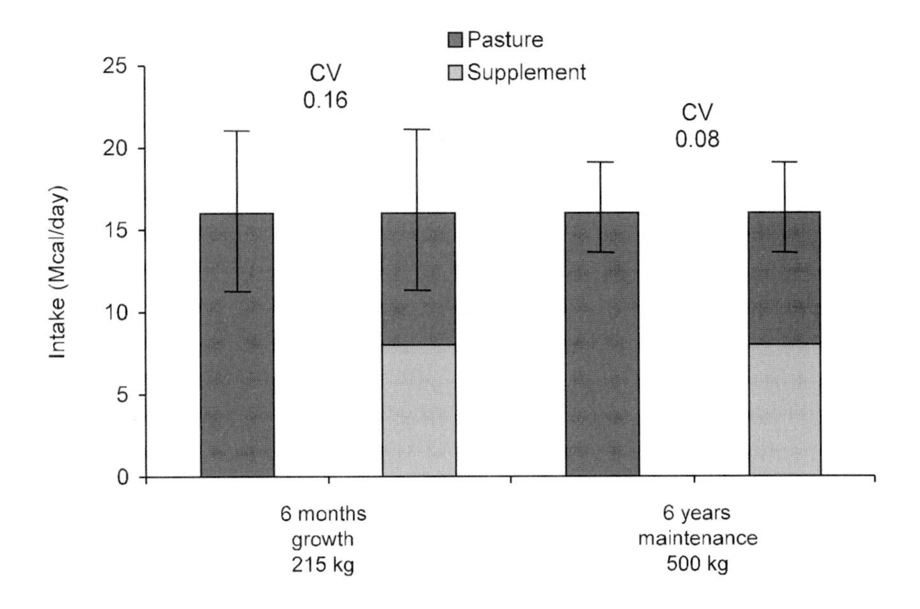

Figure 2.4 Huge variation in estimates of daily intakes of pasture by grazing horses is exacerbated by the provision of a palatable supplement that is completely consumed. It requires the application of sensitivity analysis (Kronfeld, 1998b) in ration evaluation and supplement formulation. (Modified from Kronfeld, 1998a).

basis, because as energy concentration declines, so do bioavailabilities of many nutrients, and the opposing effects of these two variables tend to cancel out.

Sensitivity analysis

For practical purposes, variation in intake, ranges of nutrient concentrations and known systematic errors can be handled by sensitivity analysis (Kronfeld, 1998b). A survey of pasture composition in central and north-central Virginia determined ranges (90% confidence intervals) of nutrient concentrations in initially 20, then 30 and later, 130 samples. Another set of ranges represented target zones for energy density and nutrient concentrations. A third set of ranges suggested DM intakes (NRC, 1989), hence approximate variation in DE intakes The range of supplement:pasture intakes of DM was 25:75 to 50:50. These ranges were used to calculate most of the nutrient contents of a supplement (Kronfeld, 1998b). The procedures are illustrated with the Pearson Square in Appendix A, and may be facilitated by ration formulation software.

Sensitivity analysis was used to formulate a trace element premix (Kronfeld, Cooper, Crandell, Gay, Hoffman, Holland, Wilson, Sklan and Harris, 1996). Target ranges for the microminerals in whole ration were 1.5- to 3-times current recommendations for growth (NRC, 1989). Endogenous trace mineral concentrations were estimated using published values for the ingredients in the supplement, and the variation in intake and the composition of the 130 forage samples was considered. Sensitivity analysis was applied to test the soundness and flexibility of the formulation. These principles could be applied in any other locale, where ranges of microminerals have been determined in the forages.

The vitamin premix was less innovative. It was assumed endogenous vitamins were zero and the supplement was fortified to provide 2-times the current recommendations for vitamins A, D, E and thiamine (NRC, 1989), assuming that the supplement provided 0.25 of the DM intake. In addition, riboflavin, niacin, biotin, folic acid, B_{12} and ascorbic acid were included on the basis of studies in other species and consultation with Dr. Ted Frye (Hoffman LaRoche, Nutley, NJ).

Pasture intake

The huge variation in estimates of pasture intake (Figure 2.4) has prompted development of more precise methods, especially using markers. In ruminants, chromic oxide and ytterbium have been perhaps the most common external markers used to estimate fecal output (Blaser, Hammes, Fontenot, Bryant, Polan, Wolf, McClaugherty, Kline and Moore, 1986). Lignin and acid insoluble ash have been used extensively as internal markers to estimate digestibility (Sunvold and Cochran, 1991). More recently, chromic oxide has been used to develop a model of faecal kinetics in the horse (Holland, Kronfeld, Sklan and Harris, 1998b). Markers currently being evaluated for their ability to estimate digestibility in the horse include alkanes (Ordakowski, 1998), ytterbium and yttrium (Holland, 1998).

Pasture composition

The energy and nutrient contents of pasture change with season and stage of growth. Relatively high sugar, protein and water contents are present in the spring and early summer (Blaser *et al.*, 1986). Fibre content increases with stem elongation, and energy, protein, phosphorus, and ß-carotene contents decrease as seeds are shed through late summer, autumn and winter. In a survey of twelve farms in northern Virginia (Kronfeld *et al.*, 1996), pastures were found marginal or deficient, compared with current recommendations (NRC, 1989), in zinc, copper and selenium throughout the year, and in phosphorus and vitamin A in the winter. The variation in nutrient composition and resultant inadequacies affect grazing animals with pasture as their primary habitat. Thus, these nutrients became the subject of field studies.

Mineral supplementation

PHOSPHORUS

Blood serum analysis indicated phosphorus depletion in weanlings but not in mares fed an all-forage diet of pasture and hay, that was marginal or deficient in phosphorus (Greiwe-Crandell, Morrow and Kronfeld, 1992). Supplementation of pasture with phosphorus via a pelleted concentrate in another group of weanlings led to adequate serum phosphorus concentrations. Unsupplemented pregnant mares had hyperphosphatemia and hyperphosphaturia, which were regarded as paradoxical to the low phosphorus in the diet. The hyperphosphaturia was accompanied by calciuria, suggesting mobilization of calcium and phosphorus from bone in pregnant mares that would ensure a sufficient supply for the foetus (Greiwe-Crandell *et al.*, 1992).

SELENIUM

Mares and weanlings fed only pasture and hay showed no clinical signs of selenium deficiency, but blood serum analysis reflected the low concentrations of selenium found in the pasture and hay (Greiwe-Crandell *et al.*, 1992). Selenium supplementation, via a pelleted concentrate, of another group maintained on pasture and hay, effectively maintained serum selenium concentration. Compared with supplemented weanlings, those fed only pasture and hay had more clinically-severe cases of an equine herpes virus type 4 infection, perhaps exacerbated by inadequate selenium intake.

ZINC AND WEANING STRESS

The social dislocative stress of weaning may affect appetite, metabolism and immune competance. Foals raised on pasture and supplemented with a concentrate coped better with weaning stress, as assessed by behavior and adrenocorticotropic hormone response tests, than foals raised on pasture and hay only (Hoffman, Kronfeld, Holland and Greiwe-Crandell, 1995). The pastures and hays were marginal or deficient in phosphorus, zinc and copper. The supplement provided two times more phosphorus, seven times more zinc and five times more copper than the forages. The advantage of feeding the concentrate may have been due to its mineral content. Serum zinc was lower during weaning in foals that had access to only pasture and hay, compared with those that were supplemented (Hoffman *et al.*, 1995). Zinc interferes with copper absorption; thus, copper deficiency may be caused by excess dietary zinc. Foals fed zinc in concentrations of 1000 or 2000 mg/kg, as compared to those fed 40 or 250 mg/kg, were severely lame after less than 10 weeks and had fractures of growth plate and articular cartilage (Bridges and Moffitt, 1990). A tentative target range for zinc may be 2- to 4- times the NRC minimum.

COPPER

Copper is an essential cofactor for the metalloenzyme lysyl oxidase, which is essential for the cross-linking of collagen. Copper concentrations in pasture of farms surveyed in northern Virginia only marginally met the NRC minimum recommendation (Kronfeld *et al.*, 1996). A negative correlation between copper concentration in foal diets and degree of osteochondrosis was reported byKnight, Gabel, Reed, Embertson, Tyznik and Bramblage (1985). In a subsequent study (Knight *et al.*, 1988), mares were fed normal and high levels of copper in late gestation and early lactation, and their foals were fed correspondingly normal and high levels of copper. The data were considered inconclusive (NRC, 1989), but were confirmed by epidemiological and experimental studies (Bridges and Harris, 1988; Williams, Stowe and Stickle, 1989; Hurtig *et al.*, 1990). Recent research has demonstrated that copper supplementation of pregnant mares may reduce the risk of skeletal abnormalities in foals (Pearce, Grace, Firth and Fennesey, 1998). Tentative target ranges may be 2- to 3-times the current mean minimum requirements of copper for pregnancy and growth.

Vitamin supplementation

VITAMIN A

Seasonal differences in vitamin A status of grazing mares was demonstrated using serum retinol concentrations and a relative dose response test (Greiwe-Crandell, Kronfeld, Gay and Sklan, 1995a). Serum retinol concentrations and the relative dose response in newborn foals were linearly correlated with corresponding data from their dams (Greiwe-Crandell, Kronfeld, Gay, Sklan and Harris, 1996). Supplementation with vitamin A at twice the NRC (1989) recommendation marginally compensated for seasonal differences (Greiwe-Crandell, Kronfeld and Sklan, 1995b). Supplementation of depleted mares with retinol acetate reduced the incidence of retained placenta and flexor deformities and increased birth weights of foals (Greiwe-Crandell, 1995b). Considering these studies and previous reports (Donoghue *et al.*, 1981; Maenpaa, Pirhonen and Koskinene, 1988), the optimal range for vitamin A intake may be between 2- and 5- times the minimum NRC recommendation.

SYNTHETICS

A water-dispersible form of ß-carotene appeared to be ineffective in improving the vitamin A status of horses (Watson, Cuddeford and Burger, 1995; Greiwe-Crandell *et al*., 1997). Moreover, the common synthetic, oxidized form of folic acid may interfere with absorption of the naturally occurring, active, reduced forms in rats, pigs and, perhaps, horses (Toribio, Bain, Mrad, Messer, Sellars and Hinchcliff, 1998).

Protein supplementation

Optimal dietary protein ranges for growth or hard work may be the most debated topic in horse nutrition. A quadratic curve applied to data from growth studies in the literature indicated maximum weight gain at a crude protein (CP) concentration of 160 g/kg DM, with 1.0 SE below the peak at 130 and 190 g/kg DM (Graham-Thiers and Kronfeld, unpublished data). Reports in the literature have indicated lysine as the first (Ott, Asquith and Feaster, 1981), and threonine as the second (Graham, Ott, Brendemuhl and TenBroeck, 1994), limiting amino acids for proper growth of young horses. Equine athletes may benefit from minimizing protein quantity and maximizing quality, which would minimize production of acid, heat and urea without compromising needs for stress and repair. In addition, dietary protein restriction would reduce ammonia in stables and nitrogen contamination of pasture.

A current pasture study on mares and foals is being conducted to compare a low protein supplement fortified with lysine and threonine with a typical (150 g CP/kg DM) supplement level (Staniar, W. B., personal communication). A similar comparison in exercising horses indicated that horses fed restricted protein fortified with lysine and threonine had a slight but significant moderation of the acidogenic effect of repeated sprints (Graham-Thiers, Kronfeld, Kline and Harris, 1998).

Carbohydrate supplementation

Proximate analysis of carbohydrates relates to plant anatomy (Henneberg and Stohmann, 1864; Van Soest, 1963). The botanical fractions fit reasonably well with the digestive physiology of ruminants, but not hind-gut fermenters such as the horse. A proximate analysis to suit the digestion and metabolism of hind-gut fermenters would determine three groups of carbohydrates (Blaxter, 1989; Gray, 1992; Kronfeld, 1996):

- Hydrolyzable carbohydrates that yield glucose and other simple sugars.
- Rapidly fermented carbohydrates that yield mainly propionate and lactate, which are metabolized via glucose and pyruvate.
- Slowly fermented carbohydrates that yield mainly acetate and butyrate, which are metabolized via acetyl-CoA.

Instead, the common American system for livestock (Van Soest, 1963) measures acid-detergent fibre (ADF, lignin and cellulose), neutral detergent fibre (NDF), which is ADF plus hemicellulose in cell walls, and non-structural carbohydrates (NSC). Ruminant nutritionists estimate NSC, as calculated by difference, and thus include fermentable components such as some hemicellulose, pectins, gums, mucilages, oligosaccharides and resistant starches, as well as hydolyzable starches and sugars (Van Soest, Robertson and Lewis, 1991). The NSC fraction may be calculated differently by subtraction of total dietary fibre (TDF; Hall, 1989). However, TDF does not include fermentable, non-hydrolyzable oligosaccharides and resistant starches, so this calculation of NSC also fails to delineate the hydrolyzable carbohydrates.

A common system used in human nutrition measures non-starch polysaccharides (NSP) and separates rapidly and slowly hydrolyzable carbohydrates (Englyst, Wiggins and Cummings, 1982; Englyst, Veenstra and Hudson, 1996). It has not been accepted widely in livestock nutrition because it makes no provision for lignin and the influence of lignification on the rate of fermentation, which affects the proportions of propionate and lactate versus acetate and butyrate.

Hydrolyzable carbohydrate has been measured in 107 forage samples and 25 concentrates (Hoffman and Kronfeld, unpublished data), using amylose and amylopectinase (Davis, 1976; Smith, 1981). The traditional method of estimating NSC by difference overestimated the fraction of the hydrolyzable carbohydrate present. Hydrolyzable carbohydrate correlated with NSC and accounted for about one-third of the NSC in forages and about two-thirds of the NSC in concentrates. Hydrolyzable carbohydrate fluctuated from month to month in pasture, but not relatively more than fermentable carbohydrate fractions (Wilson, Kronfeld, Cooper and Sklan, 1997). These partially seasonal, but also unpredictable, changes reinforced the need to provide a broad spectrum of carbohydrates in our pasture supplement.

The key feature of the pasture supplement developed is an array of fibre sources, perhaps ironically because forage is usually regarded as the main source of fibre in the equine diet. At times, however, the fibre content of pasture is low, and a supplement of slowly fermentable fibre is needed to sustain digestive health; this has been recognized for many years. The key development is the need for a full spectrum of carbohydrates, which is intended to promote a diverse microbial population. This symbiotic system should adapt readily to any abrupt change in pasture carbohydrates and sustain an adequate slow fermentation, yielding mainly acetate and butyrate, at times when overloads of hydrolyzable and rapidly fermentable carbohydrates may be exaggerating production of lactate (Figure 2.1).

Fat supplementation

Given the emphasis on fibres, the energy concentration of the supplement developed needed to be sustained by additional fat. This design is the reverse of our complete ergogenic feed for the equine athlete, which features fat first, then slowly fermentable carbohydrate in sufficient amounts to preserve the equilibrium in the lower bowel.

The substitution of fat and fibre for hydrolyzable carbohydrates has various advantages, including reduced risk of digestive disorders and improved performance (Potter, Hughes, Julen and Swinney, 1992; Kronfeld, 1996). Preference tests demonstrated higher acceptability of maize oil than other vegetable oils or animal fats (Holland, Kronfeld, Rich, Kline, Fontenot and Meachem, 1998a). Spontaneous activity and reactivity were reduced (i.e. improved tractability) by dietary supplementation with maize oil and mixtures of soya lecithin and maize oil (Holland, Kronfeld and Meachem, 1996). Fat supplementation may also be effective in controlling certain forms of exertional rhabdomyolysis (Kronfeld, 1998; Valberg, 1995). Rice bran, which contains 0.18-0.25 g/kg fat, has become a common fat supplement for horses in the last few years.

Supplementation for growth

Our fat-and-fibre supplement (FF) has been tested against a sugar-and-starch supplement (SS), which is a copy of a typical, best-selling concentrate (Omalene-200™, Purina Mills, St Louis, MO). This is important, because control diets have been sub-standard in some equine studies. Results are also compared with industry standards for reproduction and growth.

The main advantage of FF over SS was noted in growth curves of yearlings (Hoffman *et al.*, 1996). Yearlings fed SS exhibited a spring slump, evident as a decreased rate of gain in body weight and condition. Similar growth slumps were noted in Ontario and Kentucky foals, around 360 to 400 days of age (Hintz, Hintz and Van Vleck, 1979; Thompson, 1995). The slump in growth rate may be due to a metabolic change, or a result of seasonal differences in nutrient composition of pasture, easily attributed to the high sugar, starch and water content of spring pasture. This spring slump in body weight and condition was eliminated in yearlings fed FF (Hoffman *et al.*, 1996). Furthermore, compensatory growth was noted in yearlings raised on SS but not those fed FF. Compensatory growth has been associated with developmental skeletal problems, including flexural deformities (Hintz, Schryver and Lowe, 1976).

Supplementation and bone

Bone mineral content of the equine third metacarpal was estimated from radiographs standardized with aluminum wedges (Hoffman, Lawrence, Kronfeld, Dascanio, Cooper and Harris, 1997). Bone mineral content (BMC) of the third metacarpal increased with body weight and age. Two plateaus were noted: the first coincided with a change in diet from mainly milk to mainly pasture and supplement, and the second with a decrease in activity on ice and packed snow. These plateaus appeared more evident in the SS group than the FF group. Thus, the FF supplement may reduce fluctuations in the rate of increase in bone density as well as body weight and body condition during growth.

However, BMC was lower from September to May, about 5 to 13 months of age, in growing horses fed the FF supplement. Perhaps the availability of calcium may have been compromised by the formation of calcium soaps with fat (Palmquist, Jenkins and Joyner, 1986), or through cation exchange and the water holding capacity of fibre (Allen, McBurney and Van Soest, 1985) in the intestine. Perhaps additional calcium provided in the FF supplement will compensate for this effect (Wilson and Hoffman, unpublished data).

Supplementation and milk

Another advantage of the provision of a pasture supplement rich in fat and fibre was noted in milk and colostrum composition of mares (Hoffman, Kronfeld, Herbein, Swecker, Cooper and Harris, 1998). The fatty acid profile indicated consistently higher linoleic [18:*2(n-6)*] acid in milk from mares supplemented with the fat and fibre concentrate over the first six months of lactation. Research in human nutrition has indicated a role for linoleic acid in the prevention of gastric ulcers (Kearney, Kennecy, Keeling, Keating, Grubb, Kennedy and Gibney, 1989). Linoleic acid is a precursor of prostaglandin E, and stimulation by linoleic acid increased gastroduodenal prostaglandin formation (Grant, Palmer, Kelly, Wilson and Misiewicz, 1988). Prostaglandins are involved in the prevention of gastric ulcers by enhancing mucosal protective factors. Foals typically have a high incidence of gastric ulcers (Murray, Murray, Sweeney, Weld, Wingfield Digby and Stoneham, 1990), and prevention is important from both clinical and production standpoints.

Mares fed maize oil and fibre had increased protein ($P = 0.015$) and a 4.2 fold increase in immunoglobulin G ($P = 0.028$) concentration in colostrum sampled at 6 to 12 hours after foaling (Hoffman *et al.*, 1998). It was hypothesized that the maize oil contributed additional vitamin E activity that enhanced immunoglobulin synthesis in the mares. A subsequent study indicated that dietary vitamin E, supplemented during late gestation at twice the NRC minimum, increased concentrations of IgG, IgA and IgM in equine colostrum, with a resultant increase in passive transfer of IgG and IgA, but not IgM, to foals (Hoffman, unpublished data).

Conclusions

These studies have shown that mixed grass-legume pastures used for horses need supplementation with phosphorus, selenium, copper, zinc and vitamin A. Providing a fat-and-fibre supplement designed to complement the pasture avoided the spring slump in growth of yearlings and increased immunoglobulin-G and linoleic acid in mares' milk. Improvement is needed in determining optimal ranges of nutrients, in measuring pasture intake, and in differentiating physiologically oriented carbohydrate fractions. Meanwhile, the application of sensitivity analysis should enhance the flexibility, effectiveness and safety of forage supplements.

References

Allen, M. S., McBurney, M. I. and Van Soest, P. J. (1985) Cation-exchange capacity of plant cell walls at neutral pH. *Journal of Scientific Food Agriculture*, **36**, 1065–1072.

Blaser, R. E., Hammes R. C. Jr., Fontenot, J. P., Bryant, H. T., Polan, C. E., Wolf, D. D., McClaugherty, F. S., Kline, R. G. and J. S. Moore. (1986) *Forage-Animal Management Systems.* Bulletin 86-7. Virginia Agricultural Experiment Station, Virginia Tech, Blacksburg.

Blaxter, K. (1989) *Energy Metabolism in Animals and Man.* Cambridge University Press, New York.

Bridges, C. H. and Harris, E. D. (1988) Experimentally induced cartilaginous fractures (osteochondritis dissecans) in foals fed low-copper diets. *Journal of the American Veterinary Medical Association*, **193**, 215–221.

Bridges, C. H. and Moffitt, P. G. (1990) Influence of variable content of dietary zinc on copper metabolism of weanling foals. *American Journal of Veterinary Research*, **51**, 275–280.

Clarke, L. L., Roberts, M. C. and Argenzio, R. A.. (1990) Feeding and digestive problems in horses. *Veterinary Clinics of North America, Equine Practice*, **6**, 433–451.

Davis, R. E. (1976) A combined automated procedure for the determination of reducing sugars and nicotine alkaloids in tobacco products using a new reducing sugar method. *Tobacco Science,* **20**, 39–144.

Donoghue, S., Kronfeld, D. S., Berkowitz, S. J. and Copp, R. L.. (1981) Vitamin A nutrition of the equine: growth serum biochemistry and hematology. *Journal of Nutrition*, **111**, 365–374.

Englyst, H. N., Veenstra, J. and Hudson, G. J. (1996) Measurement of rapidly available glucose (RAG) in plant foods: a potenital *in vitro* predictor of the glycaemic response. *British Journal of Nutrition*, **75**, 327–337.

Englyst, H. N., Wiggins, H. S. and Cummings, J. H. (1982) Determination of the non-starch polysaccharides in plant foods by gas-liquid chromatography of constituent sugars as alditol acetates. *Analyst*, **107,** 307–318.

Food and Nutrition Board (1989) *Recommended Dietary Allowances*, National Academy Press, Washington, DC.

Graham, P. M., Ott, E. A. , Brendemuhl, J. H. and TenBroeck, S. H. (1994) The effect of supplemental lysine and threonine on growth and development of yearling horses. *Journal of Animal Science*, **72**, 380–386.

Graham-Thiers, P. M., Kronfeld, D. S., Kline, K. A. and Harris, P. A. (1998) Dietary protein influences acid-base responses to repeated sprints. *Proceedings 5th International Conference on Equine Exercise Physiology*, p 115. Japan.

Grant, H. W., Palmer, K. R., Kelly, R. W., Wilson, N. H. and Misiewicz, J. J. (1988) Dietary linoleic acid, gastric acid, and prostaglandin secretion. *Gastroenterology*, **94**, 955–959.

Gray, G. M. (1992) Starch digestion and absorption in nonruminants. *Journal of Nutrition*, **122**, 172–177.

Greiwe-Crandell, K. M., Kronfeld, D. S., Gay, L. S. and Skan, D. (1995a) Seasonal vitamin A depletion in grazing horses is assessed better by the relative dose response test than by serum retinol concentration. *Journal of Nutrition*, **125**, 2711–2716.

Greiwe-Crandell, K. M., Kronfeld, D. S., Gay, L. S., Sklan, D. and Harris, P. A. (1996) Vitamin A status of neonatal foals assessed by serum retinol concentration and a relative dose response test. *Pferdeheilkunde*, **12**, 181–183.

Greiwe-Crandell, K. M., Kronfeld, D. S., Gay, L. S., Sklan, D., Tiegs, W., and Harris, P. A. (1997) Vitamin A repletion in Thoroughbred mares with retinyl palimate or b-carotene. *Journal of Animal Science*, **75**, 2684–2690.

Greiwe-Crandell, K. M., Kronfeld, D. S. and Sklan, D. (1995b) Vitamin A and beta-carotene supplementation in horses on different forage systems. *Annales de Zootechie*, **44 (Suppl. 1)**, 308.

Greiwe-Crandell, K. M., Morrow, G. A. and Kronfeld, D. S. (1992) Phosphorus and selenium depletion in Thoroughbred mares and weanlings. *Pferdeheilkunde*,**1**, 96–98.

Hall, J. M. (1989) A review of total dietary fibre methodology. *Cereal Foods World* **34**, 526–528.

Henneberg, W. and Stohmann, F. (1864) Beitrage zur Begrundung einer rationellen Futterung der Wiederkauer II. *Braunschweig*, **29**, 48.

Hintz, H. F., Hintz, R. L., and Van Vleck, R. L. (1979) Growth rate of Thoroughbreds. Effect of age of dam, year and month of birth, and sex of foal. *Journal of Animal Science*, **48**, 480–487.

Hintz, H. F., Schryver, H. F., and Lowe, J. E. (1976) Delayed growth response and limb conformation in young horses. *Proceedings, Cornell Nutrition Conference*, Ithaca.

Hoffman, R. M., Kronfeld, D. S., Herbein, J. H., Swecker, W. S., Cooper, W. L. and Harris, P. A. (1998) Dietary carbohydrates and fat influence milk composition and fatty acid profile of mare's milk. *Journal of Nutrition*, **128**, In press.

Hoffman, R. M., Kronfeld, D. S., Holland, J. L. and Greiwe-Crandell, K. M. (1995) Preweaning diet and stall weaning method influences on stress response in foals. *Journal of Animal Science*, **73**, 2922–2930.

Hoffman, R. M., Kronfeld, D. S., Lawrence, L. A., Cooper, W. L., Dascanio, J. J. and Harris, P. A. (1996) Dietary starch and sugar versus fat and fibre: growth and development of foals. *Pferdeheilkunde*, **12**, 312–316.

Hoffman, R. M., Lawrence, L. A., Kronfeld, D. S., Dascanio, J. J., Cooper, W. L. and Harris, P. A. (1997) Dietary carbohydrates and fat influence radiographic bone mineral content of growing foals. *Proceedings of the 15th Equine Nutrition and Physiology Society*, pp 191–192. Ft. Worth, TX.

Holland, J. L. (1998) *Pasture intake, digestibility and fecal kinetics in grazing horses*. Ph.D. Dissertation, Virginia Tech, Blacksburg.

Holland, J. L., Kronfeld, D. S. and Meachem, T. N. (1996) Behavior of horses is affected by soy lecithin and corn oil in the diet. *Journal of Animal Science*, **74**, 1252–1255.

Holland, J. L., Kronfeld, D. S., Rich, G. A., Kline, K. A., Fontenot, J. P. and Meachem, T. N. (1998a) Acceptance of fat and lecithin containing diets by horses. *Applied Animal Behavioral Science*, **56**, 91–96.

Holland, J. L., Kronfeld, D. S., Sklan, D. J. and Harris, P. A. (1998b) Fecal kinetics estimated by a chromium marker in the horse. *Journal of Animal Science*, **76**, 1937–1944.

Hurtig, M. B., Green, S. L., Dobson, H. and Burton, J. (1990) Defective bone and cartilage in foals fed a low-copper diet. *Proceedings of the American Association of Equine Practitioners*, **36**, 637–643.

Kearney, J., Kennedy, N. P., Keeling, P. W. N., Keating, J. J., Grubb, L., Kennedy, M. and Gibney, M. J. (1989) Dietary intakes and adipose tissue levels of linoleic acid in peptic ulcer disease. *British Journal of Nutrition*, **62**, 699–706.

Knight, D. A., Gabel, A. A., Reed, S. M., Embertson, R. M., Tyznik, W. J. and Bramblage, L. R. (1985) Correlation of dietary mineral to incidence and severity of metabolic bone disease in Ohio and Kentucky. *Proceedings of the American Association of Equine Practitioners*, **31**, 445–461.

Knight, D. A., Weisbrade, S. E., Schmall, L. M. and Gabel, A. A. (1988) Copper supplementation and cartilage lesions in foals. *Proceedings of the American Association of Equine Practitioners*, **33**, 191.

Kronfeld, D. S. (1998a) Clinical assessment of nutritional status of the horse. In *Metabolic and Endocrine Problems of the Horse*, pp 185–217. Edited by T. D. G. Watson. Saunders, London.

Kronfeld, D. S. (1998b) A practical method for ration evaluation and diet formulation: an introduction to sensitivity analysis. In: *Advances in Equine Nutrition*, pp 77–88.Proceedings of the 1998 Equine Nutrition Conference for Feed Manufacturers, Kentucky Equine Research, Inc., Versailles.

Kronfeld, D. S. (1996) Dietary fat affects heat production and other variables of equine performance, especially under hot and humid conditions. *Equine Veterinary Journal (Supplement)*, **22**, 24–34.

Kronfeld, D. S., Cooper, W. L., Crandell, K. M., Gay, L. S., Hoffman, R. M., Holland, J. L., Wilson, J. A., Sklan, D. and Harris, P. A. (1996) Supplementation of pasture for growth. *Pferdeheilkunde*, **12**, 317–319.

Kronfeld, D. S., Ferrante, P. L. and Grandjean, D. (1994) Optimal nutrition for athletic performance, with emphasis on fat adaptation in dogs and horses. *Journal of Nutrition (Supplement)*, **124**, 2745S–2753S.

Maenpaa, P. H., Pirhonen, A. and Koskinene, E. (1988) Serum profiles of vitamins A, D and E in mares and foals during different seasons. *Journal of Animal Science*, **66**, 1418–1423.

Martin-Rosset, W. (1990) *L'alimentation des Chavaux*, INRA, Paris.

Meyer, H. (1992) *Pferdefutterung*, Verlag Paul Parcy, Hamburg.

Murray, M.L., Murray, C.M., Sweeney, H.J., Weld, J., Wingfield Digby, N.J. and Stoneham, S.J. (1990) Prevalence of gastric lesions in foals without signs of gastric disease: an endoscopic survey. *Equine Veterinary Journal*, **22**, 6-8.

NRC. (1989) *Nutrient Requirements of Horses (5th Edition)*, National Academy Press, Washington, DC.

Ordakowski, A. L. (1998) *Alkanes are potentially useful internal markers of digestive functions in the horse*, M.S. Thesis. Virginia Tech, Blacksburg.

Ott, E. A., Asquith, R. L. and Feaster J. P. (1981) Lysine supplementation of diets for yearling horses. *Journal of Animal Science*, **53**, 1496–1503.

Palmquist, D. L., Jenkins, T. C. and Joyner, A. E. (1986) Effect of dietary fat and calcium sources on insoluble soap formation in the rumen. *Journal of Dairy Science*, **69**, 1020–1025.

Pearce, S. G., Grace, N. D., Firth, E. C. and Fennesey, P. F. (1998) Effect of copper supplementation on the evidence of developmental orthopaedic disease in pasture-fed New Zealand Thoroughbreds. *Equine Veterinary Journal*, **30**, 211–219.

Potter, G. D., Hughes, S. L., Julen, T. R. and Swinney, D. L. (1992) A review of research on digestion and utilization of fat by the equine. *Pferdeheilkunde*, **1**, 119–123.

Smith, D. (1981) Removing and analyzing total nonstructural carbohydrates from plant tissue. In: *University of Wisconsin-Madison Research Report R2107*, Agricultural Bulletin Building, Madison.

Sprouse, R. F., Garner, H. E. and Green, E. M. (1987) Plasma endotoxin levels in horses subjected to carbohydrate induced laminitis. *Equine Veterinary Journal*, **9**, 25–28.

Sunvold, G. D. and R. C. Cochran. (1991) Technical note: evaluation of acid detergent lignin, alkaline peroxide lignin, acid insoluble ash, and indigestible acid detergent fibre as internal markers for prediction of alfalfa, bromegrass and prairie hay digestibility by beef steers. *Journal of Animal Science*, **69**, 4951–4955.

Thompson, K. (1995) Skeletal growth rates of weanling and yearling thoroughbred horses. *Journal of Animal Science*, **73**, 2513–2517.

Toribio, T., Bain, F. T., Mrad, D. R., Messer, N. T., Sellars, R. S. and Hinchcliff, K. (1998) Congenital defects in new-born foals of mares treated for equine protozoal myeloencephalitis during pregnancy. *Journal of the American Veterinary Medical Association*, **212**, 697–701.

Valberg, S. J. (1995) Exertional rhabdomyolysis and polysaccharide storage myopathy in Quarter Horses. *Proceedings of the American Association of Equine Practitioners*, **41**, 228–230.

Van Soest, P. J. (1963) Use of detergents in the analysis of fibrous feeds. *Association of Official Analytical Chemists*, **46**, 829.

Van Soest, P. J., Robertson, J. B. and Lewis, B. A. (1991) Methods for dietary fibre, neutral detergent fibre and nonstarch polysaccharides in relation to animal nutrition. *Journal of Dairy Science*, **74**, 3583–3597.

Vermorel, M. and Martin-Rosset, W. (1993) The French horse net energy system (UFC): concepts, scientific bases and structure. *Proceedings 44th Annual Meeting of the European Association for Animal Production*, pp 1–25. Aarhus, Denmark.

Watson, E. D., Cuddeford, D. and Burger, I. (1996) Failure of b-carotene absorption negates any potential effect on ovarian function in mares. *Equine Veterinary Journal*, **28**, 233–236.

Williams, M., Stowe, H. and Stickle, R. (1989) Relationships of nutrition and management factors to incidence of equine developmental orthopedic disease: a field survey. *Proceedings of the 7th International Conference on Production Disease in Farm Animals*, p 32. Ithaca, NY.

Wilson, J. A., Kronfeld, D. S., Cooper, W. L. and Sklan, D. (1997) Seasonal variation in the nutrient composition of northern Virginia forages. *Proceedings of the 15th Equine Nutrition and Physiology Society*, pp 336–331. Ft. Worth, TX.

APPENDIX – SENSITIVITY ANALYSIS IN RATION FORMULATION

GIVEN: Forage Cu concentration average of 9 mg/kg, with a range of 6–12 mg/kg Cu.

The concentrate:forage ratio will be between 25:75 and 50:50, with an average of 33:67.

TARGET: Total intake of 20 mg/kg Cu, with lower and upper intakes between 15 and 30 mg/kg Cu.

TASK: Determine an appropriate Cu concentration in a supplement to complement the forage.

1. Calculate using the average values and a modified Pearson's square.
   9 mg/kg Cu in Forage.
   Concentrate:Forage ratio of 33:67.
   20 mg/kg Cu TARGET.

Forage 9 z 0.67 forage

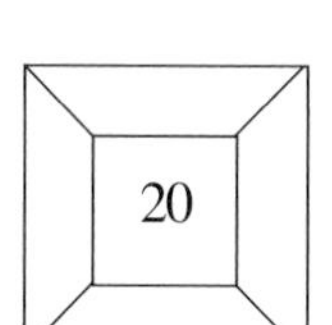

Concentrate x y 0.33 concentrate

a) $y = 20 - 9 = 11$:

Forage 9 z 0.67 forage

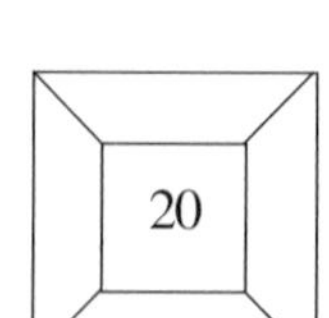

Concentrate x **11** 0.33 concentrate

b) $\frac{z}{67} = \frac{11}{33}$ $z = 22$

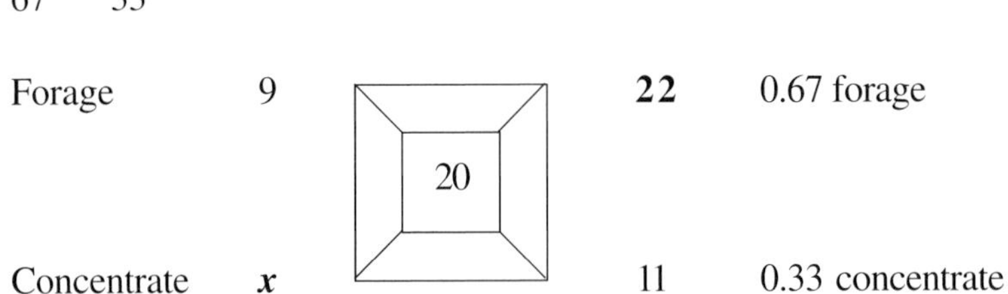

c) $x = 22+20 = 42$, appropriate concentration of Cu in the concentrate.

Forage 9 [20] 22 0.67 forage

Concentrate ***42*** 11 0.33 concentrate

2. Complete sensitivity analysis for the ranges of intakes, given 42 mg/kg Cu in the concentrate, forage ranges of 6–12 mg/kg Cu, and concentrate:forage ratios of 25:75, 33:67, and 50:50.

	Forage Cu, mg/kg	*Concentrate:Forage Ratios* 25:75	*33:67*	*50:50*
lower	6	15	18	24
middle	9	17	20	26
upper	12	20	22	27

(e.g. Forage Cu = 6; Concentrate:Forage = 25:75, then [42*0.25 + 6*0.75] = 15)

In this case, all total intake concentrations of Cu fit within the target range (15–30 mg/kg); therefore, no further iterations are needed for the concentrate's Cu content.

3

THE INFLUENCE OF NUTRITION ON FERTILITY IN DAIRY COWS

P.C. GARNSWORTHY and R. WEBB
University of Nottingham, Division of Agriculture and Horticulture, School of Biological Sciences, Sutton Bonington Campus, Loughborough, Leics LE12 5RD, UK

Introduction

Fertility is one of the major factors influencing the profitability of dairy herds. A cow will not produce milk until after she has produced her first calf, but continued regular breeding is essential for retention of a cow in the herd. Early culling due to reproductive failure represents a waste of resources and means that the cost of rearing the animal is spread over fewer productive lactations, thereby reducing profit per litre of milk. The optimum interval between one calving and the next is normally considered to be 365 days. Interest has recently been shown in extending the length of lactation, but this should be a conscious management decision and not a result of poor reproductive performance.

The objective of this paper is to consider the influence of nutrition on fertility. Many factors influence fertility, only one of which is nutrition; it is important that putative nutritional influences are seen in the context of whole-herd management. Also, nutrition is a very broad subject area, covering not only the supply of energy and nutrients to the cow, but also the animal's requirements for total and specific nutrients. This paper will review influences on fertility at the whole-animal level, consider the implications for nutrition, and then discuss possible mechanisms for the action of nutrition on reproductive processes.

Influences on fertility

Fertility is the reproductive performance of an individual cow or herd, measured as the number of calves produced per unit time. This presents problems for the researcher, since calving is a discrete event that only occurs once per year. Therefore, experimental designs, such as crossover or covariance designs, cannot be used and large numbers of replicates are needed. The production of a calf is a culmination of events involving follicle development, growth of a good-quality oocyte, ovulation, fertilisation, implantation and embryo survival. Failure at any stage in this sequence causes infertility, and failure might be caused by a number of animal, management and nutritional factors (Figure 3.1), many of which are inter-related.

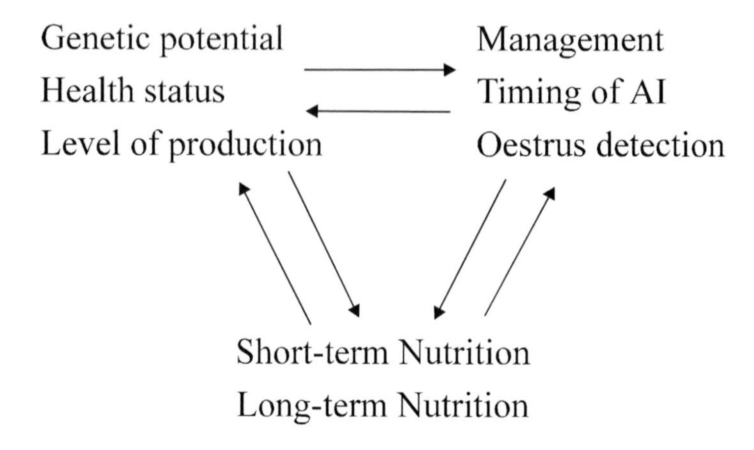

Figure 3.1 Interactions between animal, management and nutritional factors affecting fertility in dairy cows

Over the past 25 years, there have been dramatic increases in milk production in UK dairy herds, with many cows now producing over 9000 litres of milk per lactation. However, this increase in milk yield has been accompanied by a serious decline in fertility of dairy cows. Lamming, Darwash, Wathes and Ball (1998) estimated that the average conception rate to first service has declined at approximately 1% every three years.

Non-nutritional influences

Unlike other farm species, dairy cows do not exhibit strong lactational anoestrus, nor are they strongly seasonal breeders, although there is some evidence that conception rates are lower in summer (Bourchier, Garnsworthy, Hutchinson and Benson, 1987). Increased milk yields have mainly stemmed from genetic changes, with the importation of Holstein genes from North America and widespread use of bulls through artificial insemination. It is possible that single-trait selection has reduced fertility amongst dairy cows, but the move to larger herd sizes, with less attention being paid to oestrous detection, may also have contributed. Reduced fertility has been accompanied by an increased incidence of metabolic diseases and lameness.

There is no doubt that nutrition plays a key role in enabling the modern dairy cow to reach high levels of milk production. Expression of improved genetic potential has been facilitated by changes in feeding management, with increased use of complete diets fed *ad libitum*, by increased use of maize silage and better conservation of grass silage. It can probably also be claimed that our knowledge of ration formulation and the nutritive value of dietary ingredients have also improved following the publication of ARC (1980) and its successors. It could therefore be argued that the nutrition of dairy cows is better today than it was 25 years ago. However, as Agnew, Yan and Gordon. (1998) stated, the efficiency of conversion of available energy into milk energy has not increased, so high-merit animals sustain high levels of milk production partly by increased mobilisation of body reserves.

Since so many factors may have an influence on fertility, the decline in conception rates cannot be attributed solely to nutrition. It must be remembered when studying associations between two parameters, such as milk yield and conception rate, that all you can legitimately deduce is the strength of the association, not cause and effect. Both parameters may change concurrently in response to another factor. Having said that, a strong relationship shown in a survey provides a good starting point for experimental investigation.

Body reserves, food intake and milk yield

Gross measures of reproductive performance, such as conception rate to a particular service, days from calving to first oestrus and days to conception, have all been adversely associated with high milk yields, low intakes of energy and mobilisation of body reserves, which combine in the term negative energy balance. Negative energy balance is usually quoted as the main cause of poor conception rates in dairy cows.

Body reserves, milk yield and food intake are closely interrelated. If energy intake is insufficient to meets the demands of milk secretion, as it usually is in early lactation, body reserves are mobilised. If energy intake exceeds requirements, excess energy is deposited as body fat, but over the long term cows adjust their energy intake in an attempt to keep levels of body energy reserves constant (Garnsworthy, 1988). In beef cattle, calving cows in a higher body condition can improve reproductive performance (Wright, Rhind, Russel, Whyte, McBean and McMillen, 1987), but the situation is different in dairy cows. Attempts to overcome the effects of negative energy balance by increasing body condition at calving may exacerbate the problem, since there is a strong relationship between body condition at calving and the amount of condition lost in early lactation (Figure 3.2). Therefore, negative energy balance is almost inevitable in early lactation, except in cows calving with very low condition scores.

Change in body condition score is a very good indicator of energy status for a dairy cow. The University of Nottingham and ADAS conducted a survey of nearly 2000

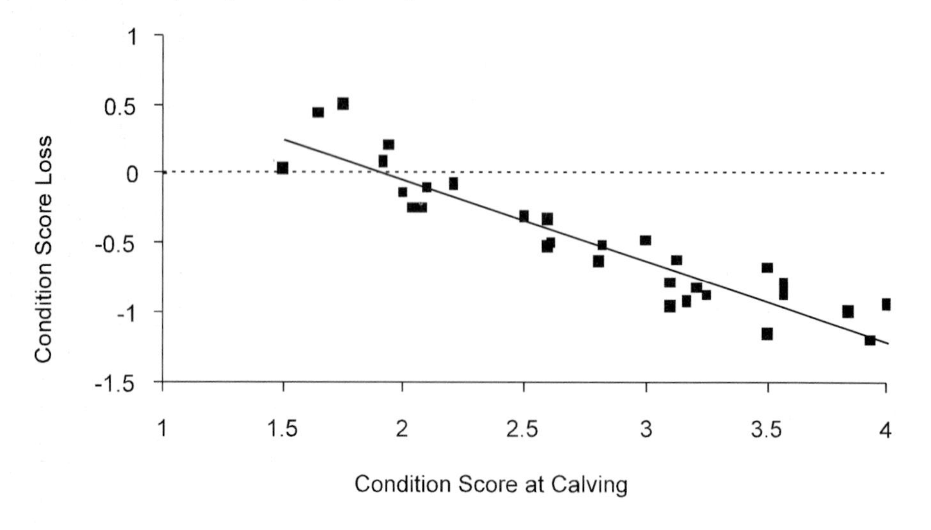

Figure 3.2 Effect of condition score at calving on loss of condition during early lactation. Each data point is a treatment mean from a published experiment (data obtained from Bourchier *et al.* (1987); Garnsworthy and Huggett (1992); Garnsworthy and Jones (1987, 1993); Garnsworthy and Topps (1982); Jones and Garnsworthy (1988, 1989); Land and Leaver (1980,1981); Treacher, Reid and Roberts (1986))

dairy cows in high-yielding herds (average annual milk yield 7210 kg/cow); preliminary findings were reported by Bourchier *et al.* (1987). Neither condition score at calving, or at service, affected conception rate to first service, except when the condition score was below 1.0 (Table 3.1). Milk yield in early lactation did not significantly affect conception rate (Table 3.2), but there was a positive relationship (P<0.001) between milk yield and the length of the period from calving to first service (Days to first service = 59 + 0.0038MY, where MY is total milk yield over the first 12 weeks of lactation).

Table 3.1 Effecct of condition score at calving or at service on conception rate to first service in 2000 high-yielding dairy cows (Bourchier *et al.,* 1987).

			Condition Score at Calving			
	<1.5	*1.5*	*2*	*2.5*	*3*	*3.5+*
Conception rate	0.47	0.56	0.57	0.57	0.56	0.56
			Condition Score at Service			
	<1.5	*1.5*	*2*	*2.5*	*3+*	
Conception rate	0.47	0.56	0.54	0.51	0.58	

Further analysis of data from this survey (Garnsworthy and Haresign, 1989) suggested that dairy cows of high genetic merit show poorest conception rates when they lose a lot of body condition between calving and service (Table 3.3). The lowest conception

Table 3.2 Effecct of milk yield during the first twelve weeks of lactation on conception rate to first service in 2000 high-yielding dairy cows (Bourchier *et al.*, 1987).

	Milk Yield (l/day)		
	<30	*30-36*	*>36*
Conception rate	0.47	0.56	0.56

rates (<0.30) were found in cows that lost more than 1.5 condition-score units between calving and service but also had relatively low milk yields (<30 litres/day). This suggests a general upset in the metabolism of a cow that affects production and fertility concurrently, but possibly independently. Further doubt for a direct effect of energy balance on fertility stems from the fact that until the early 1980s dairy cows were managed under systems that encouraged a much greater degree of mobilisation of body reserves, yet conception rates were far higher than today.

Table 3.3 Effecct of condition score loss from calving to first service on conception rate to first service in 2000 high-yielding dairy cows (Bourchier *et al.*, 1987).

	Condition score loss			
	>1.5	*1*	*0.5*	*>0*
Conception rate	47	56	53	63

Requirements for specific nutrients

In addition to the general effects of energy and protein balance on fertility, dairy cows have specific requirements for certain nutrients, such as vitamins and minerals. Vitamins A, D and E have been shown to have direct effects on fertility, but others, such as thiamine, niacin, vitamin B12 and choline may have act indirectly through their effects on metabolism (Bieber-Wlaschny, 1988). Many minerals can also have direct and indirect effects on fertility, often in association with vitamins (McDonald, Edwards, Greenhalgh and Morgan, 1995).

Influence of nutrition on metabolic hormones

High milk yields and/or undernutrition cause an increase in circulating concentrations of growth hormone (GH), accompanied by decreases in circulating levels of insulin and liver GH receptors. These factors reduce the production of insulin-like growth factor-I (IGF-I) and IGF binding protein in the liver, so the net effect is that plasma concentrations of IGF-I falls, despite elevated levels of GH (Pell and Bates, 1990; Thissen, Ketelslegers and Underwood, 1994; Armstrong, Cohick, Harvey, Heimer and Campbell, 1993; Spicer, Crowe, Prendiville, Goulding and Enright, 1992). In contrast, when insulin concentration is high during overfeeding, IGF-I production increases. This is despite a decline in GH secretion and is probably due to enhanced binding of GH to its receptor, caused by insulin.

The IGF system has many components. There are two peptides, IGF-I and IGF-II, that both act on the type 1 IGF receptor (IGF-1R). The type 2 IGF receptor is thought to act as a clearance molecule for IGF-II. There is also a family of six IGF binding proteins (IGFBPs 1-6). These can either inhibit IGF activity (by reducing its bioavailability to bind to the type 1 receptor) or enhance it (by targeting IGFs to a particular tissue and acting as a local reservoir within it) (Jones and Clemmons, 1995). All of these factors may act as signals of metabolic status to the reproductive system, as discussed below. As well as being produced by the liver, these members of the IGF system are also expressed in the ovaries and the reproductive tract (Armstrong and Webb, 1997; Webb and Armstrong, 1998).

Mechanistic influences on fertility

At the mechanistic level, several reproductive stages have been identified where high milk yields, inadequate nutrition or excessive mobilisation of body reserves may have detrimental effects. These include the hypothalamus/pituitary gland, follicular growth, corpus luteum function, oocyte quality, uterine environment and embryo survival.

EFFECT OF DIET ON GONADOTROPHINS

As well as affecting metabolic hormones, diet can also influence the production of luteinising hormone (LH) by the pituitary gland, although follicle-stimulating hormone (FSH) does not seem to be affected by diet (Rhodes, Fitzpatrick, Entwistle and De'ath, 1995). Significant changes in body weight can alter the pattern of LH secretion (Gonzales-Padilla, Wiltban and Niswender, 1975; Yelich, Wettemann, Marston and Spicer, 1996). Canfield and Butler (1990) observed that baseline and mean concentration of LH and LH pulse frequency were higher in cows after the energy balance nadir compared with before the energy balance nadir. Extremes of nutrition appear to affect pituitary gonadotrophin secretion - although in most cases the pituitary gland is capable of releasing adequate amounts and patterns of gonadotrophins, the response to gonadotrophin-releasing hormone (GnRH) is reduced once a compromising body condition score is reached (Roberson, Stumpf, Wolfe, Cupp, Kojima, Werth, Kittok and Kinder, 1992). In beef cows body condition at parturition was found to be directly related to LH pulse frequency, particularly in thin cows (Bishop, Wettemann and Spicer, 1994; Wright, Rhind, Smith and Whyte, 1992).

EFFECT OF DIET ON OVARIAN FUNCTION

Follicular development in cattle is primarily controlled by the co-ordinated action of hypothalamic GnRH and pituitary gonadotrophins (Campbell, Scaramuzzi and Webb, 1995; Webb, Gong, Law and Rusbridge, 1992; Webb, Gong and Bramley, 1994; Gong and Webb, 1996). These authors have demonstrated that without pulsatile release of LH, dominant follicle development could not proceed past 7-9 mm in diameter (Gong, Campbell, Bramley, Gutierrez, Peters and Webb, 1996). When FSH concentrations were subsequently suppressed, after GnRH-agonist treatment, follicle growth stopped at 4 mm in diameter. Therefore, since follicle development depends on sustained support by gonadotrophins, nutritionally induced changes in gonadotrophin secretion will affect follicular development. Folliculogenesis in cattle is characterised by a continuous turnover of follicular waves (see Webb *et al.* 1992; Gong and Webb, 1996). Each wave of follicle development involves the simultaneous growth of a cohort of 5-7 antral follicles from the growing pool to >5 mm in diameter and then the selection of one of these follicles to grow rapidly while other cohort follicles regress. The selected follicle continues to grow to become a dominant follicle with a diameter of about 15 mm and remains at maximum size for 2-3 days before regression, followed by a new wave of follicle development. It appears that approximately 80% of cattle have three waves of follicular growth and development during each oestrous cycle and it is the dominant follicle produced during the third wave that usually ovulates. However, follicular waves have also been shown to occur during the post-partum anoestrus period (Boland, Goulding and Roche, 1991; Beam and Butler, 1997). Following ovulation and fertilisation optimum progesterone, from the newly formed corpus luteum is essential in order to provide optimal conditions in the oviduct and uterus for embryonic growth and development.

Waves of follicular development in cows are normally initiated soon after calving (Boland *et al*, 1991; Beam and Butler, 1997). During this period, the degree of negative energy balance can affect follicular development. The lower energy balance of lactating cows is accompanied by fewer follicles of <15 mm in diameter (de la Sota, Lucy, Staples and Thatcher, 1993) but higher numbers of small (3-5 mm) and medium-sized (6-9 mm) follicles (Lucy, Staples, Michel and Thatcher, 1991a, 1991b). Furthermore, when cows are fed on low energy diets, the growth rate of preovulatory follicles is slower than that of follicles from cows fed on high energy diets (Murphy, Enright, Crowe, McConnell, Spicer, Boland and Roche, 1991; Lucy, Beck, Staples, Head, de la Sota and Thatcher, 1992a).

In beef cattle, the preovulatory follicle has a smaller diameter when animals are losing weight (Murphy *et al.,* 1991; Grimard, Humblot, Ponter, Mialot, Sauvant and Thibier, 1995; Rhodes *et al.,* 1995). The reduction in follicle diameter was positively correlated with weight loss until the animals ceased ovulating (Rhodes *et al.,* 1995). In contrast, an increase in diameter of the largest follicle was seen when heifers were gaining weight (Rhodes *et al.,* 1995; Murphy *et al.,* 1991; Spicer, Enright, Murphy and Roche, 1991). These follicles were also dominant, since the number of subordinate follicles was reduced as the first ovulation postpartum approached (Gutierrez, Galina, Zarco and Rubio, 1994).

The ovarian IGF system is a key component of the regulatory system that controls folliculogenesis within the ovary (Armstrong and Webb, 1997; Webb and Armstrong, 1998). It appears that dietary energy and protein can also regulate the expression of mRNA encoding components of the ovarian IGF system. For example, diets shown previously to increase the number of small ovarian follicles in cattle (Gutierrez, Oldham, Bramley, Gong, Campbell and Webb, 1997a), decreased the expression of mRNA encoding IGFBP-2 in similar size small (<4 mm) healthy bovine follicles (Armstrong, Baxter, Sinclair, Robinson, McEvoy and Webb, unpublished observations). This is similar to the case in the development of the dominant follicle (>8 mm) in cattle, where granulosa cell IGFBP-2 mRNA expression is reduced (Armstrong, Baxter, Gutierrez, Hogg, Glazyria, Campbell, Bramley and Webb, 1998). Reduced production of IGFBP-2 probably results in upregulating the bioavailability of IGFs, leading to enhanced gonadotrophic stimulation of ovarian follicular cells (Gutierrez, Campbell and Webb, 1997b).

Factors such as GH, insulin and IGF can have a pronounced influence on ovarian follicle development in cattle (Armstrong and Webb, 1997; Webb and Armstrong, 1998). Therefore, it is likely that changes in these factors during early lactation will alter the pattern of ovarian follicle growth and development, thereby influencing reproductive function. To test this working hypothesis, possible links between metabolic changes and patterns of ovarian follicular growth and development during the early postpartum period, were studied in dairy cows on intensive and extensive systems (Gong, Logue, Crawshaw and Webb, unpublished data). No obvious differences in the pattern of ovarian follicular development between either the high and low input systems or autumn and spring calving seasons were observed. These results suggest that reproductive function need not be compromised by high milk output, providing nutritional inputs are adequate. This emphasises that it is not the absolute input or output, but the balance between input and output, which is really important.

Treatment of lactating cows with bovine somatotrophin (BST), which increases milk production and reduces energy balance, stimulated an increase in the number of follicles reaching the larger size categories and decreased the number of small (<5 mm) follicles (de la Sota *et al.,* 1993; Lucy *et al.,* 1992a). In heifers, administration of BST increased the number of small follicles (<5 mm), but did not alter the number of large follicles (>5 mm) or the dynamics of follicle turnover (Gong, Bramley and Webb, 1991; 1993). BST had no effect on circulating concentrations of FSH and LH or gonadotrophin binding to theca and granulosa cells (Gong *et al.*, 1991). The action of BST therefore appears to be via circulating insulin and IGF-I concentrations (Gong *et al.* 1993; de la Sota *et al.* 1993; Lucy, Collier, Kitchell, Dibner, Hauser and Krivi, 1993). It is likely that insulin and/or IGF-I alter the response of follicular cells to gonadotrophins, as shown by recent *in vitro* studies culturing bovine follicular cells (Gutierrez *et al.*, 1997b) and *in vivo* dose-response studies (Gong, Baxter, Bramley and Webb, 1997). This also indicates that a change in circulating gonadotrophin concentrations is not essential.

Immunisation of prepubertal heifers against GH releasing hormone reduces the circulating concentrations of GH. The proportion of heifers developing follicles above 7 mm in diameter was only 0.22, compared with 0.77 of control heifers (Cohick, Armstrong, Withacre, Lucy, Harvey and Campbell, 1996). Recently, increased dietary intake has been associated with increased small follicle recruitment during the first follicular wave of the oestrous cycle in Holstein-Friesian heifers (Gutierrez *et al.*, 1997a). The number of small follicles returned to normal as soon as dietary treatments were terminated. The number of medium-size follicles increased in all groups, at the time when numbers of small follicles were decreasing. Taken together, these results indicate that nutrition appears to affect mainly the recruitment of small follicles.

Metabolic hormones and reproductive function were recently measured in dairy cows of High and Low genetic merit during early lactation (Gutierrez, Gong and Webb, unpublished data). Resumption of normal oestrous cycles postpartum occurred approximately 8 days later in the High-line cows, and this was associated with lower plasma insulin concentrations. A further study was designed to investigate if feeding diets to increase circulating insulin concentrations during the early postpartum period can overcome the delay in the first ovulation postpartum in animals selected for increased milk yield. Two isoenergetic diets were formulated to either stimulate or depress plasma insulin concentrations, and these were fed to cows of High or Low genetic merit. Again, the initiation of the first ovulation and the resumption of normal oestrous cycles postpartum were delayed in the cows of high genetic merit. This was associated with lower circulating insulin concentrations, but did not involve an alteration in basal plasma gonadotrophin concentrations and patterns of ovarian follicular development during the early postpartum period. Feeding the diet designed to increase circulating insulin concentrations advanced the initiation of the first ovulation postpartum so that fewer cows failed to ovulate within the first 50 days of lactation (Table 3.4; Gong, Garnsworthy and Webb, unpublished data).

Increasing the fat content of the diet increases both the number and size of follicles present in the ovary and can also shorten the interval from parturition to first ovulation (Lucy *et al.*, 1991b; Lucy, Savio, Badinga, de la Sota and Thatcher, 1992b; Hightshoe,

Table 3.4 Number of animals that did not ovulate within 50 days of calving in High and Low genetic merit dairy cows fed on diets that induced high or low concentartions of plasma insulin (Gong, Garnsworthy and Webb; unpublished data).

		Diet	
		High Insulin	*Low Insulin*
Genetic merit	High	2/10	5/10
	Low	0/10	4/10

Cochran, Corah, Kiracofe, Harmon and Perry, 1991; Ryan, Spoon and Williams, 1992; Lammoglia, Willard, Hallford and Randel, 1997; Beam and Butler, 1997; Thomas and Williams, 1996). Raised blood cholesterol could increase precursor availability for follicular steroid synthesis. Since oestradiol-17b induces proliferation of granulosa cells this might, in turn, increase follicular progesterone (Talavera, Park and Williams, 1985; Carroll, Jerred, Grummer, Combs, Pierson and Hauser, 1990; Hawkins, Niswender, Oss, Moeller, Odde, Sawyer and Niswender, 1995; Burke, Carroll, Rowe, Thatcher and Stormshak, 1996). Another possibility is that clearance rates of progesterone from plasma may be reduced (Hawkins *et al*., 1995). This elevation in progesterone could have a beneficial effect on fertility, as suboptimal progesterone concentrations are associated with high return rates in cows.

EFFECTS OF DIET ON OOCYTE QUALITY AND EMBRYO SURVIVAL

In sheep, it has been shown that energy intake by oocyte donors, but not dietary protein concentration, influences blastocyst production from oocytes cultured *in vitro* (McEvoy, Robinson, Aitken, Findlay and Robertson., 1997a). It has also been found that dietary energy affects blastocyst production from oocytes collected from 2-4 mm bovine follicles (McEvoy, Sinclair, Staines, Robinson, Armstrong and Webb, 1997b). This fits with the observations discussed above that nutrition affects the growth of small follicles and IGF expression.

Dietary PUFA composition may alter progesterone output via changes in the production of eicosanoids within the corpus luteum. Altering the balance of endogenous prostaglandin production within the corpus luteum could potentially alter both the production of progesterone and the overall length of the luteal phase. Maternal progesterone concentrations during the early luteal phase appear to affect development of the embryo and its ability to inhibit luteolysis through secretion of antiluteolytic trophoblast interferon (IFNL) (Mann, Lamming and Fisher, 1998).

Diets high in rumen degradable protein lead to high concentrations of rumen ammonia, which may alter uterine pH and endometrial Na and K fluxes, with reduced embryo survival. The detrimental effect of this increase in ammonia on protein synthesis in day 4 ovine embryos has been clearly shown (Robinson and McEvoy, 1996). In cattle, high ammonia levels have been associated with reduced oocyte cleavage rates (Sinclair, Kuran, Sorohan, Staines and McEvoy, 1999). Slowly degraded protein leads to lower ammonia concentrations, which may explain why fishmeal sometimes improves fertility (Armstrong, Goodall, Gordon, Rice and McGaughey, 1990). Fishmeal also supplies essential fatty acids, which may enhance embryo survival through their anti-luteolytic properties (Coelho, Ambrose, Binelli, Burke, Staples, Thatcher and Thatcher, 1997).

EMBRYO DEVELOPMENT

Maintenance of pregnancy is dependent on the embryo prolonging the lifespan of the corpus luteum by preventing luteolysis. This is achieved by secretion of IFNt by the conceptus (Roberts, Cross and Leaman, 1992) which in turn inhibits the development of oxytocin receptors (OTR) in the endometrium (Wathes and Lamming, 1995). The development of OTR is essential for pulsatile secretion of $PGF_2\mu$ to occur. Several studies in both cattle and sheep have shown that lipid infusions can alter the ability of the uterus to respond to a challenge of oxytocin in the late luteal phase in terms of $PGF_2\mu$ release and may affect cycle length (Thatcher, Staples, Danet-Desnoyers, Oldick and Schmitt, 1994; Burke *et al.*, 1996). An endometrial cyclo-oxygenase inhibitor isolated from the bovine endometrium was also identified as linoleic acid (Danet-Desnoyers, Johnson, O'Keefe and Thatcher, 1993; Thatcher, Meyer and Danet-Desnoyers, 1995) and was present at a higher concentration in pregnant compared with non-pregnant uteri. It is therefore possible that the ability of a particular cow to maintain a pregnancy by inhibiting luteolysis may be influenced by her dietary PUFA intake.

There is considerable evidence that the IGF system is important in embryo development (Wathes, Reynolds, Robinson and Stevenson, 1998). IGFs could potentially influence the embryo indirectly by altering the metabolic and secretory activity of the reproductive tract. It has been shown that many components of the IGF system are expressed in the bovine and ovine reproductive tracts (Geisert, Lee, Simmen, Zavy, Fliss, Bazer and Simmen, 1991; Stevenson, Gilmour and Wathes, 1994; Reynolds, Stevenson and Wathes, 1997; Kirby, Thatcher, Collier, Simmen and Lucy, 1996, Robinson, Mann, Lamming and Wathes, 1997). As discussed above, the main regulator of the maternal IGF system is nutrition (McGuire, Vicini, Bauman and Veenhuizen, 1992; Thissen *et al.*, 1994).

Conclusions

In conclusion, further knowledge is required before the complete relationships between milk yield, nutrition and fertility can be explained. Poor fertility is not an inevitable consequence of high genetic merit, but may be due to a combination of factors that include genetic susceptibility, management, disease, milk yield, energy balance, body reserves and specific nutritional circumstances.

The absolute requirements of an oocyte or developing embryo for energy and protein are minuscule, compared with the cow's requirements for milk production. However, nutrition can have major effects on metabolic hormones, which can cascade through the IGF system to have repercussions at the tissue or cellular level.

Because so many factors influence reproductive function in dairy cows, it is very difficult to predict whether an individual cow will be fertile under a given set of circumstances. However, it is possible to identify certain risk factors that may predispose cows to infertility (Figure 3.3). These include excesses or deficiencies of energy and protein, severe negative energy balance and excessive weight loss. The susceptibility of a cow to these risks is influenced by her genetic merit, general health and management conditions.

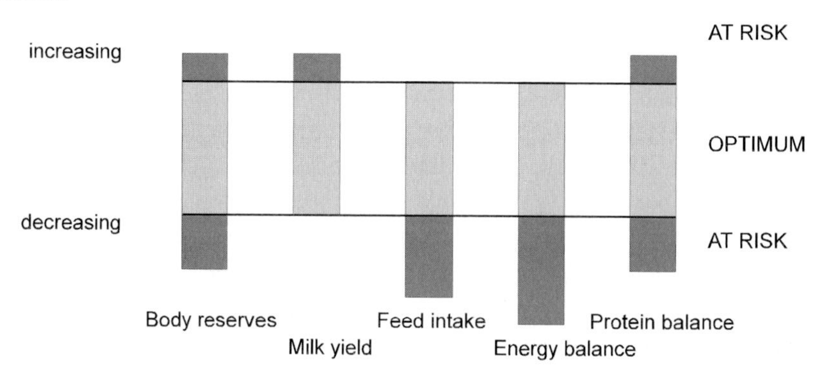

Figure 3.3 Possible risk factors causing infertility in dairy cows. For optimum fertility, all production measures should be in the central region; high or low values for these measures increase the risk of infertility

References

Agricultural Research Council (1980) *The Nutrient Requirements of Ruminant Livestock*, Commonwealth Agricultural Bureaux, Slough.

Agnew, R.E., Yan, T. and Gordon, F.J. (1998) Nutrition of the high genetic merit dairy cow – energy metabolism studies. In *Recent Advances in Animal Nutrition – 1998* (eds P.C. Garnsworthy and J. Wiseman), pp 181-208, Nottingham University Press, Nottingham.

Armstrong, D. G., Baxter, G., Gutierrez, C. G., Hogg, C. O., Glazyria, A. L., Campbell, B. K., Bramley, T. A. and Webb, R. 1998. Insulin-like growth factor binding protein-2 and –4 messenger ribonucleic acid expression in bovine ovarian follicles:

effects of gonadotrophins and developmental states. *Endocrinology* **139:** 2146-2154.

Armstrong, J. D., Cohick, W. S., Harvey, R. W., Heimer, E. P. and Campbell, R. M. 1993. Effect of feed restriction on serum somatotrophin, insulin-like growth factor-1 (IGF-1) and binding proteins in cyclic heifers actively immunised against growth hormone releasing factor. *Domestic Animal Endocrinology* **10:** 315-324.

Armstrong, J. D., Goodall, E. A., Gordon, F. J., Rice, D. A. and McGaughey, W. J. 1990. The effect of levels of concentrate offered and inclusion of maize gluten or fishmeal in the concentrate on reproductive performance and blood parameters of dairy cows. *Animal Production* **50:** 1-10.

Armstrong, D. G. and Webb, R. 1997. Ovarian follicular dominance: novel mechanisms and protein factors. Reviews of Reproduction **3:** 134-146.

Beam, S. W. and Butler, W. R. 1997. Energy balance and ovarian follicle development prior to the first ovulation postpartum in dairy cows receiving three levels of dietary fat. *Biology of Reproduction* **56:** 133-142.

Bieber-Wlashny, M. (1988) Vitamin requirements of the dairy cow. In *Nutrition and Lactation in the Dairy Cow* (ed P.C. Garnsworthy), pp 135-156, Butterworths, London.

Bishop, D. K., Wettemann, R. P. and Spicer, L. J. 1994. Body energy reserves influence the onset of luteal activity after early weaning of beef cows. *Journal of Animal Science.***72:** 2703-2708.

Boland, M. P., Goulding, D. and Roche, J. F. 1991. The use of ultrasound to monitor ovarian function in farm animals. *Ag. Biotechnology News Information* **2:** 841-844.

Bourchier, C. P., Garnsworthy, P. C., Hutchinson, J. M. and Benson, T. A. 1987. The relationship between milk yield, body condition and reproductive performance in high-yielding dairy cows. *Animal Production* **44:** 460 (abstract).

Burke, J. M., Carroll, D. J., Rowe, K.E., Thatcher, W.W. and Stormshak, F. 1996. Intravascular infusion of lipid into ewes stimulates production of progesterone and prostaglandin. *Biology of Reproduction* **55:** 169-175

Campbell, B.K., Scaramuzzi, R. J. and Webb, R. 1995. Control of antral follicle development and selection in sheep and cattle. *Journal of Reproduction and Fertility Supplement* **49:** 335-350.

Canfield, R. W. and Butler, W. R. 1990. Energy balance and pulsatile LH secretion in early postpartum dairy cattle. *Domestic Animal Endocrinology* **7:** 323-330.

Carroll, D. J., Jerred, M. J., Grummer, R. R., Combs, D. K., Pierson, R. A. and Hauser, E. R. 1990. Effects of fat supplementation and immature alfalfa to concentrate ratio on plasma progesterone, energy balance and reproductive traits of dairy cattle. *Journal of Dairy Science* **73:** 2855-2863.

Coelho, S., Ambrose, J. D., Binelli, M., Burke, J., Staples, C. R., Thatcher, M. J. and Thatcher, W. W. 1997. Menhaden fish meal attenuates estradiol- and oxytocin-induced uterine secretion of $PGF_2\mu$ in lactating dairy cattle. *Theriogenology* **47:** 143.

Cohick, W. S., Armstrong, J. D., Withacre, M. D., Lucy, M. C., Harvey, R. W. and Campbell, R. M. 1996. Ovarian expression of insulin-like growth factor-I (IGF-I), IGF binding proteins, and growth hormone (GH) receptor in heifers actively immunized against GH-releasing factor. *Endocrinology* **137:** 1670-1677.

Danet-Desnoyers, G., Johnson, J. W., O'Keefe, S. F. and Thatcher, W. W. 1993. Characterization of a bovine endometrial prostaglandin synthesis inhibitor (EPSI). *Biology of Reproduction* **48:** Supplement 115.

de la Sota, R. L., Lucy, M. C., Staples, C. R. and Thatcher, W. W. 1993. Effects of recombinant bovine somatotropin (Sometribove) on ovarian function in lactating and nonlactating dairy cows. *Journal of Dairy Science* **76:** 1002-1013.

Garnsworthy, P.C. (1988) The effect of energy reserves at calving on performance of dairy cows. In *Nutrition and Lactation in the Dairy Cow* (ed P.C. Garnsworthy), pp 157-170, Butterworths, London.

Garnsworthy, P. C. and Haresign, W. 1989. Fertility and nutrition. In *Dairy Cow Nutrition – The Veterinary Angles.* (Ed. A. T. Chamberlain) pp23-34. The University of Reading, Department of Agriculture.

Garnsworthy, P. C. and Huggett, C. D. 1992, The influence of the fat concentration of the diet on the response by dairy cows to body condition at calving. *Animal Production* **54**: 7-13.

Garnsworthy, P. C. and Jones, G. P. 1987. The influence of body condition at calving and dietary protein supply on voluntary food intake and performance in dairy cows. *Animal Production* **44:** 347-353.

Garnsworthy, P. C. and Jones, G. P. 1993. The effects of dietary fibre and starch concentrations on the response by dairy cows to body condition at calving. *Animal Production* **57:** 15-21.

Garnsworthy, P. C. and Topps, J. H. 1982. The effect of body condition of dairy cows at calving on their food intake and performance when given complete diets. *Animal Production* 35: 113-119.

Geisert, R. D., Lee, C., Simmen, F. A., Zavy, M. T., Fliss, A. E., Bazer, F. W. and Simmen, R. C. M. 1991. Expression of messenger RNAs encoding insulin-like growth factor-I, -II and insulin-like growth factor binding proteins-2 in bovine endometrium during the estrous cycle and early pregnancy. *Biology of Reproduction* **45:** 975-983.

Gong, J. G., Baxter, G., Bramley, T. A. and Webb, R. 1997. Enhancement of ovarian follicle development in heifers by treatment with recombinant bovine somatotrophin: a dose-response study. *Journal of Reproduction and Fertility* **110:** 91-97.

Gong, J. G., Bramley, T. A. and Webb, R. 1991. The effect of recombinant bovine somatotropin on ovarian function in heifers: Follicular populations and peripheral hormones. *Biology of Reproduction* **45:** 941-949.

Gong, J. G., Bramley, T. A. and Webb, R. 1993. The effect of recombinant bovine somatotrophin on ovarian follicular growth and development in heifers. *Journal of Reproduction and Fertility* **97:** 247-254.

Gong, J. G. and Webb, R. 1996. Control of ovarian follicle development in domestic ruminants: its manipulation to increase ovulation rate and improve reproductive performance. *Animal Breeding,* Abstract **64:** 195-204.

Gong, J. G., Campbell, B. K., Bramley, T. A., Gutierrez, C. G., Peters, A. R. and Webb, R. 1996. Suppression in the secretion of follicle-stimulating hormone and luteinizing hormone, and ovarian follicle development in heifers continuously infused with a gonadotropin-releasing hormone agonist. *Biology of Reproduction* **55:** 68-74.

Gonzalez-Padilla, E., Wiltban, J. N. and Niswender, G. D. 1975. Puberty in beef heifers. 1. The interrelationship between pituitary, hypotalamic and ovarian hormones. *Journal of Animal Science* **40:** 1091-1104.

Grimard, B., Humblot, P., Ponter, A. A., Mialot, J. P., Sauvant, D. and Thibier, M. 1995. Influence of postpartum energy restriction on energy status, plasma LH and oestradiol secretion and follicular development in suckled beef cows. *Journal of Reproduction and Fertility* **104:** 173-179.

Gutierrez, C. G., Campbell, B. K. and Webb, R. 1997b. Development of a long-term bovine granulosa cell culture system: induction and maintenance of estradiol production, response to follicle-stimulating hormone and morphological characteristics. *Biology of Reproduction* **56:** 608-616.

Gutierrez, C. G., Galina, C. S., Zarco, L. and Rubio, I. 1994. Pattern of follicular growth during prepuberal anoestrous and transition from anoestrous to oestrous cycles in *Bos indicus* heifers. *Advanced Agricultural Research* **3:** 1-11.

Gutierrez, C. G., Oldham, J., Bramley, T. A., Gong, J. G., Campbell, B. K. and Webb, R. 1997a. The recruitment of ovarian follicles is enhanced by increased dietary intake in heifers. *Journal of Animal Science* **75:** 1876-1884.

Hawkins, D. E., Niswender, K. D., Oss, G. M., Moeller, C. L., Odde, K. G., Sawyer, H. R. and Niswender, G. D. 1995. An increase in serum lipids increases luteal lipid content and alters disappearance rate of progesterone in cows. *Journal of Animal Science* **73:** 541-545.

Hightshoe, R. B., Cochran, R. C., Corah, L. R., Kiracofe, G. H., Harmon, D. L. and Perry, R.,C. 1991. Effects of calcium soaps of fatty acids on postpartum reproductive function in beef cows. *Journal of Animal Science* **69:** 4097-5004.

Holmes, C. W. 1988. Genetic merit and efficiency of milk production by the dairy cow. In *Nutrition and Lactation in the Dairy Cow*, Ed. Philip C Garnsworthy, published by Butterworths.

Jones, J. I. and Clemmons, D. R. 1995. Insulin-like growth factors and their binding proteins. *Endocrine Reviews* **16:** 3-34.

Jones, G. P. and Garnsworthy, P. C. 1988. The effects of body condition at calving and dietary protein content on dry-matter intake and performance in lactating dairy cows given diets of low energy content. *Animal Production* **47:** 321-333.

Jones, G. P. and Garnsworthy, P. C. 1989. The effects of dietary energy content on the response by dairy cows to body condition at calving", *Animal Production*, **49**, 183-191

Kirby, C. J., Thatcher, W. W., Collier, R. J., Simmen, F. A. and Lucy, M. C. 1996. Effects of growth hormone and pregnancy on expression of growth hormone receptor, insulin-like growth factor-1, and insulin-like growth factor binding protein-2 and -3 genes in bovine uterus, ovary and oviduct. *Biology of Reproduction* **55:** 996-1002.

Lamming, G.E., Darwash, A.O., Wathes, D.C. and Ball, P.J. (1998) The fertility of dairy cattle in the UK: current status and future research. *Journal of the Royal Agricultural Society of England*, **159:** 82-93.

Lammoglia, M. A., Willard, S. T., Hallford, D. M. and Randel, R. D. 1997. Effects of dietary fat on follicular development and circulating concentrations of lipids, insulin, progesterone, estradiol-17ß, 13,14-dihydro-15-keto-prostaglandin F2α and growth hormone in estrous cyclic Brahman cows. *Journal of Animal Science* **75:** 1591-1600.

Land, C. and Leaver, J.D. 1980 The effect of body condition at calving on the milk production and feed intake of dairy cows. *Animal Production*, **30**, 449 (abstract)

Land, C. and Leaver, J.D. 1980 The effect of body condition at calving on the production of Friesian cows and heifers. *Animal Production*, **32**, 362-363 (abstract)

Lucy, M. C., Beck, J., Staples, C. R., Head, H. H., de la Sota, R. L. and Thatcher, W. W. 1992a. Follicular dynamics, plasma metabolites, hormones and insulin-like growth factor (IGF-1) in lactating cows with positive or negative energy balance during the preovulatory period. *Reproduction Nutrition Development* **32:** 331-341.

Lucy, M. C., Collier, R. J., Kitchell, M. L., Dibner, J. J., Hauser, S. D. and Krivi, G. G. 1993. Immnunohistochemical and nucleic acid analysis of somatotropin receptor populations in the bovine ovary. *Biology of Reproduction* **48:** 1219-1227.

Lucy, M. C., Savio, J. D., Badinga, L., de la Sota, R. L. and Thatcher, W. W. 1992b. Factors that affect ovarian follicular dynamics in cattle. *Journal of Animal Science* **70:** 3615-3626.

Lucy, M. C., Staples, C. R., Michel, F. M. and Thatcher, W. W. 1991a. Energy Balance and size and number of ovarian follicles detected by ultrasonography in early postpartum dairy cows. *Journal of Dairy Science* **74:** 473-482.

Lucy, M. C., Staples, C. R., Michel, F. M. and Thatcher, W. W. 1991b. Effect of feeding calcium soaps to early postpartum dairy cows on plasma prostaglandin

$F_2\alpha$, luteinizing hormone, and follicular growth. *Journal of Dairy Science* **74:** 483-489.

Mann, G. E., Lamming, G. E. and Fisher, P. A. 1998. Progesterone control of embryonic interferon tan production during early pregnancy in the cow. *Journal of Reproduction and Fertility,* Abstract Series **21:** Abstract 37.

McDonald, P., Edwards, R.A., Greenhalgh, J.F.D. and Morgan, C.A. (1995) *Animal Nutrition*, 5th Edition, pp 97-127, Longman Scxientific and Technical, Harlow.

McEvoy, T. G., Robinson, J. J., Aitken, R. P., Findlay, P. A. and Robertson, I. S. 1997a. Dietary excesses of urea influence the viability and metabolism of preimplantation sheep embryos and may affect fetal growth among survivors. *Animal Reproduction Science* **47:** 71-90.

McEvoy, T. G., Sinclair, K. D., Staines, M. E., Robinson, J. J., Armstrong, D. G. and Webb, R. 1997b. *In vitro* blastocyst production in relation to energy and protein intake prior to oocyte collection. *Journal of Reproduction and Fertility,* Abstract Series **19**, Abstract **132:** page 51.

McGuire, M. A., Vicini, J. L., Bauman, D. E. and Veenhuizen, J. J. 1992. Insulin-like growth factors and binding proteins in ruminants and their nutritional regulation. *Journal of Animal Science* **70:** 2901-2910.

Murphy, M. G., Enright, W. J., Crowe, M. A., McConnell, K., Spicer, L. J., Boland, M. P. and Roche, J. F. 1991. Effect of dietary intake on pattern of growth of dominant follicles during the oestrous cycle in beef heifers. *Journal of Reproduction and Fertility* **92:** 333-338.

Pell, J. M. and Bates, P. C. 1990. The nutritional regulation of growth hormone action. *Nutrition Research Reviews* **3:** 163-192.

Reynolds, T. S., Stevenson, K. R. and Wathes, D. C. 1997. Pregnancy-specific alterations in the expression of the insuli-like growth factor system during early placental development in the ewe. *Endocrinology* **138:** 886-897.

Rhodes, F. M., Fitzpatrick, L. A., Entwistle, K. W. and De'ath, G. 1995. Sequential changes in ovarian follicular dynamics in *Bos indicus* heifers before and after nutritional anoestrus. *Journal of Reproduction and Fertility* **104:** 41-49.

Roberson, M. S., Stumpf, T. T., Wolfe, M. W., Cupp, A. S., Kojima, N., Werth, L. A., Kittok, R. J. and Kinder, J. E. 1992. Circulating gonadotrophins during a period of restricted energy intake in relation to body condition in heifers. *Journal of Reproduction and Fertility* **96:** 461-469.

Roberts, R. M., Cross, J. C. and Leaman, D. W. 1992. Interferons as hormones of pregnancy. *Endocrine Reviews* **13:** 432-452.

Robinson, R. S., Mann, G. E., Lamming, G. E. and Wathes, D. C. 1997. The localization of IGF-I,-II and IGF type 1 receptor in the bovine uterus on day 16 of pregnancy. *Journal of Endocrinology* **155:** Supplement p64.

Robinson, J. J. and McEvoy, T. G. 1996. Feeding level and rumen degradable nitrogen effects on embryo survival. In "Techniques for Gamete Manipulation and Storage". ICAR Satellite Symposium, Hamilton, New Zealand.

Ryan, D. P., Spoon, R. A. and Williams, G. L. 1992. Ovarian follicle characteristics, embryo recovery and embryo viability in heifers fed high fat diets and treated with follicle stimulating hormone. *Journal of Animal Science* **70:** 3505-3513.

Sinclair, K. D., Kuran, M., Sorohan, P., Staines, M. E. and McEvoy, T. G. (1999). Dietary nitrogen metabolism, folliculogenesis and IVP in cattle. *International Embryo Society* (in press).

Spicer, L., Crowe, M. A., Prendiville, D. J., Goulding, D. and Enright, W. J. 1992. Systemic but not intraovarian concentrations of insulin-like growth factor-I are affected by short-term fasting. *Biology of Reproduction* **46:** 920-925.

Spicer, L. J., Enright, W. J., Murphy, M. G. and Roche, J. F. 1991. Effect of dietary intake on concentrations of insulin-like growth factor-1 in plasma and follicular fluid, and ovarian function in heifers. *Domestic Animal Endocrinology* **8:** 431-437.

Stevenson, K. R., Gilmour, R. S. and Wathes, D. C. 1994. Localization of insulin-like growth factor-I and -II messenger ribonucleic acid and type 1 IGF receptors in the ovine uterus during the estrous cycle and early pregnancy. *Endocrinology* **134:** 1655-1664.

Talavera, F., Park, C. S. and Williams, G. L. 1985. Relationships among dietary lipid intake, serum cholesterol and ovarian function in Holstein heifers. *Journal of Animal Science* **60\;** 1045.

Thatcher, W. W., Meyer, M. D. and Danet-Desnoyers, G. 1995. Maternal recognition of pregnancy. *Journal of Reproduction and Fertility (Suppl.)* **49:** 15-28.

Thatcher, W. W., Staples, C. R., Danet-Desnoyers, G., Oldick, B. and Schmitt, E. P. 1994. Embryo health and mortality in sheep and cattle. *Journal of Animal Science* **72** (suppl. 3)**:** 16-30.

Thissen, J-P., Ketelslegers, J-M. and Underwood, L. E. 1994. Nutritional regulation of the insulin-like growth factors. *Endocrinology Reviews* **15:** 80-101.

Thomas, M. G. and Williams, G. L. 1996. Ovarian follicular characteristics, embryo recovery and embryo viability in heifers fed high-fat diets and treated with follicle-stimulating hormone. *Journal of Animal Science* **70:** 3505.

Treacher, R.J., Reid, I.M. and Roberts, C.J. 1986 Effect of body condition at calving on health and performance of dairy cows. *Animal Production*, **43**, 1-6.

Wathes, D. C. and Wooding, F. B. P. 1980. An electron microscopic study of implantation in the cow. *American Journal of Anatomy* **159:** 285-306.

Wathes, D. C. and Lamming, G. E. 1995. The oxytocin receptor, luteolysis and the maternal recognition of pregnancy. *Journal of Reproduction and Fertility Supplement* **49:** 53-67.

Wathes, D. C., Reynolds, T. S., Robinson, R. S. and Stevenson, K. R. 1998. Role of the insulin-like growth factor system in uterine function and placental development in ruminants. *Journal of Dairy Science* **81:** 1778-1789.

Webb, R. and Armstrong, D. G. 1998. Control of ovarian function; effect of local interaction and environmental influence on follicular turnover in cattle: a review. *Livestock Production Science* **53:** 95-112.

Webb, R., Gong, J. G. and Bramley, T. A. 1994. Role of growth hormone and intrafollicuar peptides in follicle development in cattle. *Theriogenology* **41:** 25-30.

Webb, R., Gong, J. G., Law, A. S. and Rusbridge, S. M. 1992. Control of ovarian function in cattle. *Journal of Reproduction and Fertility Supplement* **45:** 141-156.

Wright, I. A., Rhind, S. M., Smith, A. J. and Whyte, T. K. 1992. Effects of body condition and estradiol on luteinizing hormone secretion in post-partum beef cows. *Domestic Animal Endocrinology* **9:** 305-312.

Yelich, J. V., Wettemann, R. P., Marston, T. T. and Spicer, L.J. 1996. Luteinizing hormone, growth hormone, insulin-like growth factor-I, insulin and metabolites before puberty in heifers fed to gain at two rates. *Domestic Animal Endocrinology* **13:** 325-338.

4

THE INFLUENCE OF NUTRITION ON LAMENESS IN DAIRY COWS

J.E. Offer, R.J. Berry and D.N. Logue
SAC Veterinary Science Division, Dairy Health Unit, SAC Auchincruive, KA6 5AE

Introduction

Lameness in dairy cows is widespread in all the major dairying countries and the recent Farm Animal Welfare Council report identifies lameness, reduced reproductive efficiency and mastitis as the three main causes of poor welfare in UK dairy cattle. A heightened awareness of pain in lame animals has been demonstrated (Whay, Waterman, and Webster, 1997; Whay, Waterman, Webster, and O'Brien, 1998), reinforcing the suffering that arises with the condition. Lameness is also a source of economic loss from treatment and milk loss, and through its effect on reproductive efficiency (Collick, Ward and Dobson, 1989; Sprecher, Hostetler and Kaneene 1997). It has been estimated by Logue, Offer, Chaplin, Knight, Hendry, Leach, Kempson, and Randall (1995), updating Esslemont's calculations of 1990, that lameness causes an overall loss of £100 million per year to the UK dairy industry. However it must be remembered that these estimates are based on the total losses, not the avoidable losses that may be half of this (McInerney, Howe, and Schepers, 1990). Lameness seems to be a greater problem in dairy cows than in beef suckler cows, although there are interpretative problems. There has been an apparent increase in levels of lameness in the UK over the last 20 years (Clarkson, Downham, Faull, Hughes, Manson, Merrit, Murray, Russell, Sutherst, and Ward, 1996; Esslemont and Kossaibati, 1996) that may be linked with the 15% improvement in milk production per cow per year (Dairy Facts and Figures 1980 and 1996) since a number of studies have found a negative relationship between milk production and dairy cow health and welfare (Emanuelson, 1988; Lyons, Freeman, and Kuck,, 1991; Brotherstone and Hill, 1991; Hoekstra, Lugt, Werf, and Ouweltjes, 1994; Gröhn, Hertl and Harman, 1994; Pryce, Veerkamp, Thompson, Hill and Simm, 1998). The widespread introduction of the Holstein to the UK, coupled with determined selection for milk production, has been recognised as a contributory factor (Veerkamp,

Hill, Stott, Brotherstone and Simm 1995; Webster, 1995a). However this may only be a partial explanation, for there have been concomitant changes in general nutrition and other management factors over this period. Thus, before discussing the role of nutrition in this multifactorial disease it is necessary to consider some of the wider issues involved in understanding one of the most intractable problems of dairy cow welfare.

Prevalence of lameness

Estimates of the level of lameness in the UK vary widely depending how the data were collected. Veterinary based surveys (Eddy and Scott, 1980; Russell, Rowlands, Shaw, and Weaver, 1982) reported lameness incidence figures of less than 10 cases per 100 cows per year, but these underestimated the problem, since figures based on farmer data (Whitaker, Smith, daRosa, and Kelly, 1983; Booth, 1989) or more intensive surveys (Prentice and Neal, 1972; Arkins 1981, Esslemont and Peeler, 1993) have shown an incidence of approximately 33 cases per 100 cows. Indeed it is likely that both the veterinary and the farmer based surveys have been underestimates as Wells, Trent, Marsh and Robinson, (1993) reported farmers and herd managers only classified 40% of "lame" cows diagnosed by investigators. Thus the apparently high incidence of 55 cases per 100 cows per year reported by Clarkson, *et al.*, (1996) in England is probably a more accurate reflection of the extent of the condition in the UK (Table 4.1).

Table 4.1 Estimates of the incidence of lameness in dairy cows

		British Isles	*Rest of Europe*	*Austalia and New Zealand*
Number of publications		8	5	5
Total cows affected (estimate)		25766	3841	2070
Total cows records (estimate)		202877	13286	40598
Proportion affected	*(mean of data)*	*0.13*	*0.29*	*0.05*
	(mean of studies)	*0.22*	*0.29*	*0.08*
	(range)	*0.07-0.55*	*0.16-0.47*	*0.02-0.14*

The incidence in Australia and New Zealand is somewhat lower than that recorded in the UK and the rest of Europe but while this probably reflects increased grazing time it is suspected that, in some cases, there has been unwitting censorship of the data by the recording farmers.

Types of lameness

There are a variety of descriptive names for lesions associated with lameness and these have been accurately noted in some surveys but not in others. Greenough, Weaver, Esslemont and Galindo (1997) comment that there is now a need for accuracy of recording of incidence and lesion identification. Without this, accurate epidemiological analysis and interpretation is impossible. It is recognised that for many widely based studies the categories have to be simplified. Based on present understanding of the pathogenesis of the various lesions and their risk factors, data were amalgamated into four categories; claw, interdigital, "non-foot" (upper limb, back etc.) and uncertain (Table 4.2). Using these criteria it can be seen that it is conditions of the foot that are most often observed in lameness cases and that of these claw horn lesions are the most important (Table 4.2). For this reason much lameness research has focused on understanding the development of conditions affecting the claw (Ossent, Greenough and Vermunt, 1997).

Table 4.2 Frequency of causes of lameness from world-wide studies

	British Isles	*Rest of Europe*	*North America*	*Australia and NewZealand*
Number of publications	*6*	*6*	*1*	*5*
Number of observations	19714	4869	245	427
Total I/D or Digital lesions	0.20	0.27	0.09	0.18
Total Claw	0.71	0.67	0.87	0.79
White Line	*0.22*	*0.10*	*0.12*	*0.40*
Sole	*0.40*	*0.22*	*0.60*	*0.35*
Foot	*0.07*	*0.23*	*0.13*	*0.05*
Uncertain	0.04	0.07	0.01	0.01
Limb lameness	0.05	n/a	0.04	0.04

Pathogenesis

In the 1950s and '60s our understanding of claw lameness in cattle was heavily influenced by analogy with laminitis in the horse. Nilsson (1963) suggested that the appearance of haemorrhages in the sole and white line of the cow was associated with a loss of integrity of the hoof horn and that this occurred as a result of laminitis. He and later authors proposed that, although this was often insufficient to cause clinical signs (hence "subclinical laminitis"), it was nevertheless a major predisposing factor in the pathogenesis of claw lesions that caused lameness (Nilsson, 1963). This theory has been steadily refined by a series of authors (Peterse, Korver, Oldenbroek, and Talmon,

1984; Bradley, Shannon and Neilson., 1989; Greenough and Vermunt, 1991; Ossent and Lischer, 1994 and 1998). Ossent and Lischer have proposed a three-stage pathogenesis to explain the sequence of sole lesion development suggesting that changes in the laminar region of the dermis lead to separation (or at least a distortion) of the dermal and epidermal layers (Stage 1). Consequently, the position of the third phalanx, the final bone of the weight-bearing column, changes relative to the softer tissues of the claw causing pressure-induced haemorrhage and necrosis in the corium of the sole so disrupting hoof horn formation (Stage 2). The predominant sign of this is the appearance of blood and cell debris within the claw horn of the sole and white line. This then leads to a loss of horn integrity and the more severe lesions of the horny capsule (Stage 3). The shape of the distal surface of the third phalanx and the angle at which it drops within the hoof capsule determines the distribution of claw lesions seen in the sole and explains the 'typical sole ulcer site' - the junction between the sole and heel caused by pressure resulting from the flexor process (the most distal point of the sunken bone) - and the development of toe ulcers from the rotation of the third phalanx. They also suggest that white line lesions are caused directly by the initial laminar damage leading to accumulation of the blood and debris in the horn. This part of the hypothesis is supported by the slightly different patterns of development between sole and white line lesions observed by Leach, Logue, Kempson, Offer, Ternent, and Randall (1997) and Offer, McNulty and Logue (1998). If disruption of the dermal-epidermal junction is severe then the change in position of the distal phalanx may be irreversible explaining how, once a cow presents with a severe claw condition such as sole ulceration, it is much more likely to present with the same condition in a subsequent lactation (Enevoldsen, Grohn, and Thysen, 1991; Alban, 1995).

Aetiology

While this hypothesis fits observations of the development of claw horn lesions, especially sole and white line haemorrhages, the exact aetiology of the changes that occur in the foot that lead to the initial formation of these lesions is still unproven. Some believe that these claw horn lesions result from the physical, concussive trauma from the cow walking on hard surfaces such as concrete or rough tracks and walkways. Others favour the theory of the release of an unidentified vasoactive substance(s) that acts on the intricate blood capillary network of the corium causing vessel wall paralysis, vasodilation, venous stagnation and oedema. It is thought that this causes opening of ateriovenous anastomoses or shunts which direct the blood away from the capillary bed leading to hypoxia and further damage to capillary walls. The combination of oedema and reduced blood supply interrupts the normal oxygen and nutrient supply to the tissues underlying the basal epidermal cells. This leads to formation of poor quality horn which would be more prone to physical insult and to loss of structural integrity in the epidermal dermal junction. Once this bond is weakened the weight of the cow can cause the changes already outlined in Stage 1 of Ossent and Listener's pathogenesis.

A third hypothesis for claw lesion aetiology has recently been proposed by Webster (1998) who suggested that hormonal changes associated with calving, particularly the increased levels of relaxin, are a major contributory factor. Since one effect of relaxin is to increase the ability of collagen and other connective tissues to deform (and so allow movement of the pelvis during birth), it is suggested that this same effect adds to the instability of the dermal-epidermal junction by allowing unusually large movement (or even breakage) in the network of collagen attachments between the distal phalanx and the basement membrane of the laminae.

In fact all of these effects may contribute to epithelial disruption since all fit our present state of knowledge. However the identity, even the existence, of such 'vasoactive substances' is still to be confirmed as is any effect of relaxin upon the collagen fibres of the foot. Because of the analogy with the horse and the observations that nutrition can influence lesion formation and lameness, most authors suggest that these "vasoactive substances" are of dietary origin: perhaps 'normal' rumen metabolites, or substances resulting from rumen dysfunction. These have included histidine or histamine (Nilsson, 1963), ammonia or other products of protein breakdown and metabolites resulting from (subclinical) rumen acidosis; D-lactate, endotoxins (Boosman, 1990; Mortensen, Hesselholt and Basse, 1986).

Nutrition and risk factors for lameness

Since rumen function or dysfunction were suspected as the main factors influencing lameness, the concentrate part of the diet was initially the subject of most research. However, these studies have not confirmed any of the theories of aetiology and pathogenesis and were reviewed by Vermunt and Greenough (1994). There have been a number of comparative experiments investigating the interrelationship between nutrition, management and lameness (Table 4.3). Most have involved a simple continuous feeding design. These studies confirmed what was until then anecdotal evidence for the effects of nutrition on lameness. They showed the importance of maintaining a stable rumen environment in limiting lameness. When high levels of protein or readily fermentable carbohydrate were fed in the concentrate, or high levels of concentrate were fed as large meals, hoof health was compromised (Table 4.3).

However many of these experiments had a number of deficiencies of design. A particular concern was that some lesions were the consequence of earlier disruption of the claw. Most recent studies therefore have been longitudinal observations of lesion development in first calving cows (heifers) and this approach has been generally accepted as the most suitable monitor of the extent of insult seen in the peri-parturient period

Table 4.3 Studies of the relationship between nutrition and lameness

	Study	*Year*	*Effect on lameness*
Level of Concentrate:	Leaver and Webster	1983	flat rate> fed to yield
	Peterse *et al.*,	1984	high>low
	Manson and Leaver	1988a	high>low
	Kelly and Leaver	1990	rapid>slowly fermentable
	Logue Offer and Hyslop	1994	NS
Forage:Concentrate ratio:	Livesey and Fleming	1984	high>low
	Manson and Leaver	1989	high>low
Protein:in Concentrate:	Manson and Leaver	1988b	high>low
	Offer Roberts and Logue	1997	no sig. diff in RDP/UDP proportion

and first half of first lactation (Bradley *et al.*, 1989; Chesterton Pfeiffer, Morris, and Tanner, 1989; Greenough and Vermunt, 1991; Kempson and Logue, 1993; Leach, *et al.* 1997). These studies, along with the epidemiological analysis of survey data (Clarkson *et al.*, 1996), have not to date defined the aetiology of claw horn lesions, but they have shown that such lameness is associated with a range of risk factors that includes: calving and lactation, housing in cubicles, season and higher rainfall (winter), higher yields, flat rate feeding. While various forage characteristics have been implicated (type of forage, DM, protein content) this is poorly understood and is under experimental investigation at present.

In summary these studies show that there is no single cause of lameness and that the problem is multifactorial, with many different risk factors that interact and confound the effects of other factors. The effects of calving, lactation, housing, environment and season are the major influences in lameness and lesion formation. While nutrition is a factor its effects can only be properly assessed in association with other risk factors and its relative importance is now slowly being elucidated.

NUTRITION AND CALVING

Virtually all these studies have established a clear link between the time of calving, when nutrition is changed in order to support the demands of lactation, and the formation of clinical or subclinical claw horn lesions 3 months after calving (Arkins 1981; Russell, *et al.,* 1982; Kempson and Logue, 1993). Observational studies in first calving heifers have confirmed this and shown that calving is possibly the most important risk factor for lameness associated with claw lesion formation. Figure 4.1 summarises five studies carried out at SAC and the Hannah Research Institute in which claw lesions have been measured in various groups of heifers after calving.

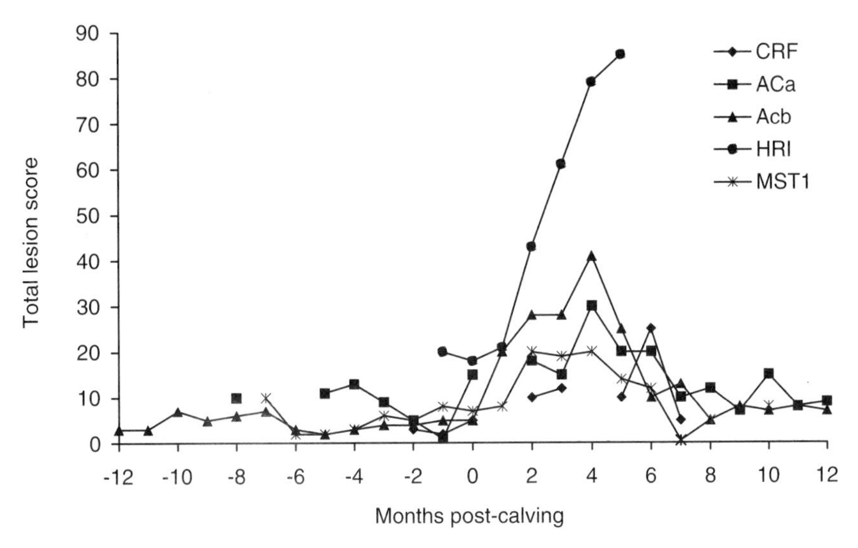

Figure 4.1 Mean total lesion score for 5 cohort studies

The majority of these lesions affected the sole/heel junction of the outer hind claw and rose to a peak at around 3 to 4 months post-calving (Figure 4.1) regardless of the environmental and management conditions. These even occurred in a small group housed on straw over winter and calved at grass in the summer (Logue, *et al.,* 1994). Analysis has shown that the pattern of lesion development after calving is similar in all claws but approximately 0.75 of all claw lesions are seen on the outer hind claw and most of the remaining 0.25 lesions most were seen in the inner front claw. Force plate studies have shown that these two claws carry the greater proportion of the cows weight (Toussaint-Raven, 1985). Other theories for the preponderance of lesions in the lateral hind claw include the way that cows place the foot on the ground (Burgi, 1998) or the conformation of the udder in a newly calved cow forcing the cow to take a wider stance than normal and causing increased weight-bearing in the outer claw (Webster, 1995b).

An observational study of a cohort of autumn calving cattle managed similarly either on a clover based system, receiving no artificial nitrogen, or on conventionally managed ryegrass swards showed no significant effect on the incidence of lameness. Modelling of the pooled data showed the pattern development of sole and white line lesions in the outer hind claw (Figures 4.2 and 4.3) is similar to that shown in Figure 4.1, reaching a peak at around 3 - 4 months (later in subsequent lactations) after calving

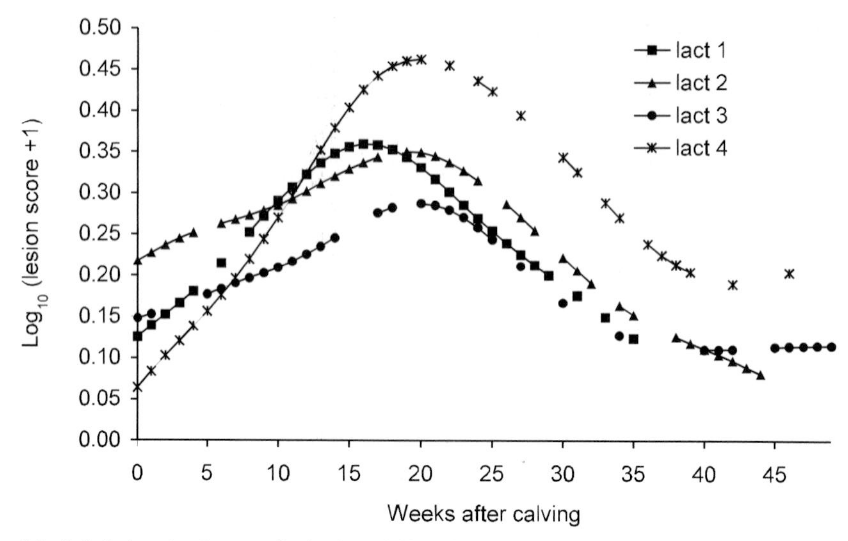

Figure 4.2 Sole lesion development in the lateral hind claws (lactations 1-4)

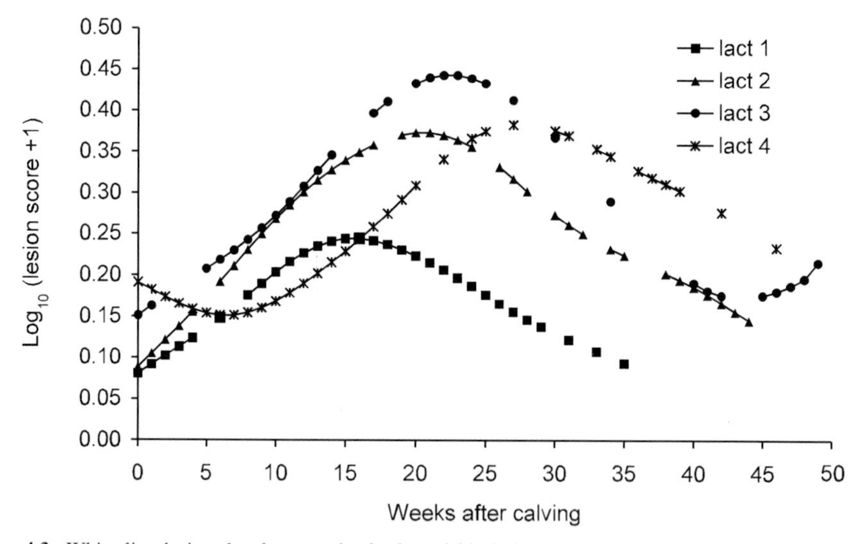

Figure 4.3 White line lesion development in the lateral hind claw (lactations 1-4)

and declining thereafter. Lesions were significantly worse in older animals and there was a significant interaction between time after calving and parity on lesion formation. The difference in the shape of the curves for sole and white line lesions suggests that although the aetiology may be similar, the pathogenesis may be different, as proposed by Ossent and Lischer (1998). The increased severity of lesions in older animals is supported by the increase in risk odds for lameness in older animals shown from the Liverpool Survey (Ward and French, 1997). It was suggested that the shift in timing of the peak lesion score to later in lactation was due to repeated lesion formation and animals' calving sometime after housing in later lactations instead of simultaneously as in the first lactation (Offer *et al.,* 1998). Thus already we are seeing some of the complexity of these interrelationships.

Results from a number of more recent studies involving heifer groups have been more equivocal about the importance of nutrition than some of the earlier feeding experiments. At SAC, as part of an investigation into Metabolic Stress (Logue, Berry Offer, Chaplin, Crawshaw, Leach, Ball and Bax, 1999) the two Acrehead herds (each of 70 cows, housed in adjacent halves of the same shed and milked by the same person, through the same milking parlour) were fed considerably different concentrate inputs (Table 4.4).

Table 4.4 Diets of animals in on Low Input (LI) and High Output (HO) management systems

	1994/5		*1995/6*		*1996/7*	
Composition (kg FW)	*LI*	*HO*	*LI*	*HO*	*LI*	*HO*
Silage	40	40	26	35	32	36
Grainbeet	10	15	7.5	10	4	8
Barley straw	-	-	3	-	-	-
Maize gluten	4.2	-	-	-	-	-
Wholecrop barley	-	-	-	-	7.5	6.5
Fodderbeat	-	-	-	-	17.5	20
Fishmeal	0.7	-	-	-	-	-
Blended concentrate	-	4	-	-	-	-
Parlour concentrate	1	2	1.5	3	2	3
Fresh weight mix	55.9	61	38	48	63	73.5
Mix ration						
kg DM	19.5	21.2	12.9	16.7	15.9	21
ME (MJ/kg DM)	12.0	11.3	10.5	11.7	11.7	11.8
CP (g/kgDM)	171	184	139	160	131	135
OIL (g/kgDM)	53	51	47	53	40	43
NDF (g/kgDM)	442	402	540	476	365	404
Starch and Sugar (g/kgDM)	94	81	85	92	221	207
Ash (g/kgDM)	82	102	77	84	99	80

Figure 4.4 summarises the claw lesion scores for first lactation heifer groups during the first 2 months pre-calving and first 6 months post-calving of either a low input herd (LI, yielding 5,000kg/cow/year) or high output herd (HO, yielding 8,500kg/cow/year) during the years 1994-97. In summary, while there were significant differences between the 3 years and between the two herds in 96/97, the general picture is one of little major difference between herds despite quite dramatic differences in concentrate input (0.5

vs. 2.0 tonnes/cow/year). It is suggested that the increased lesion scores for both herds in 96/97 relate to the higher levels of starch/sugar and lower NDF in the ration compared with previous years. Interestingly, claw lesions in those receiving only 0.5 tonnes/cow/year concentrate were significantly worse than for those receiving 2.0 tonnes/cow/year in that year. Two factors may have influenced this: firstly, LI started with a higher lesion score at calving; secondly, the metabolic challenge for LI was greater than for HO and they were consequently less able to withstand the additional challenges of calving and the higher starch/sugar rations. Put simply, this is the cumulative results of the interaction of several risk factors for lesion formation (calving, nutrition and metabolic challenge).

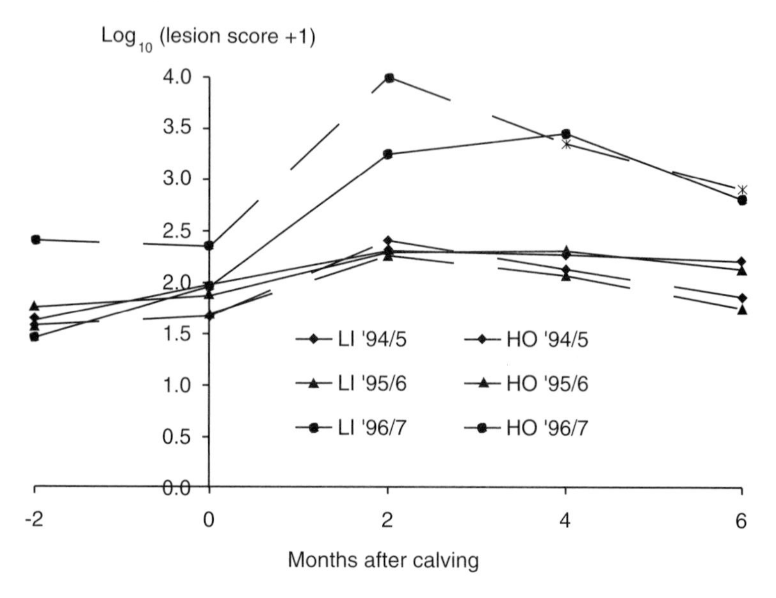

Figure 4.4 Total lesion score in first lactation heifers on low input (LI) or high output (HO) management systems 1994-1997

These data seem to follow a somewhat similar pattern to that described by Bergsten (1995), Bergsten and Frank (1996) and Olsson, Bergsten and Wiktorsson (1998) in Sweden. They were unable to show any significant effect of concentrate level on lameness or lesion formation. Conversely, the CVL/ADAS/RCVS group at Bridgets have reported a highly significant effect of concentrate input upon heifer claw lesion score development (Livesey, Harrington, Johnston, May and Metcalf, 1998). However in their 2×2 factorial experiment comparing the effects on lameness of high with low levels of concentrate and straw yards with cubicles, they found much greater differences due to housing type than to level of concentrate; where cubicles gave a higher lesion score than a straw yard. Thus the results from these more recent studies involving heifer groups have been more equivocal about the importance of nutrition than some of the earlier feeding experiments.

A further aspect of the effects of nutrition on lameness is supplementation with mineral and vitamins. Recently there has been an upsurge of interest in the effects of biotin supplementation on dairy cow lameness. In a very well controlled experiment Midla, Hoblet, Weiss and Moeschberger (1998a and b) monitored claw lesion development using heifer groups supplemented or not with biotin. Although no significant difference in overall lesion score was found between treatments (Midla *et al.*, 1998a) there was significantly less white line separation in supplemented animals (Midla *et al.*, 1998b). There is also some tentative evidence that biotin can increase the rate at which claw lesions heal (Koller, Lischer, Geyer, Ossent , Schulze and Auer, 1998) and supplementation with organic forms of zinc may improve the tensile strength of hoof horn and thus increase the resistance of the hoof capsule to physical trauma (Stern and Guyer, 1998). Generally these additives may have some positive qualities but their value to the farm depends on their cost.

NUTRITION AND HOUSING

When the clinical cases of lameness in all cows (not just first lactation animals) on the 'Metabolic Stress' project were analysed, the results were not as would be expected from the data on lesion score in heifers. Overall HO had a higher incidence of clinical lameness ($p<0.05$) despite there being no difference in lesion scores for first lactation heifers (Table 4.5). In 1995-96 a major contributor to this difference was an outbreak of foul in the foot in HO during the autumn. Although both herds were affected, HO was much more severely affected than LI and this was considered as the major contributor to the poorer mobility of cows in HO. Over the three years there was a higher incidence of lameness caused both by digital and interdigital dermatitis and by claw lesions in HO.

Table 4.5 Number of lesions/year identified at clinical examinations for lameness

	Low input	*High output*
(Inter) digital dermatitis	18	36
Claw Horn Lesion	27	38
Other	1	2
Total	46	76

We believe that, thanks to the use of both extended grazing in the late autumn and daylight grazing in the spring and early autumn, the lower time spent indoors (25% less) by the LI herd and also the more broken nature of this housing, reduced the challenge to the feet and the prevalence of both digital and interdigital and claw horn lesions in this group. In addition there may have been a cumulative effect of the management and feeding regimes on cow hoof health since lesions are known to be more severe and are also repeated in later lactations. While we also believe that the environmental challenge was a principle component in the greater development of claw horn lesions in the older cattle we cannot rule out an indirect nutritional interaction.

Nutrition can indirectly affect lameness by influencing the quantity and characteristics of the slurry and underfoot conditions within the housing. There is now ample evidence that by far the most important risk factor for these digital and interdigital lesions is a dirty environment (Rodriguez-Lainz, Hird, Carpenter and Read, 1996). Furthermore it is known that slurry greatly increases the permeability of the intertubular horn of the sole and the heel (Kempson, Langridge and Logue 1998), leaving the horn of the sole and particularly the heel vulnerable to microbial invasion and physical trauma. Heel erosion is ubiquitous amongst housed cattle and although not a primary cause of lameness, can, if severe, change the conformation and biomechanics of the foot in such a way as to make lesion formation more likely. Figure 4.5 shows how heel erosion in autumn calving dairy cattle housed on concrete develops over the winter months and is at a peak by April/May when the animals are turned out for spring grazing again. Once out at grass and away from the influence of slurry and concrete the heels begin to recover until the next housing in autumn.

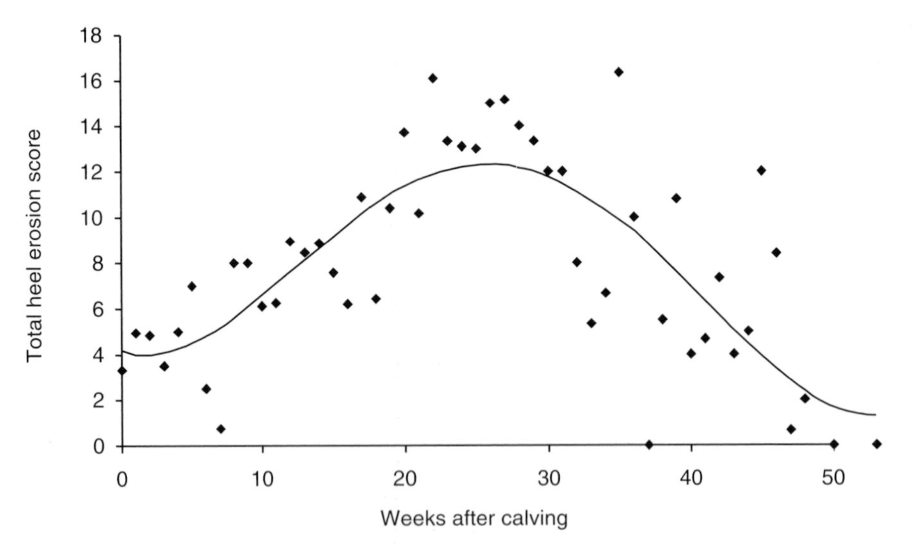

Figure 4.5 Heel erosion score in autumn calving cattle housed from mid October to late April in a concrete based cubicle house

There is often a net hoof wear in autumn calved cattle, and hooves become steeper over the winter housing period, but no direct relationship with diet has been established.

Measurements of hoof hardness, which is notoriously difficult to measure, have shown little if any effect of nutrition. The *in vitro* measurement of claw horn epithelial cell proliferation and keratinization from small groups of mid-lactation cattle has shown that the initiation of a silage based diet and introduction to housing (cubicles and concrete) causes a cumulative reduction in DNA synthesis by the laminar epithelium (MacCallum, Knight, Wilde, Kim, and Logue, 1996). This fall in the rate of epidermal growth may reduce the ability of the claw horn to withstand damage and again demonstrates the multifactorial nature of lameness.

A further aspect of nutrition which may influence lameness and lesion formation is the quality of the diet and the way that this interacts with cow behaviour. On the Metabolic Stress project animals in LI were fed a diet with a large proportion of forage. As a result the animals had to spend a greater proportion of their time feeding and sacrificed time lying down in cubicles (as shown in Figure 4.6). Lack of adequate cubicle comfort and a consequent reduction of lying time have been considered by some as the major aetiological factors influencing severe claw lesions. (Colam-Ainsworth, Lunn, Thomas and Eddy 1989; Singh, Ward, Lautenbach, Hughes and Murray,1993 Singh, Ward and Murray, 1993). However, since in this study there was a substantial difference in lying time in 1995/6, but no obvious difference in the effect on claw lesion development in the first calving heifer groups over their first housed period, it is clear that the matter is not that simple! (Berry, Logue, Waran, Appleby and Offer, 1997). The even more extreme heifer cohort studies of Leonard, O'Connell and O'Farrell (1994 and 1995) confirm this view point.

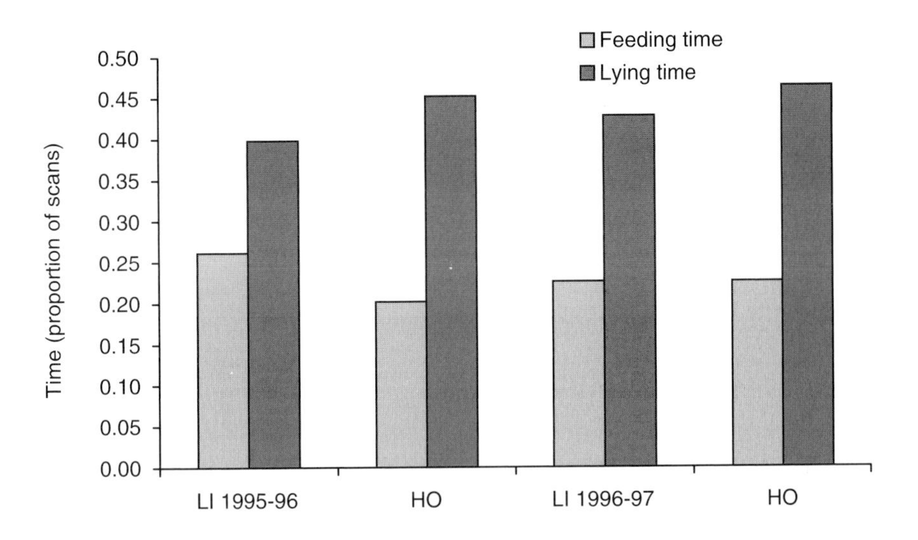

Figure 4.6 Summary time budget for low input (LI) and high output (HO) cohorts in Metabolic Stress study (1995-1997)

Without more detailed investigations it is only possible to speculate about the various inter-relationships which might explain this further.

NUTRITION AND SEASON

Season has a big influence on lameness and lesion formation closely related to many of the other management factors, nutrition and housing that are changed because of the season and outdoor environmental conditions. Season has other effects, largely outwith management control, such as rainfall, humidity and day length. Whilst rainfall has been shown by epidemiological evidence to influence lameness (Russell, *et al.,* 1982), the direct effect of day length is currently under investigation. It has been shown that the interaction with season and calving can have a major influence on lesion formation.

Animals from the Metabolic Stress project mentioned earlier were split into autumn-calving and spring-calving groups for analysis (i.e. calving before and after Christmas respectively). Autumn-calving animals had significantly higher total lesion scores than spring-calvers at 2, 4, and 6 months post calving ($p<0.01$) (Figure 4.7)

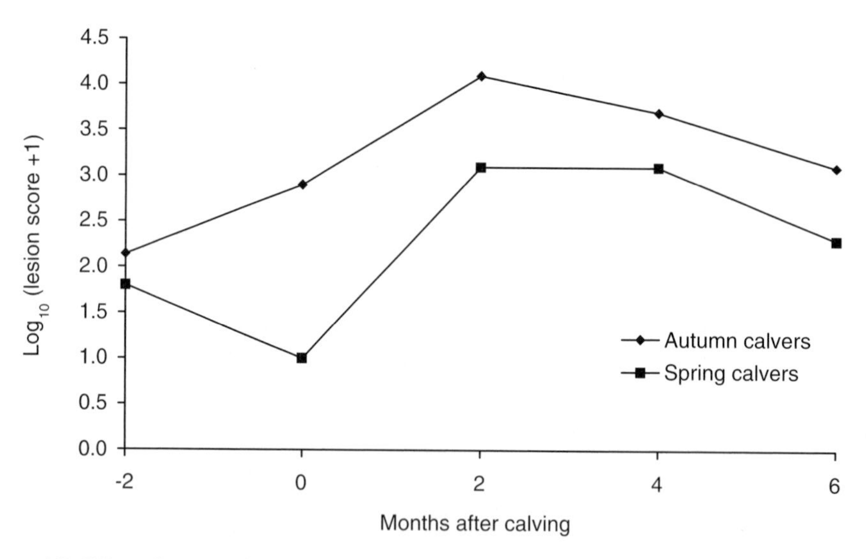

Figure 4.7 Effect of season of calving on total lesion score (1996-1997)

Overall, autumn-calving cows had a higher incidence of lameness than spring calving animals ($p<0.05$). Cows calving in the autumn, at or close to housing for the winter, experience two known risk factors (both housing and calving) together and lesion scores rise higher than they otherwise would. This effect is exacerbated by the reduction in laminar epithelium DNA synthesis rates during winter (MacCallum, Knight, Wilde and Hendry, 1998). This suggests that cows calving in spring are able to recover from the initial challenge of housing before they calve and consequently lesion scores tend not to increase to the levels seen in autumn-calving animals. Additionally, autumn-calving animals are at increased risk of infectious causes of lameness whilst at peak milk production, whereas spring-calving animals have the added advantage of being out at grass for at least half of their lactation when lesions may be expected to be at their worst.

Summary

The general nutrition of dairy cows has improved over the last 20 years as our understanding of the problems facing the high yielding cow increases. Improvements in ration formulation and feeding practice has allowed dairy cows to meet the metabolic demands of calving and lactation despite being asked to work harder as profit margins are squeezed. We have shown in our studies of metabolic stress that cows can meet the demands of either low input or high output systems (towards the extremes of dairy cow management in Scotland) but that high levels of production may be detrimental to hoof health because of the increased environmental challenge rather than the level of feeding *per se* (provided that nutrition is well formulated). The increased interest in welfare and lameness in particular has meant that we are getting better at monitoring lameness. We believe that measurement of lameness must be standardised in order to assess this apparent increase more fully. However if, as epidemiology suggests, there is an increase in the prevalence and incidence of lameness, it seems that nutrition is not as important as a risk factor as was once thought and that other, in some cases related, factors already discussed such as quality and length of housing may be responsible.

We are still far from understanding the full intricacies of this problem but several matters have been clarified. We now know that calving, housing and season are the primary risk factors for lameness and claw lesion formation. When animals are well managed, fed balanced rations and the rumen is functioning optimally, nutrition is can vary quite widely without adversely affecting their feet provided that the environmental challenge is also limited. However when the animal is exposed to known nutritional risks (rapid change in diet, large meals of rapidly fermentable carbohydrate, high levels of protein) and at the same time is experiencing other risk factors, (calving, housing on concrete) then hoof health can be compromised. Ideally the number of risk factors faced by a dairy cow at any particular time should be minimised. This is especially important in the dairy heifer; the lower the baseline lesion score prior to calving the better for hoof health. All the evidence is that this should be obtained by steady growth rates and training the animals to the various management regimes they are likely to encounter in later life. Care is also needed with the introduction of newly calved cows, especially first calving heifers, into a cubicle house. Good cubicle comfort and adequate access to forage is essential. Care of the young cow will mean fewer lesions in older ones!

Acknowledgement

We gratefully acknowledge help and advice from SAC, Edinburgh University and BioSS colleagues. One of us (RB) was in receipt of a MAFF studentship and SAC receives funding from SOAEFD.

References

Alban, L. (1995) Lameness in Danish dairy cows- frequency and possible risk factors. *Preventative Veterinary Medicine* **22,**: 213-225.

Anon. (1996) Dairy facts and figures 1996. A National Dairy Council Publication, London.

Anon. (1997) Report on the welfare of dairy cattle. FAWC, Tolworth, Surrey (MAFF publication PB3426).

Anon. (1980) United Kingdom dairy facts and figures 1980. The Federation of United Kingdom Milk Marketing Boards, Milk Marketing Board of England and Wales, Surrey.

Arkins, S. (1981). Lameness in dairy cows. *Irish Veterinary Journal* **35,**: 135 - 140 and 163-170.

Bergsten, C. and Frank, B. (1996) Sole haemorrhages in tied primiparous cows as an indicator of periparturient laminitis: effects of diet, flooring and season. *Acta Veterinaria Scandinavia* **37,** 383- 394.

Bergsten, C. (1995) Management and nutritional effects on foot health in dairy cows. *Proceedings of the 46th Annual Meeting of the European Association of Animal Production* 1995, Prague, 4 - 7th Sept Abstr. p82

Berry R.J., Logue, D.N., Waran, N.K., Appleby, M.C. and Offer, J.E. (1997) Effect of high and low production regimes on the behaviour of dairy cows. *Proceedings of the 51st Scientific meeting of the Association of Veterinary Research Workers* Abst. 1B20 p 20

Boosman, R. (1990) Bovine laminitis: Histopathologic and arteriographic aspects, and its relation to endotoxaemia. Proefschrift, Utrecht, 1990

Booth, J.M. (1989) Lameness and mastitis losses. *Veterinary Record* **125**, 161

Bradley, H.K., Shannon, D. and Neilson, D.R. (1989) Subclinical laminitits in dairy heifers. *Veterinary Record*, **125**, 177 - 179

Brotherstone, S. and Hill, W. G. (1991) Dairy herd life in relation to linear type traits and production 1. Phenotypic and genetic analyses in pedigree type classified herds. *Animal Production* **53**, 279-287

Burgi (1998) Determine maintenance hoof trimming by observing movement *Proceedings of the Xth International Symposium on Lameness in Ruminants, Lucerne, Switzerland.* (Eds. Lischer, Ch. J. and Ossent, P.) 20 - 22

Chesterton, R.N., Pfeiffer, D.U., Morris, R.S. and Tanner, C.M. (1989) Environmental and behavioural factors affecting the prevalence of foot lameness in NewZealand Dairy Herds - a case control study. *New Zealand Veterinary Journal*, **37**, 135 - 142

Clarkson, M. J., Downham, D. Y., Faull, W. B., Hughes, J. W., Manson, F. J., Merrit, J. B., Murray, R. D., Russell, W. B., Sutherst, J. and Ward, W. R. (1996) Incidence and prevalence of lameness in dairy cattle. *Veterinary Record* **138**, 563-567

Colam-Ainsworth, P., Lunn, G. A., Thomas, R. C. and Eddy, R. G. (1989) Behaviour of cows in cubicles and its possible relationship with laminitis in replacement dairy heifers. *Veterinary Record* **125**, 573-575

Collick, D. W., Ward, W. R. and Dobson, H. (1989) Association between types of lameness and fertility. *Veterinary Record* **125**, 103-106.

Eddy, R.G. and Scott, C.P. (1980) Some observations on the incidence of lameness in dairy cattle in Somerset *Veterinary Record* **106**, 140-144.

Emanuelson, U. (1988) Recording of production diseases in cattle and possibilities for genetic improvements: a review. *Livestock Production Science* **76**, 89-106

Enevoldsen, C., Grohn, Y. T. and Thysen, I. (1991) Sole ulcers in dairy cattle: associations with season, cow characteristics, disease and production. *Journal of Dairy Science* **74**, 1284-1298

Esslemont, R.J. and Peeler, E.J. (1993) The scope for raising margins in dairy herds by improving fertility and health. *British Veterinary Journal* **149**, 537-547.

Greenough, P. R. and Vermunt, J. J. (1991) Evaluation of subclinical laminitis in a dairy herd and observations on associated nutritional and management factors. *Veterinary Record* **128**, 11-17

Greenough, P. R., Weaver A. D., Esslemont R. J. and Galindo F. A. (1997) Basic concepts of bovine lameness. In: *Lameness in cattle* (3rd edition). (Eds., Greenough, P. R. and Weaver, A. D.) London. p3-13

Gröhn Y. T., Hertl J. A. and Harman J. L. (1994) Effect of early lactation milk yield on reproductive disorders in dairy cows. *American Journal of Veterinary Research* **55**, 1521-1528

Hoekstra, J., Lugt, A. W. van der, Werf, J. H. J. ven der and Ouweltjes, W. (1994) Genetic and phenotypic parameters for milk production and fertility traits in upgraded dairy cattle. *Livestock Production Science* **12**, 225-232

Kelly, E.F. and Leaver, J.D. (1990) Lameness in dairy cattle and the type of concentrate given. *Animal Production*, **51**, 221 - 227

Kempson, S.A., Langridge, A. and Logue, D.N. (1998) Laminitis versus papillaryitis. *Proceedings of the 10th International Symposium on Lameness in Ruminants*. Sept 7 - 10, 1998 Lucerne, Switzerland pp 153 - 154

Kempson, S.A. and Logue, D.N. (1993) Ultrastructural observations of hoof horn from dairy cows: changes in the white line during the first lactation. *Veterinary Record*, **132**, 524 - 527

Koller, U., Lischer, C.J., Geyer, H., Ossent, P., Schulze, J. and Auer J.A.(1998) The effect of biotin in the treatment of uncomplicated sole ulcers in cattle. A controlled study. *Proceedings of the Xth International Symposium on Lameness in Ruminants, Lucerne, Switzerland.* (Eds. Lischer, Ch. J. and Ossent, P.) 230 -232

Leach, K. A., Logue, D. N., Kempson, S. A., Offer, J. E., Ternent, H. E. and Randall, J. M. (1997) Claw horn lesions in dairy cattle: development of sole and white line haemorrhages during the first lactation. *The Veterinary Journal* **154**, 215-225

Leaver, J.D and Webster, D.M (1983) Assessment of lameness in dairy cattle on different systems of concentrate feeding. In *SAC Crichton Royal Annual Report (1983)* pp 27 - 29

Leonard F. C., O'Connell J. and O'Farrell K. (1994) Effect of different housing conditions on behaviour and foot lesions in Friesian heifers. *Veterinary Record* **134**, 490-494

Leonard F. C., O'Connell J. and O'Farrell K. (1995) Effect of overcrowding on claw health in first-calved Friesian heifers. *British Veterinary Journal* **152**, 459-472

Livesey, C. T. and Fleming, F. L. (1984) Nutritional influences on laminitis, sole ulcer and bruised sole in Friesian cows. *Veterinary Record* **114**, 510-512

Livesey, C. T., Harrington, T., Johnston, A. M., May, S. A. and Metcalf, J. A. (1998) The effect of diet and housing on the development of sole haemorrhages, white line haemorrhages and heel erosions in Holstein heifers. *Animal Science* **67**, 9-16

Logue, D.N., Offer, J.E., Leach, K.A. Kempson, S.A. and Randall. J.M. (1994) Lesions of the hoof in first-calving dairy heifers. *Proceedings of the 8th International Symposium on Disorders of the Ruminant Digit*, Banff, Canada, Ed. Greenough P.R. 272

Logue, D.N., Berry, R.J., Offer, J.E., Chaplin. S.J., Crawshaw, W.M., Leach, K.A., Ball, P.J.H. and Bax, J. (1999) Consequences of 'metabolic load' for lameness and disease. *Proceedings of the International Symposium on Metabolic Stress in Dairy Cows* 28 - 30 Oct 1998, Edinburgh (In Press)

Logue, D.N., Offer, J.E. and Hyslop, J.J. (1994) Relationship of diet, hoof type and locomotion score with lesions of the sole and white line. *Animal Production,* **59**, 173 - 181

Logue, D.N., Offer, J.E., Chaplin, S.J., Knight, C.H., Hendry, K.A.K., Leach, K.A., Kempson, S.A. and Randall, J.M. (1995) Lameness in dairy cattle. *Proceedings of the 46th Annual Meeting of the European Association of Animal Production* 1995, Prague, 4 - 7th Sept Abstr. p218

Lyons, D. T., Freeman, A. E., and Kuck, A. L. (1991) Genetics of health traits. *Journal of Dairy Science* **74,** 1092-1100

MacCallum, A.J., Knight, C.H., Wilde, C.J., Kim, J.S. and Logue, D.N. (1996) Environmental and nutritional influences on bovine laminitis. *Hannah Research Institute Report* (1996) Abst

MacCallum, A.J., Knight, C.H., Wilde, C.J. and Hendry, K.A.K. (1998) Cell proliferation and keratinization in bovine hoof during the development of the cow and during lameness challenge. *Proceedings of the Xth International Symposium on Lameness in Ruminants, Lucerne, Switzerland.* (Eds. Lischer, Ch. J. and Ossent, P.) p 236 - 238

Manson, F. J. and Leaver, J. D. (1988a.)The influence of concentrate amount on locomotion and clinical lameness in dairy cattle. *Animal Production* **47,** 185-190

Manson, F. J. and Leaver, J. D. (1988b) The influence of dietary protein intake and of hoof trimming on lameness in dairy cattle. *Animal Production* **47**, 191-199

Manson, F. J. and Leaver, J. D. (1989) The effect of concentrate:silage ratio and of hoof trimming on lameness in dairy cattle. *Animal Production* **49,** 15-22

McInerney, J.P., Howe, K.S. and Schepers, J.P. (1990) A framework for the economic analysis of disease in farm livestock. *Report of a research project (ref CSA 873) funded by MAFF.* The University of Exeter, Agricultural Economics Unit. 87pp

Midla, L.T., Hoblet, K.H., Weiss, W.P. and Moeschberger, M.L. (1998a) Supplemental dietary biotin for prevention of lesions associated with aseptic subclinical laminitis (*pododermatitis aseptica diffusa*) in primiparous cows. *American Journal of Veterinary Research*, **59**, 733 - 737

Midla, L.T., Hoblet, K.H., Weiss, W.P. and Moeschberger, M.L. (1998b) Biotin feeding trial to prevent lesions associated with pododermatitis aseptica diffusa in lactating cattle). *Proceedings of the Xth International Symposium on Lameness in Ruminants, Lucerne, Switzerland.* (Eds. Lischer, Ch. J. and Ossent, P.) p 225 - 226

Mortensen, K., Hesselholt, M and Basse, A. (1986) Pathogenesis of bovine laminitis (*pododermatitis aseptica diffusa*). Experimental models. *Proceedings of the 14th World Congress on Diseases in Cattle*, Dublin, Ireland, 1025 - 1030

Nilsson, S.A. (1963) Clinical, morphological and experimental studies of laminitis in cattle. *Acta Veterinaria Scandinavica*, **4**, Suppl 1, 1 - 276

Offer, J. E., McNulty, D. And Logue, D. N. (1998) Modelling lesion development in a cohort of Holstein-Friesian cows (lactations 1 to 4). *Proceedings of the Xth International Symposium on Lameness in Ruminants, Lucerne, Switzerland.* (Eds. Lischer, Ch. J. and Ossent, P.) p 159-160

Offer, J.E., Logue, D.N. and Roberts, D.J. (1997) The effect of protein source on lameness and solear lesion formation in dairy cattle. *Animal Science*, **65**, 143 - 149

Olsson, G., Bergsten, C. and Wiktorsson, H. (1998) The influence of diet before and after calving on the food intake, production and health of primiparous cows, with special reference to sole haemorrhages. *Animal Science*, **66**, 77 - 86

Ossent, P., Greenough P.R. and Vermunt J.J. Laminitis (1997) In *Lameness in cattle* (3rd edition). (Eds., Greenough, P. R. and Weaver, A. D.) London. 277 - 292

Ossent, P. and Lischer, C. J. (1994) Theories on the pathogenesis of bovine laminitis. *Proceedings of the 8th International Symposium on disorders of the Ruminant Digit*, Banff, Canada. 207-209.

Ossent, P. and Lischer, C.J. (1998) Bovine laminitis: the lesions and their pathogenesis. *In Practice* **20**, 415 - 427

Peterse, D. J., Korver, S., Oldenbroek, J. K. and Talmon, F. P. (1984) Relationship between levels of concentrate feeding and the incidence of sole ulcers in dairy cattle. *Veterinary Record* **115**, 629-630.

Prentice, D.E. and Neal, P.A. (1972) Some observations on the incidence of lameness in dairy cattle in West Cheshire *Veterinary Record* **91**, 1 - 7

Pryce, J. E. Veerkamp, R. F. Thompson, R. Hill W.G. and Simm, G. (1998) Genetic parameters of common health disorders and measures of fertility in Holstein Friesian dairy cattle. *Animal Science* **65**, 353-360.

Rodriguez-Lainz, A. J., Hird, D. W., Carpenter, T. E. and Read, D. H. (1996) Case control of papillomatous digital dermatitis in Southern California dairy farms. *Preventive Veterinary Medicine* **28**, 117-131.

Russell, A. M., Rowlands, G. J., Shaw, S. R. and Weaver, A. D. (1982) Survey of lameness in British dairy cattle. *Veterinary Record* **111**, 155-160.

Singh, S. S., Ward, W. R., Lautenbach, K., Hughes, J. W. and Murray, R. D. (1993a) Behaviour of first lactation and adult dairy cows while housed and at pasture and its relationship with sole lesions. *Veterinary Record* **133**, 469-474.

Singh, S. S., Ward, W. R. and Murray, R. D. (1993b) Aetiology and pathogenesis of sole lesions causing lameness in cattle: a review. *Veterinary Bulletin* **63**, 303-315.

Sprecher, D. J., Hostetler, D. E. and Kaneene, J. B. (1997) A lameness scoring system that uses posture and gait to predict dairy cattle reproductive performance. *Theriogenology* **47**, 1179 -1187

Stern, A., Guyer, H., Morel, I. and Kessler, J. (1998) Effect of organic zinc on horn quality in beef cattle *Proceedings of the Xth International Symposium on Lameness in Ruminants, Lucerne, Switzerland.* (Eds. Lischer, Ch. J. and Ossent, P.) 233 - 235

Toussaint-Raven, E. (1985) *Cattle footcare and claw trimming*. Farming Press, Ipswich.

Veerkamp, R. F., Hill, W.G., Stott A. W., Brotherstone S. and Simm, G. (1995) Selection for longevity and yield in dairy cows using transmitting abilities for type and yield. *Animal Science* **61**, 189-198.

Vermunt, J.J. and Greenough, P.R. (1994) Predisposing factors of laminitis in cattle. *British Veterinary Journal*, **150**, 151 - 164

Ward, W.R. and French N.P. (1997) Foot lameness in cattle. *Proceedings of the 51st Scientific meeting of the Association of Veterinary Research Workers* Abst. 1B17 p 19

Webster, A.J.F. (1998) *Report on the biology of lameness workshop.* Funded by SOAEFD and MDC, SAC Report (*In press*)

Webster, A. J. F. (1995a) Welfare strategies in future selection and management strategies. In *Breeding and feeding the high genetic merit dairy cow. Occasional publication No 19 British Society of Animal science 1995* (Eds T J L Lawrence, Gordon FJ and Carson A.) p 87- 93

Webster A.J.F. (1995b) Animal Welfare - a cool eye towards Eden. Blackwell Science, Oxford

Wells, S. J., Trent, A. M., Marsh, W. E. and Robinson, R. A. (1993) Prevalence and severity of lameness in lactating dairy cows in a sample of Minnesota and Wisconsin herds. *Journal of the American Veterinary Medical Association* **202**, 78-82.

Whay, H.R., Waterman, A. E., Webster, A. J. F. and O'Brien, J. K. (1998) The influence of lesion type on the duration of hyperalgesia associated with hindlimb lameness in dairy cattle. *The Veterinary Journal* **156**, 23-29.

Whay, H.R., Waterman, A. E. and Webster, A. J. F. (1997) Associations between locomotion, claw lesions and nociceptive threshold in dairy heifers during the peri-partum period. *Veterinary Journal* **154**, 155-161.

Whitaker, D. A., Smith, E. J. daRosa, G. L., and Kelly, J. M. (1993) Some effects of nutrition and management on the fertility of dairy cattle. *Veterinary Record* **133**, 61-64.

5

MANIPULATION OF MILK FAT IN DAIRY COWS

M. DOREAU[1], Y. CHILLIARD[1], H. RULQUIN[2] and D.I. DEMEYER[3]
[1] *Institut National de la Recherche Agronomique, Unité de Recherches sur les Herbivores, Theix, 63122 Saint-Genès Champanelle, France*
[2] *Institut National de la Recherche Agronomique, Station de Recherches sur la Vache Laitière, 35590 Saint-Gilles, France*
[3] *University of Ghent, Department of Animal Production, Proefhoevestraat 10, 9090 Melle, Belgium*

Introduction

The first attempts to modify milk fat by nutritional means date from the last century. Throughout the twentieth century until the 1970s, the aim was to improve butter production by an increase in milk fat content. Numerous reviews have been devoted to the low milk fat syndrome caused by an excess of concentrates. A large change has taken place during the two past decades, with an increasing concern for milk protein, especially in countries where cheese-making is important. The aim is to increase the protein:fat ratio in milk, or at least to decrease the fat content. The present situation is motivated by the establishment of milk fat quotas, independent of milk yield quotas, or by penalties for high fat content of milk in many countries of the European Community.

The 1970s also mark the beginning of manipulation of milk fat composition. Two reasons can be evoked : 1) in developed countries milk production increased enough to ensure self-sufficiency in dairy products. For this reason interest in quality took precedence over increase in production; 2) medical research suggested a negative role of saturated fatty acids (FA) on cardiovascular diseases and the consumption of margarines was promoted at the same time. Techniques have been proposed to increase the polyunsaturated content in milk FA. More recently, other modifications of milk FA composition have been proposed, based on new research : increasing n-3 FA, decreasing trans FA, increasing conjugated linoleic acid (CLA) (Demeyer and Doreau, 1999). Unfortunately, as expressed by Kennelly (1996a), “the ideal milk fatty acid profile is a moving target”. This is a consequence of the difficulty in establishing a scale between the different dietary factors acting positively or negatively on human health, and of accounting for the multifactorial origin of human diseases. For example, epidemiological data have shown risks of cardiovascular diseases due to trans FA (Willett, Stampfer, Manson, Colditz, Speizer, Rosner, Sampson and Hennekens, 1993), but other studies found no evidence of their toxicity (Wolff, 1994). In the same way, the interest in decreasing dietary saturated fats, always in effect in the United States (Havel, 1997) has been questioned in France (Apfelbaum, 1990).

Besides an effect of FA on human health, increasing attention is paid to milk organoleptic and marketing qualities. In particular, the increase in polyunsaturated FA (PUFA) in milk has negative consequences on milk quality by the development of oxidative processes and off-flavours (see Doreau and Chilliard, 1997a). The use of vitamin E to prevent oxidation, which had been found inefficient, may be positive in some cases (Focant, Mignolet, Marique, Clabots, Breyne, Dalemans and Larondelle, 1998). On another hand, different chemical constituents have been identified in the fat fraction : aldehydes, methyl ketones, esters, lactones and terpenoids (Joblin and Hudson, 1997). Although the effect of these molecules on flavour has not been clearly established, and the threshold of their concentration in milk needed to develop flavour, is unknown, attempts are made to relate their concentration in food to milk flavours. For example, it has been shown that the content of lactones may be enhanced by cereals (Urbach, 1990). In the same way, close relationships have been established between terpenoids in forages and in milk (Viallon, Verdier-Metz, Denoyer, Pradel, Coulon and Berdagué, 1999). Although fragmentary results are available in this area, it appears that in the near future a better knowledge of aromatic compounds in feeds and fat could lead to specific flavours in milk.

A new stage was reached 2-3 years ago. The BSE crisis emphasised the anxiety of the consumer over the health value of their food, and irrational behaviours appeared : blind belief in the media, especially when they predict disasters. To take into account this new behaviour, public authorities established the "precautionary principle" i.e. the quest for maximal safety, to minimize the risks of food for public health. For this reason, it is likely that the ideal milk fat composition will change in the next years, and the ideal cow's diet also.

Nevertheless, scientists must go on trying to manipulate milk fat to provide a rapid response to changes in market demand. Increased PUFA is always an objective, but human consumption of these FA can be achieved by vegetable or marine oils. At the moment, the main objective is to reduce the amount of possibly "negative" compounds in milk, such as saturated and trans unsaturated FA, and to increase "positive" compounds, such as CLA, which has the great advantage of being present mainly in ruminant meat and milk. It can be argued that progress in the technology of products and in transgenesis removes the need for developing feeding strategies to modify milk composition. The authors believe that if agricultural products are priced according to their quality, a part of the increase in value should benefit the farmer, especially if the increase in quality can be achieved by the use of "natural" diets, i.e. of vegetable origin.

Different methods of decreasing butterfat

There are two effective ways to strongly decrease butterfat : concentrate-rich diets and fish oil. Ground grass has the same effect (Clapperton, Kelly, Banks and Rook, 1980), but its use is not discussed here, due to the lack of practical use of this type of feeding. It is also possible to decrease fat in milk using sources of fat other than fish oil, or propionate-enhancing additives, but results are inconsistent.

DIETS RICH IN CEREALS

It has long been known that consumption of a large amount of concentrates decreases milk butterfat content. Many reviews have been devoted to this subject, so this section only considers the main ideas. The decrease in butterfat is generally significant (down to less than 30 g/kg) when the proportion of concentrates in the diet is more than 0.6. Milk fat content lower than 25 g/kg is almost always reached with diets containing 0.7-0.8 concentrates. However fat concentrations lower than 20 g/kg can sometimes be obtained with diets containing no more than 0.65 concentrates (Kennelly, 1996b). No threshold can be determined. Figure 5.1 illustrates the linear relationships between milk fat and proportion of concentrates in the diet on the one hand (Journet and Chilliard, 1985), and between milk fat and dietary acid detergent fibre content on the other (Sutton, 1989). From these reviews and other more recent reviews (Grummer, 1991, Palmquist, Beaulieu and Barbano, 1993, Kennelly, 1996b, Fredeen, 1996), the main factors affecting the response of milk fat to increasing concentrates can be defined as below, provided that the proportion of concentrate is high enough to induce a significant decrease in butterfat :

1. The decrease is higher in mid-lactation than in early lactation (Jenny, Polan and Thye, 1974) because of the supply of mobilised lipids in early lactation.
2. The decrease is greater when concentrates are given 1 or 2 times a day, compared to mixed diets or concentrate distribution in multiple meals (Sutton, Hart, Broster, Elliott and Schuller, 1986).
3. The decrease is greater with cereals than with concentrates rich in fibre (Sutton, Bines, Morant, Napper and Givens, 1987) or soluble carbohydrates, such as molasses, but not compared with sucrose (Sutton, 1989).
4. The decrease is greater when cereals are ground instead of cracked or rolled.
5. The decrease is greater when dietary starch is highly degradable (Sutton, Oldham and Hart, 1980).
6. The decrease requires the absence of a buffer in the diet (Kalscheur, Teter, Piperova and Erdman, 1997a)

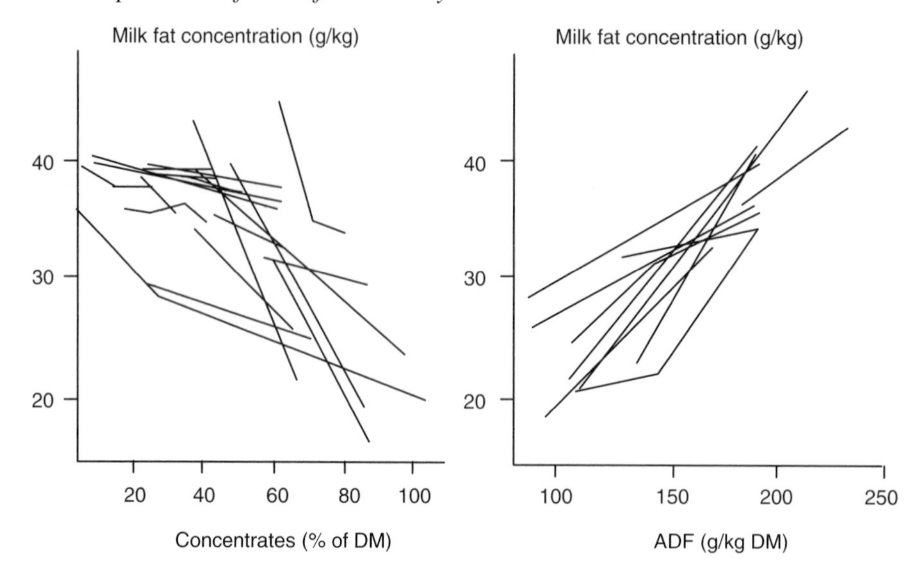

Figure 5.1 Relationship between milk fat concentration and the percentage of concentrates in the diet (Journet and Chilliard, 1985) or the ADF content of the diet (Sutton, 1989) © Journal of Dairy Science, 1989

7. A high individual variation is observed for the same diet (Gaynor, Waldo, Capuco, Erdman, Douglass and Teter, 1995).

Diets with a high proportion of concentrate modify milk fat composition. Short- and medium-chain fatty acids, including palmitic acid, are reduced. Stearic acid is also reduced and C18:1 isomers are increased. Among them, vaccenic acid (trans-11 C18:1) is generally increased. In some experimentss the proportion of palmitic acid did not decrease (Rémond and Journet, 1971), but output of this FA still decreased markedly.

Increasing the proportion of concentrates generally increases the energy concentration of a diet and thus milk protein content. If, in addition, milk yield increases, fat content may decrease by a dilution effect. Generally, high-concentrate diets decrease butterfat to a greater extent (often by more than 10 g/kg) than they increase protein (generally by less than 4 g/kg). Therefore, the protein:fat ratio increases.

DIETS ENRICHED WITH FISH OIL

Fish oils (and oils from marine mammals) are characterised by the presence of highly unsaturated 20- and 22-carbon FA, of which eicosapentaenoic (20:5 n-3) and docosahexaenoic (22:6 n-3) are the most important (Ackman, 1982). Among possible feeds for ruminants, only plankton and algae are also rich in these FA (Givens, Cottrill, Davies, Lee, Mansbridge and Moss, 1997).

Fish oils decrease butterfat consistently. This effect was first shown in 1894 by Sebelien, quoted by Opstvedt (1984), then largely demonstrated in the 1930s mainly with cod liver oil and recently with other oils, such as menhaden. Some experiments with specific oils, such as from salmon or shark failed to depress butterfat (review by Jarrige and Journet, 1959), but this cannot be explained by the FA composition of the oils and is perhaps due to unsuitable designs.

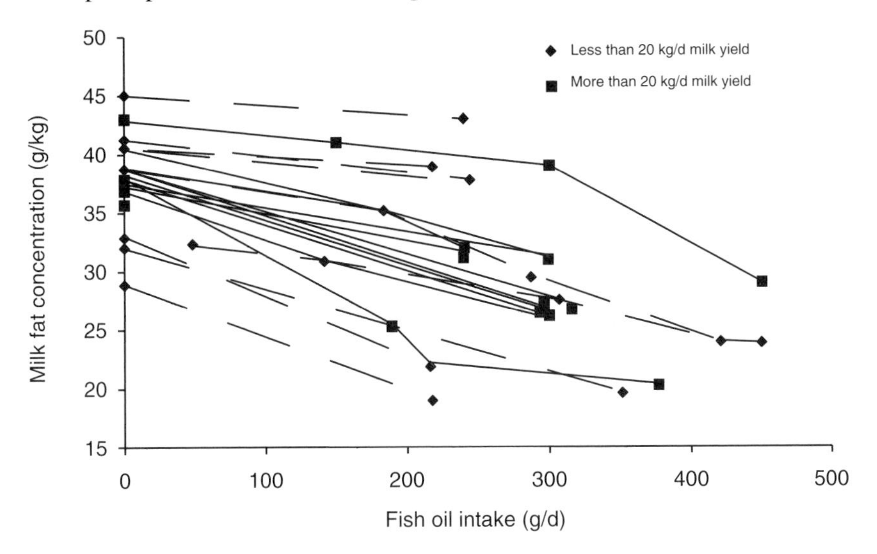

Figure 5.2 Influence of level of inclusion of fish oil on milk fat concentration: review of the literature

The minimal daily supply of fish oil needed to significantly decrease milk butterfat is low. It has been estimated as 55 g by Jarrige and Journet (1959), 75-100 g by Pike, Miller and Short (1994) and 100-150 g from results summarised by Opstvedt (1984). The decrease is frequently observed with fish meal supplements which induce rather low FA supplies (Hussein and Jordan, 1991). Most experiments with fish oil have been carried out with daily supplies of 200 to 400 g. The effect of the proportion of fish oil in the diet appears to be linear (Figure 5.2), but too few experiments have been made with two or more levels of oil inclusion for the range of linearity to be confirmed. Figure 5.2 also shows that the extent of the decrease is independent of the butterfat content of the control diet. On the other hand, the decrease could be more important when milk yield is higher : for cows producing less or more than 20 kg milk daily, it reaches 6.3 and 10.1 g/kg for a mean inclusion of 250 and 305 g fish oil, respectively. This effect has to be confirmed by direct measurements, because results with low-producing cows are generally ancient and obtained with cod liver oil, whereas results with high-producing cows have been obtained more recently with menhaden-type or mixed oils. Moreover it has been observed by Chilliard and Doreau (1997) that the decrease in butterfat is greater for primiparous than for multiparous cows. The magnitude of the decrease in butterfat may increase in diets with a high proportion of concentrates (Pike *et al.*, 1994), but this needs to be confirmed since the response was not found in the study of Nicholson and Sutton (1971).

The decrease in butterfat has been shown to be lower when fish oil is protected by proteins (Storry, Brumby, Hall and Tuckley, 1974) or infused into the abomasum or duodenum (Pennington and Davis, 1975, Chilliard, Chardigny, Chabrot, Ollier, Sébédio and Doreau, 1999). When the added fish oil is hydrogenated, no decrease is observed (Brumby, Storry and Sutton, 1972, Sundstøl, 1974). This shows that the decrease is related mainly to the products of biohydrogenation (which are not the same FA as those produced by industrial hydrogenation), and secondly to the naturally occurring FA.

Most of the decrease in butterfat is due to a large decrease in short- and medium chain FA, that is not compensated for by an increase in long-chain FA. In particular, the incorporation of dietary total 20- and 22-carbon FA into milk is low, ranging from 0.15 (Brumby *et al.*, 1972) to 0.3-0.4 (Storry *et al.*, 1974, Hagemeister, Precht and Barth, 1988). With unprotected oils, the natural C20:5 and C22:6 are in very low concentrations in milk; the main 20- and 22-carbon FA are monounsaturated isomers.

Fish oil, like other lipid sources, decreases milk protein content. The feeding of fish oils with protected amino acids restores milk protein content to a similar level found in a control diet, but no interaction between fish oil and amino acid is observed (Chilliard and Doreau, 1997). On the other hand, fish oil has an adverse effect on milk taste (Lacasse, Kennelly and Ahnadi, 1998), which has to be taken into account for its use.

Algae have been recently used in dairy cow feeding. The inclusion of 910 g algae daily significantly decreased milk butterfat (Franklin, Schingoethe and Baer, 1998).

OTHER SOURCES OF FAT

Among natural or protected fats and oils of any type, except fish oil, only vegetable oils reduce milk fat, by 2.8 g/kg on average (Table 5.1). This mean value masks a high variability : Kalscheur, Teter, Piperova and Erdman (1997b) observed a decrease by 4 g/kg with soyabean oil. However, this decrease is too low to obtain low-fat milk. Feeding vegetable oils and seeds decreases the proportion and output of short- and medium-chain FA and increases the proportion and output of long-chain FA (Kennelly, 1996a).

Most vegetable oils and seeds (including rapeseed, soyabean, cottonseed, sunflower, safflower, maize), are rich in oleic and/or linoleic acid. Linseed oil is characterised by a high content in linolenic acid. Few experiments have been carried out with linseed oil.

According to Kennelly (1996a) linseed oil does not cause a decrease in butterfat although fat composition is modified; the decrease in short- and medium-chain FA being compensated for by an increase in long-chain FA ; moreover low amounts of trans-11 C18:1 are found. However, linseed oil decreased milk fat in mid-lactation but

Table 5.1 Effect of lipid supplementation on milk fat content according to the nature of dietary lipids (reviews by Chilliard, 1993 and Chilliard, Doreau, Gagliostro and Elmeddah, 1993)

Dietary lipids	*No of supplemented groups*	*Milk fat content (supplemented - control, g/kg)*
Animal and blended fats	22	-1.4
Protected tallow	26	+4.0**
Saturated fats	10	+0.5
Calcium salts of palm oil	29	+0.4
Vegetable oils	8	-2.8*
Oilseeds	34	-0.9*
Protected vegetable oils	26	+6.4**

* significantly different from zero, $P<0.05$
** significantly different from zero, $P<0.01$

not in early lactation in a trial by Brunschwig, Kernen and Weill (1997), and increased milk fat in a trial by Nevens, Alleman and Peck (1926). A mixture of rapeseed and linseed oil decreased fat content in a trial by Focant *et al.* (1998). More information is needed on the effect of linseed oil on fat content. In rare cases, other vegetable oils provoke a strong drop in fat concentration. An example is given in Table 5.2 (Doreau and Chilliard, unpublished data). Curiously, the output of long-chain FA was not increased by oils or seeds. Other examples of very low fat contents have been observed by Palmquist (1984) with Ca salts of blended fat or soyabean, and Chilliard and Doreau (unpublished data) with animal fats. In the same way, a reduction of milk fat by 6 g/kg has been observed with palm oil FA (Bremmer, Ruppert, Clark and Drackley, 1998). These atypical results cannot be explained at the moment.

FEED ADDITIVES

Ionophore antibiotics, which are propionate enhancers, are used mainly for beef cattle and very few experiments have been carried out with dairy cows. A moderate negative effect on milk fat is generally observed (reviews by Sprott, Goehring, Beverly and Corah, 1988 and Van Amburgh, 1997 ; Cant, Fredeen, McIntyre, Gunn and Crowe, 1997). Significant decreases may occur at high ionophore supplies: more than 4 g/kg milk with 450 mg/d monensin in a trial by Van der Werf, Jonker and Oldenbroek (1998). No variation in milk protein content was observed.

Monopropylene glycol, a precursor of propionate used for the treatment or prevention of ketosis, decreases milk fat when fed in large amounts (-4 g/kg for an intake of 1.2

Table 5.2 An atypical effect of 44 g/kg inclusion of oil as soyabean oil or whole soyabeans on milk fat (Doreau and Chilliard, unpublished data)

	Control	*Whole soya beans*	*Soyabean oil + soyabean meal*
Milk yield (kg/d)	21.7	21.4	20.6
Milk fat (g/kg)	35.7[a]	25.4[b]	25.5[b]
FA output (g/d)			
Total	798[a]	532[b]	528[b]
C4 to C14	212[a]	107[b]	98[b]
C16:0	207[a]	123[b]	117[b]
C18:0	90	93	81
C18:1	233	223	213
C18:2	27	30	19

Diet based on 0.6 maize silage and 0.4 concentrates.
Means on the same row with different superscripts significantly differ ($P < 0.01$)

kg/d, Hurtaud, Vérité and Rulquin, 1991). This decrease is mainly due to a decrease in 18-carbon FA (Hurtaud, unpublished data).

Manipulation of milk fat composition

The ideal milk fat composition would improve human health, organoleptic quality of dairy products and technological value for processing. Some of these objectives are sometimes contradictory. An increase in n-3 PUFA reduces the risk of cardiovascular diseases (Kinsella, Lokesh and Stone, 1990); PUFA increase the spreadability of butters, but also the tendency for oxidation (reviews by Palmquist *et al.*, 1993 and Doreau and Chilliard, 1997a). On the other hand, the negative effects of trans FA could be counterbalanced by the positive effect of CLA, which is anticarcinogenic (Kelly and Bauman, 1996), both FA being positively related (Jiang, Bjoerck, Fondén and Emanuelson, 1996). The aim of this section is to summarise methods of modifying milk fat composition, without taking a stand about the ideal composition of milk.

INCREASE IN PUFA

With natural feedstuffs the concentration of PUFA in milk fat is always low, even when high amounts of PUFA are fed, due to the high level of ruminal biohydrogenation. With diets rich in concentrates, PUFA (mainly C18:2 and C18:3), which are generally lower than 0.04 of total FA reach 0.10 of total FA (Clapperton *et al.*, 1980). Until now, the only way to increase PUFA in milk fat significantly was to use oils or oilseeds protected by encapsulation with a matrix of protected proteins. It is thus possible to obtain milk containing up to 0.35 linoleic acid with safflower oil and up to 0.22 linolenic acid with linseed oil (McDonald and Scott, 1977, Kennelly, 1996a). An increase in PUFA with 20 or 22 carbons can be achieved by the use of protected fish oils (Storry *et al.*, 1974). It must be noted that a high level of protection requires an excellent control of the process, which is not always achieved in commercial products (Ashes, Gulati, Cook, Scott and Donnelly, 1979). Other techniques for protection of oils fail to increase PUFA in significant proportions : calcium salts of PUFA (Ferlay, Chilliard and Doreau, 1992, Enjalbert, Nicot, Bayourthe, Vernay and Moncoulon, 1997) are extensively hydrogenated, fatty acyl amides from soyabean oil increase linoleic acid to a low extent, up to 6% (Jenkins, Bateman and Block, 1996). Feeding unprotected oilseeds increases PUFA in milk to a moderate extent, depending on nature of seeds : it has been observed with soya beans, especially when roasted (Tice, Eastridge and Firkins, 1994), but no effect was found in some experiments (Chouinard, Girard and Brisson, 1997a).

DECREASE IN SATURATED FA

Another objective in manipulation of milk fat is to decrease the proportion of saturated FA, i.e. to increase unsaturated FA, especially C18:1. The addition of fat to the diet generally reduces the amount of short- and medium-chain saturated FA, due to lower de novo synthesis, but does not modify the amount of saturated long-chain FA. However, substitution of palmitic acid by stearic acid is often observed (e.g. Banks, Clapperton, Kelly, Wilson and Crawford, 1980). The most efficient was to lower the ratio C16:0 : C18:0, to supply a diet with oils poor in palmitic acid, such as rapeseed or sunflower which contain less than 10% of this FA (Grummer, 1991). The reduction in de novo synthesis is thus combined with a low incorporation of exogenous C16:0. A decrease in saturated FA can also be obtained by incomplete hydrogenation of oleic acid. This has been achieved by feeding oleamide, resistant to biohydrogenation (Jenkins, 1998).

VARIATIONS IN TRANS C18:1 AND CONJUGATED C18:2 FA

Feeding oleaginous oils generally involves an increase in trans FA, which can be either moderate (from 3 to 6% in a trial by Banks *et al.*, 1980, from 1 to 8% in trials by Banks, Clapperton and Ferrie, 1976, and Casper, Schingoethe, Middaugh and Baer, 1988) or more pronounced (from 1 to 13% in a trial by Wonsil, Herbein and Watkins, 1994). When oleaginous seeds are fed instead of oils, the amount of trans FA decreases (Banks *et al.*, 1980). Calcium salts of rapeseed oil increase moderately (Enjalbert *et al.*, 1997) or do not increase (Kowalski, Pisulewski and Spanghero, 1999) trans C18:1.

Milk fat produced from fish oil is characterised by a high content in trans-11 C18:1 (Wonsil *et al.*, 1994, Mansbridge and Blake, 1997, Lacasse *et al.*, 1998). The technological process of hydrogenation of fat produces trans FA. For this reason, feeding partially hydrogenated fats increases milk trans FA (Banks, Clapperton, Girdler and Steele, 1984). It has been recently shown that in this case the main isomers are trans-9 and trans-10, and that trans-11 represents only 15% of total trans C18:1 whereas trans-11 is the major isomer when oleaginous oils are fed (Griinari, Chouinard and Bauman, 1997a).

The CLA content of milk has been recently studied by several research groups. Various conjugated isomers of linoleic acid are present in milk, but the most important (ca. 90 - 95 %) is cis-9 trans-11 (Jiang *et al.*, 1996, Chilliard *et al.*, 1999). The main dietary effects on CLA content have been studied in particular by Chouinard, Corneau, Bauman, Butler, Chilliard and Drackley (1998) and reviewed by Demeyer and Doreau (1999). It appears that the highest content of CLA in milk (more than 16 g/kg) is found with dietary supplies of fish oil and vegetable oils rich in linoleic (sunflower, soyabean) and linolenic (linseed) acids. Oils rich in oleic acid (peanut) are less rich in CLA ; with seeds (soyabeans, cottonseeds) no increase in CLA is shown compared with control diets, at levels lower than 8 g/kg of total milk FA.

It appears that dietary factors that increase trans FA also increase CLA. A positive relationship has been established by Jiang *et al.* (1996), but Lin, Boylston, Chang, Luedecke and Shultz (1995) suggest a more complex relationship involving several other FA.

Modifications of ruminal metabolism related to milk fat

The different means of modifying milk fat content or composition – increasing concentrates, supplying dietary lipids and, to a lesser extent, ionophores – induce modifications of carbohydrate fermentation and lipid metabolism in the rumen that may act on mammary lipid synthesis. All contribute to the shift of ruminal volatile FA (VFA) towards propionate and the increase in intermediate long-chain FA originating from biohydrogenation. On the other hand, concentrates and ionophores are dietary factors that reduce biohydrogenation of dietary FA.

INCREASING PROPIONATE PROPORTION

The main factors leading to an increase in propionate when concentrates are fed in large amounts are summarised on Figure 5.3. From this figure it is possible to understand the different factors that modify the VFA pattern: amount of concentrates, amount and

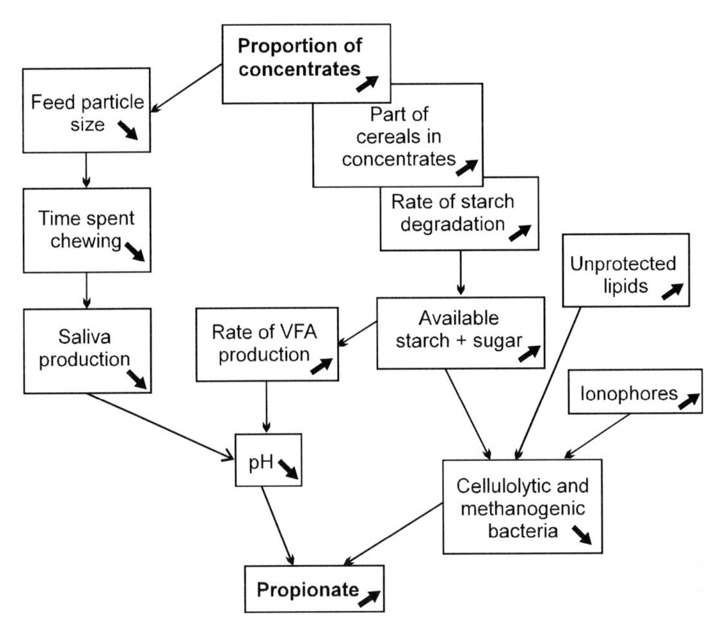

Figure 5.3 Factors modifying propionate proportion in ruminal VFA mixture

degradability of starch, presentation of grain and number of meals, factors modifying the ruminal pH.

Fat supplementation is also known to increase propionic acid in the VFA mixture. This is mainly a consequence of a decrease in cellulolytic and methanogenic microorganisms in the ruminal ecosystem (Prins, Van Nevel and Demeyer, 1972, Maczulak, Dehority and Palmquist, 1981). The proportion of propionate is increased more by oils than by their corresponding seeds, and more by linseed oil than by oils rich in linoleic acid (Jouany, Michalet-Doreau and Doreau, 1998). Fish oils also increase propionic acid to a large extent, although cellulolytic activity in the rumen is not depressed (Doreau and Chilliard, 1997b).

Ionophores antibiotics also enhance propionic acid production by a toxic effect on cellulolytic bacteria (Nagaraja, Newbold, Van Nevel and Demeyer, 1997). However the combination of ionophores and fat does not lead to an extra production of propionate (Zinn, 1988) whereas the addition of ionophores is less effective with a concentrate-rich diet than with a forage-rich diet (Nagajara, Newbold, Van Nevel and Demeyer, 1997).

RUMINAL BIOHYDROGENATION OF PUFA

Biohydrogenation can be defined as the sum of mechanisms related to the modification of dietary unsaturated FA, leading to saturated and intermediate unsaturated compounds.

Biohydrogenation occurs only in free FA, after hydrolysis of triglycerides, phospholipids and galactolipids from the diet. The extent and the factors causing variation in biohydrogenation of 18-carbon FA have been described by Doreau and Ferlay (1994). Linolenic acid is hydrogenated to a greater extent than linoleic acid. Linoleic acid is partially taken up by bacteria (Bauchart, Legay-Carmier, Doreau and Gaillard, 1990) so that its biohydrogenation is incomplete with diets poor in linoleic acid. However, the quantity of precursors for biohydrogenation is not a major cause of variation in biohydrogenation. The disappearance of linoleic and linolenic acids in the rumen is usually reduced by high-concentrate diets, due to a low ruminal pH which limits the first step, hydrolysis of triglycerides (Van Nevel and Demeyer, 1996a). However, factors other than pH may slow down biohydrogenation with high starch intakes, suggesting that cellulolytic microorganisms are more able to hydrolyse lipids and/or to hydrogenate FA than amylolytic ones (Latham, Storry and Sharpe, 1972, Gerson, John and King, 1985). On the other hand, diurnal variations in ruminal propionic acid and in duodenal C18:1 are concomitant (Doreau, Batisse and Bauchart, 1989), suggesting a link due to microbial activity. Biohydrogenation, which has been mainly studied for 18-carbon FA, also occurs to a large extent on 20- and 22-carbon FA (Doreau and Chilliard, 1997b, Van Nevel, Fievez and Demeyer, 1999).

Biohydrogenation may be decreased by protection of lipids. The natural protection of oilseeds by their coat is of limited extent. Technological processes have thus been evaluated. The encapsulation with formaldehyde-treated proteins is the most effective (Gulati, Scott and Ashes, 1997). On the contrary, formation of Ca salts does not prevent biohydrogenation of PUFA (Ferlay *et al*., 1992, Van Nevel and Demeyer, 1996b, Doreau, Demeyer and Van Nevel, 1997).

Antibiotics, especially ionophores, have been shown to decrease biohydrogenation (Kobayashi, Wakita and Hoshino, 1992). This is due to a limitation of lipolysis, independently of a modification of pH (Van Nevel and Demeyer, 1995).

C18:1 TRANS AND CLA

Biohydrogenation of PUFA in the rumen results in a large array of monounsaturated FA, including positional and geometric isomers (Bickerstaffe, Noakes and Annison, 1972). The most important is vaccenic acid (trans-11 C18:1) which is produced from hydrogenation of linoleic acid. The biochemical pathways have been described by Harfoot and Hazlewood (1997). For example, linoleic acid (cis-9, cis-12 C18:2) is rapidly isomerised to a CLA (cis-9 trans-11 C18:2) which is reduced to vaccenic acid (trans-11 C18:1) then partially to stearic acid (C18:0). Hydrogenation of linolenic acid results in the production of trans-11 and trans-15 C18:1. Although the main pathways described in the literature are simple, in practice the high number of isomers (at least 8 trans C18:1 according to Fellner, Sauer and Kramer, 1995) suggests that minor pathways have not yet been elucidated. Moreover, hydrogenation of trans C18:1 isomers to stearic acid has been shown to be high for trans-8, 9 and 10 and low for trans-11 (Kemp, Lander and Gunstone, 1984). The kinetics of production of the different intermediates of PUFA biohydrogenation suggests that the rate of conversion of vaccenic acid into stearic acid is slow (Singh and Hawke, 1979).

Compared with studies of milk composition, few experiments have been published with separation between cis and trans C18:1 isomers at the ruminal or duodenal level, most of them being obtained using imprecise methods. It is thus rather difficult to relate the amount of trans FA to characteristics of the lipid supplement. Nevertheless, it appears that the highest levels of trans FA are obtained with diets rich in linoleic acid or in fish oil (Wonsil *et al.*, 1994, Kalscheur *et al.*, 1997b). Compared with triglycerides, the corresponding calcium salts do not decrease the extent of trans FA formation (Kankare, Antila, Väätäinen, Miettinen and Setälä, 1989, Enjalbert *et al.*, 1997). Conversely, fatty acyl amides, in which linoleic acid is largely hydrogenated, do not lead to a high production of trans FA, for unknown reasons (Jenkins, 1995). From trials by Wonsil *et al.* (1994) and Chilliard *et al.* (1999) it appears that fish oil, although being rich in 20- and 22-carbon FA, results in large amounts of trans-11 C18:1. Some data are not consistent with the general trend. Steele, Noble and Moore (1971) and Børsting, Weisbjerg and Hvelplund (1992) did not observe any increase in duodenal trans FA when vegetable or fish oils were fed. A problem of separation between cis and trans isomers could have occurred in these experiments.

Increasing the proportion of concentrates in the diet is known to lead to incomplete saturation of FA, shown by increases in total C18:1. It has recently been shown that high-concentrate diets increased specifically trans C18:1 (Kalscheur *et al.*, 1997a). In the same way, ionophores involve incomplete biohydrogenation, with an increase in trans C18:1 and, only when linoleic acid is added to the medium, of CLA (Fellner, Sauer and Kramer, 1997). In addition, pasture induces high levels of trans FA (Rulquin, unpublished data).

The positive relationship between trans-11 C18:1 and CLA at the ruminal level was observed by Noble, Moore and Harfoot (1974). Recently, Jiang *et al.* (1996) deduced from this strong correlation that isomerisation and the first reduction of linoleic acid are not rate-limiting. The progressive increase in trans C18:1 observed when PUFA are incubated in vitro shows that the second reduction takes place at a slower rate than the first one. On the other hand, it can be observed that factors which lead to an increase in trans C18:1 are either factors limiting lipolysis, and thus the start of biohydrogenation (concentrates, ionophores, fish oil, etc), or high amounts of linoleic acid.

Control of milk fat secretion

The three theories described below were suggested 30 years ago and progressively enriched or invalidated as research progressed. In fact each of them does not exclude the others; a global theory remains to be proposed.

AMOUNT OF ABSORBED NUTRIENTS

The first theory elaborated to explain the decrease in butterfat is the limitation of nutrients available to mammary gland. In particular, the shift in VFA pattern towards propionate could involve a limitation to the amount of acetate and ß-hydroxybutyrate available for de novo synthesis. This explanation apparently works for all feeds which decrease butterfat : cereals, fish oil and, to a lesser extent, other sources of fat and ionophores. The potential importance of this phenomenon is stressed on Figure 5.4 which plots from available literature arterio-venous differences at the mammary level against arterial concentration (Rulquin, 1997). This figure, which confirms in particular data of Annison, Bickerstaffe and Linzell (1974), shows the constancy of the uptake rate of acetate and b-hydroxybutyrate, even at low arterial concentrations. A limitation of de novo FA synthesis could thus arise from low arterial concentration of precursors. A positive relationship between the ruminal production of the different VFA and arterial concentration of acetate and ß-hydroxybutyrate can be established (Nozière, Martin, Rémond, Kristensen, Bernard and Doreau, to be published, Table 5.3) confirming that high-concentrate diets reduce acetate concentrations (Sutton *et al.*, 1986).

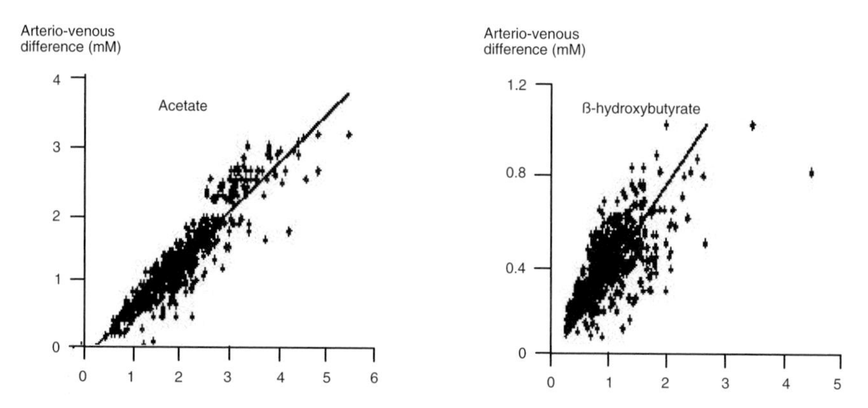

Figure 5.4 Relationship between mannary arterio-venous differences and arterial concentrations of acetate and ß-hydroxybutyrate in dairy cows: review of the literature (Rulquin, 1997)

However, the significance of this statement is limited by the fact that, in many cases, the increases in concentrates result, in terms of VFA production, in an increase in propionate without a decrease in acetate (Annison *et al.*, 1974, review by France and Siddons, 1993). In the same way, Hurtaud, Rulquin and Vérité (1993) and Casse, Rulquin and Huntington (1994), who infused propionate in the rumen, obtained a decrease in milk fat probably related to a decrease in acetate and ß-hydroxybutyrate uptake by the udder, whereas ruminal production of acetate and butyrate was not modified. Annison *et al.* (1974) suggested that the decrease in arterial concentration of acetate with high-concentrate diets was due to a decrease in endogenous production of acetate. On the other hand, propionate may play a direct role in milk fat synthesis. Indeed by ruminal infusions of propionate Hurtaud, Rulquin and Vérité (1998a, Table 5.4) decreased milk fat, the proportion of odd-numbered FA being increased. This confirms that when propionate flow towards the mammary gland is very high, it may contribute to fat synthesis (Massart-Leën, Roets, Peeters and Verbeke, 1983).

Table 5.3 Effect of ruminal infusions of VFA to sheep receiving a hay-based diet on arterial concentrations of precursors for de novo milk fat synthesis (Noziére, Martin, Rémond, Kristensen, Bernard and Doreau, to be published)

VFA ruminal infusion (C2:C3:C4, mmol/h)	*0:0:0*	*43:14:7*	*7:45:6*	*0:9:34*
Composition of ruminal VFA mixture (C2:C3:C4)	72:19:7	70:20:9	48:42:8	47:21:31
Acetate (m*M*)	0.59	0.81	0.59	0.53
ß-hydroxybutyrate (m*M*)	0.24	0.24	0.11	0.43
Propionate (m*M*)	0.01	0.02	0.05	0.02
Glucose (m*M*)	3.47	3.52	3.62	3.41

Table 5.4 Effect of propionate or glucose infusions on milk fat (Hurtaud *et al.*, 1998a)

	Control	*Propionate infusion*	*Glucose infusion*
Milk yield (kg/d)	26.7[a]	28.1[b]	26.3[a]
Milk fat (g/kg)	35.8[a]	31.5[b]	31.8[b]
FA output (g/d)			
Total	936[a]	867[b]	818[c]
C4 to C14	218[a]	196[b]	215[a]
C16:0+C16:1	287	285	286
C18:0+C18:1+C18:2	269[a]	239[b]	176[c]

Diet based on 0.6 maize silage, 0.1 hay and 0.3 concentrates.
Means on the same row with different superscripts significantly differ ($P < 0.05$)

Fat supplementation of diets results in an increase in milk long-chain FA due to incorporation of FA in milk triglycerides. These FA arise mainly from triglycerides of VLDL and chylomicrons, to a lesser extent from non-esterified FA and are generally taken up in proportion to their concentration (Gagliostro, Chilliard and Davicco, 1991, Cant, DePeters and Baldwin, 1993, Rulquin, 1997, Enjalbert, Nicot, Bayourthe and Moncoulon, 1998). It is noticeable that for low non-esterified FA concentrations their extraction rate is negative (Rulquin, 1997, Thivierge, Chouinard, Lévesque, Girard, Seoane and Brisson, 1998), due to the release of FA after partial desesterification by lipoprotein lipase. The possible differential uptake rate from one FA to another has not been clearly shown, results being often inconsistent because of the differences between experiments in lipid composition. For example, with diets not supplemented with lipids, vaccenic acid seems to be taken up to a greater extent than stearic acid or oleic acid (Thompson and Christie, 1991, Enjalbert *et al.*, 1998). With diets supplemented with lipids, the differences between FA become lower, limiting the consequences of the position of FA on the glycerol molecule (Enjalbert *et al.*, 1998).

ROLE OF INSULIN

The effect of increased insulin secretion was first suggested by McClymont and Vallance (1962) from the observation that intravenous infusion of glucose decreased milk fat. Insulin may act by partition of metabolites towards lipogenesis or muscle oxidation rather than towards the mammary gland. It has been recently confirmed that duodenal infusions of glucose decreased milk fat without inducing insulin resistance (Lemosquet, Rideau, Rulquin, Faverdin, Simon and Vérité, 1997, Hurtaud *et al.*, 1998a, Table 5.4). According Hurtaud *et al.*, 1998a and Hurtaud, Rulquin and Vérité, 1998b, glucose infusions strongly decrease milk long-chain FA and slightly increase short- and medium chain FA. Although glucose and propionate both stimulate insulin secretion, only glucose leads to a decrease in lipid mobilisation. It seems, therefore, that the effect of insulin on milk fat is related to the nutrient balance of the animal.

The insulin hypothesis has been questioned by McGuire, Griinari, Dwyer and Bauman (1995) and Griinari, McGuire, Dwyer, Bauman and Palmquist (1997b), who did not decrease milk fat yield (but decreased milk fat percentage in the latter experiment) by raising plasma insulin level using a hyperinsulinemic – euglycemic clamp for several days. These results suggest that insulin level alone is not responsible for decreases in milk fat. On the other hand, feeding fat sometimes decreases plasma insulin level (Palmquist and Moser, 1981, Choi and Palmquist, 1996) whereas in other trials plasma insulin was not modified (review by Chilliard, 1993). Moreover, the response of milk fat to a high-concentrate diet appears independent of plasma insulin level (Gaynor *et al.*, 1995). It can be hypothesized that the increase in insulin is a consequence of the diet, when rich in concentrates and increasing propionate production, that could act on de novo FA synthesis, but that insulin may not play any role in milk fat depression independently of nutrient supply to the mammary gland.

ROLE OF LONG-CHAIN FA

The effect of long-chain FA on the decrease in de novo FA synthesis and their incorporation into triacylglycerols in the mammary gland was shown first with stearic, oleic and linoleic acid and more recently with trans C18:1 FA. The mode of action of FA, especially trans isomers, on the secretion of short- and medium-chain FA by the udder has been shown at two levels : 1) reduction of acetyl CoA carboxylase activity for de novo synthesis, suggested by Palmquist and Jenkins (1980) from the statement that butyrate secretion in milk, which does not depends on the malonyl CoA pathway, often increases when the other short- and medium-chain FA are depressed, and 2) inhibition of FA esterification (Askew, Emery and Thomas, 1971, Hansen and Knudsen, 1987).

The negative effect of trans FA on milk fat was suggested first by Davis and Brown (1970), then by Pennington and Davis (1975). Close negative relationships between the levels of duodenal and milk trans FA on the one hand, and milk fat content on the other, have then been established (Wonsil *et al.*, 1994, Griinari, Bauman and Jones, 1995). The specific role of trans FA has been recently proved by abomasal infusions of FA mixtures rich in cis or trans (250 - 300 g/d) C18:1 isomers (Gaynor, Erdman, Teter, Sampugna, Capuco, Waldo, Hamosh, 1994, Romo, Casper, Erdman and Teter, 1996) so that trans isomers are now considered as the main causes of milk fat depression.

In some cases, an increase in trans FA does not result in a milk fat depression. This has been shown with soyabeans (Chouinard *et al.*, 1997a) and calcium salts of rapeseed oil (Chouinard, Girard and Brisson, 1997b). The increase in trans FA, from 20 to 50-60 g/kg of total milk FA, could have been too low to decrease fat concentration, but did not avoid a decrease in short- and medium-chain FA, compensated for by an increase in long-chain FA. The absence of a decrease in milk fat, when trans FA was markedly increased, has been observed when partially hydrogenated fats are fed (Wonsil *et al.*, 1994, Kalscheur *et al.*, 1997b); a large supply of trans FA (225 g/d) was necessary to decrease milk fat in a trial by Selner and Schultz (1980).

It happens that a strong decrease in milk fat concentration corresponds to a moderate increase in trans FA. This has been observed by Gaynor *et al.* (1995) with concentrate-rich diets and may suggest that factors other than trans FA are involved in milk fat depression. The multifactorial character of milk fat depression is revealed in a trial by Griinari *et al.* (1995) who separated the effects of increasing the proportion of concentrate and the unsaturation of dietary FA in the diet. The former increases ruminal propionate:acetate ratio and plasma insulin level ; the latter raises trans FA in milk, but the decrease in butterfat is obtained by the combination of high concentrate and unsaturated FA in the diet. Surprising results of Focant *et al.* (1998) also question the hypothesis of the pre-eminent role of trans FA in milk fat depression : the inclusion of vitamin E in a diet supplemented with rapeseed and linseed oil restored milk fat to a control level without changes in FA composition, especially trans 18:1.

These apparent limits to the role of trans FA have been recently countered by Griinari *et al.* (1997a). They suggested a specific effect of trans-10 C18:1 instead of the well-known trans-11 C18:1. This hypothesis, which remains to be verified, could explain the results mentioned above. Indeed, the increase in concentrates particularly enhances trans-10 C18:1. On the other hand, partially hydrogenated FA in the diet contain numerous trans 18:1 isomers, of which trans-9 C18:1 is the most abundant. However, when partially hydrogenated fats are fed, the absence of a decrease in milk fat could also be related to the absence of modification of ruminal VFA pattern (Drackley and Elliott, 1993).

Apart from trans FA, it is likely that very long-chain FA have a specific negative action on milk fat. It has been shown that infusions of fish oil into the abomasum or duodenum decrease butterfat, although to a lesser extent than fish oil directly fed or infused into the rumen (Pennington and Davis, 1975, Léger, Sauvant, Hervieu and Ternois, 1994, Chilliard *et al.*, 1999). As neither propionic acid nor trans FA are increased, a direct effect of 20- and 22-carbon FA is assumed.

To a lesser extent, other sources of FA have a direct effect on milk fat. Sources of fats such as saturated fats or calcium salts from palm oil, which do not lead to the production of trans FA or to an increase in propionate, reduce de novo FA synthesis. By duodenal infusion of rapeseed oil, Chilliard, Gagliostro, Fléchet, Lefaivre and Sébastian (1991) reduced de novo FA synthesis, especially 10- to 16-carbon FA, suggesting that acetyl CoA carboxylase activity may decrease. Such a result was found by Drackley, Klusmeyer, Trusk and Clark (1992) with a mixture of unsaturated FA, and by Christensen, Drackley, LaCount and Clark (1994) with rapeseed, soyabean and sunflower FA. On the contrary, duodenal infusions of palmitic, stearic and oleic acid did not decrease de novo synthesis in a trial by Enjalbert *et al.* (1998).

Trans FA are not synthesized by the udder. However, a mammary synthesis of cis-9 trans-11 C18:2 from trans-11 C18-1 by the stearoyl CoA desaturase has been recently shown (Corl, Chouinard, Bauman, Dwyer, Griinari and Nurmela, 1998). This action may increase milk CLA content, but produced very little decrease in the proportion of milk trans FA, which is present in greater proportion than CLA in milk. By contrast, trans-11 C18:1 could have a negative effect on desaturation of stearic acid to oleic acid (Enjalbert *et al.*, 1998).

TOWARDS A GLOBAL THEORY

Among these mechanisms, it appears that the direct action of insulin is highly questionable. The action of trans FA is now sometimes considered as sufficient to explain low milk fat (Griinari *et al.*, 1997a). Although the most important decrease in milk fat appears to be due to an increase in trans FA, this affirmation has to be moderated by the above mentioned results obtained with infusions of propionate and glucose, and of cis FA present in lipid supplements. The comparative action of these FA could be achieved by developing techniques of abomasal or duodenal infusions of individual FA. Moreover, since factors leading to the production of trans FA also led to an increase in propionic acid, it is possible to find excellent positive correlations between ruminal propionate and decreases in milk fat. The specific role of trans FA is now evident, but the absence of a role for the limitation of precursors of de novo synthesis remains to be proved, for example by ruminal infusions of acetate and butyrate with high-concentrate diets.

Conclusion

Nutritional means to modify milk fat amount and composition exist. However they have to be considered together with the other consequences of their use. For example, the proportion of concentrates in dairy diets is generally determined by factors other than decreasing milk fat: economical consideratoins, risks of acidosis . . . In the same way, the use of fish oil and some other additives may adversely affect the taste of dairy products and have a negative image for the consumer. Finally, it can be considered that the decrease in milk fat content is associated with increases in trans FA and CLA.

Regarding composition of the fat fraction in milk, several questions have to be answered in the near future. Some of them depend on the medical profession : is the simultaneous increase in trans FA and CLA globally beneficial or detrimental for human health? Besides this, a large area of research remains for animal nutritionists. Knowledge of the control of ruminal lipid metabolism and of the mammary milk fat synthesis has to be increased. New techniques of protection of PUFA against ruminal microorganisms could be developed, provided they are accepted by the consumer and do not lead to problems of oxidation or off-flavours. Finally, ways to improve the organoleptic quality of milks have to be explored : maximal incorporation of specific FA in the diet, association of fats with antioxidants, development of specific flavours in milk.

References

Ackman, R.G. (1982) Fatty acid composition of fish oils. In *Nutritional Evaluation of Long-Chain Fatty Acids in Fish Oil*, pp. 25-88 Edited by S.M. Barlow and M.E. Stansby. Acad. Press, London, UK

Annison, E.F., Bickerstaffe, R. and Linzell, J.L. (1974) Glucose and fatty acid metabolism in cows producing milk of low fat content. *Journal of Agricultural Science, Cambridge*, **82**, 87-95

Apfelbaum, M. (1990) Cholestérol, c'est fou ! In *Cholestérol et Prévention Primaire*, pp. 153-157, CIDIL, Paris, France

Ashes, J.R., Gulati, S.K., Cook, L.J., Scott, T.W. and Donnelly, J.B. (1979) Assessing the biological effectiveness of protected lipid supplements for ruminants. *Journal of the American Oil Chemists' Society*, **56**, 522-527

Askew, E.W., Emery, R.S. and Thomas, J.W. (1971) Fatty acid specificity of glyceride synthesis by homogenates of bovine mammary tissue. *Lipids*, **6**, 777-782

Banks, W., Clapperton, J.L. and Ferrie, M.E. (1976) Effect of feeding fat to dairy cows receiving a fat-deficient basal diet. II. Fatty acid composition of the milk fat. *Journal of Dairy Research,* **43**, 219-227

Banks, W., Clapperton, J.L., Kelly, M.E., Wilson, A.G. and Crawford, R.J.M. (1980) The yield, fatty acid composition and physical properties of milk fat obtained by feeding soya oil to dairy cows. *Journal of the Science of Food and Agriculture*, **31**, 368-374

Banks, W, Clapperton, J.L., Girdler, A.K. and Steele, W. (1984) Effect of inclusion of different forms of dietary fatty acid on the yield and composition of cow's milk. *Journal of Dairy Research*, **51**, 387-395

Bauchart, D., Legay-Carmier, F., Doreau, M. and Gaillard, B. (1990) Lipid metabolism of liquid-associated and solid-adherent bacteria in rumen contents of dairy cows offered lipid-supplemented diets. *British Journal of Nutrition*, **63**, 563-578

Bickerstaffe, R., Noakes, D.E. and Annison, E.F. (1972) Quantitative aspects of fatty acid biohydrogenation, absorption and transfer into milk fat in the lactating goat, with special reference to the cis and trans-isomers of octadecenoate and linoleate. *Biochemistry Journal,* **130**, 607-617

Børsting, C.F., Weisbjerg, M.R. and Hvelplund, T. (1992) Fatty acid digestibility in lactating cows fed increasing amounts of protected vegetable oil, fish oil or saturated fat. *Acta Agriculturae Scandinavica, Section A, Animal Science,* **42**, 148-156

Bremmer, D.R., Ruppert, L.D., Clark, J.H. and Drackley, J.K. (1998) Effects of chain length and unsaturation of fatty acid mixtures infused into the abomasum of lactating dairy cows. *Journal of Dairy Science,* **81**, 176-188

Brumby, P.E., Storry, J.E. and Sutton, J.D. (1972) Metabolism of cod-liver oil in relation to milk fat secretion. *Journal of Dairy Research*, **39**, 167-182

Brunschwig, P., Kernen, P. and Weill, P. (1997) Effets de l'apport d'un concentré enrichi en acides gras polyinsaturés sur les performances de vaches laitières à l'ensilage de maïs. *Rencontres Recherches Ruminants*, **4**, 361

Cant, J.P., DePeters, E.J. and Baldwin, R.L. (1993) Mammary uptake of energy metabolites in dairy cows fed fat and its relationship to milk protein depression. *Journal of Dairy Science*, **76**, 2254-2265

Cant, J.P., Fredeen, A.H., MacIntyre, T., Gunn, J. and Crowe, N. (1997) Effect of fish oil and monensin on milk composition in dairy cows. *Canadian Journal of Animal Science*, **77**, 125-131

Casper, D.P., Schingoethe, D.J., Middaugh, R.P. and Baer, R.J. (1988) Lactational responses of dairy cows to diets containing regular and high oleic acid sunflower seeds. *Journal of Dairy Science*, **71**, 1267-1274

Casse, E.A., Rulquin, H. and Huntington, G.B. (1994) Effect of mesenteric vein infusion of propionate on splanchnic metabolism in primiparous holstein cows. *Journal of Dairy Science,* **77**, 3296-3303

Chilliard, Y., Gagliostro, G., Fléchet, J., Lefaivre, J. and Sebastian, I. (1991) Duodenal rapeseed oil infusion in early and midlactation cows. 5. Milk fatty acids and adipose tissue lipogenic activities. *Journal of Dairy Science*, **74**, 1844-1854

Chilliard, Y. (1993) Dietary fat and adipose tissue metabolism in ruminants, pigs, and rodents: a review. *Journal of Dairy Science*, **76**, 3897-3931

Chilliard, Y., Doreau, M., Gagliostro, G. and Elmeddah, E. (1993) Addition de lipides protégés (encapsulés ou savons de calcium) à la ration de vaches laitières. Effets sur les performances et la composition du lait. *INRA Productions Animales*, **6**, 139-150.

Chilliard, Y. and Doreau, M. (1997) Influence of supplementary fish oil and rumen-protected methionine on milk yield and composition in dairy cows. *Journal of Dairy Research,* **64**, 173-179

Chilliard, Y, Chardigny, J.M., Chabrot, J, Ollier, A, Sebedio, J.L. and Doreau, M. (1999) Effects of ruminal or postruminal fish oil supply on conjugated linoleic acid (CLA) content of cow milk fat. *Proceedings of the Nutrition Society,* in press

Choi, B.R. and Palmquist, D.L. (1996) High fat diets increase plasma cholecystokinin and pancreatic polypeptide, and decrease plasma insulin and feed intake in lactating cows. *Journal of Nutrition*, **126**, 2913-2919

Chouinard, P.Y., Girard, V. and Brisson, G.J. (1997a) Performance and profiles of milk fatty acids of cows fed full fat, heat-treated soybeans using various processing methods. *Journal of Dairy Science*, **80**, 334-342

Chouinard, P.Y., Girard, V. and Brisson, G.J. (1997b) Lactational response of cows to different concentrations of calcium salts of canola oil fatty acids with or without bicarbonates. *Journal of Dairy Science*, **80**, 1185-1193

Chouinard, P.Y., Corneau, L., Bauman, D.E., Butler, W.R., Chilliard, Y. and Drackley, J.K. (1998) Conjugated linoleic acid content of milk from cows fed different sources of dietary fat. *Journal of Dairy Science,* **81** (Suppl.) 1, 233

Christensen, R.A., Drackley, J.K., LaCount, D.W. and Clark, J.H. (1994) Infusion of four long-chain fatty acid mixtures into the abomasum of lactating dairy cows. *Journal of Dairy Science,* **77**, 1052-1069

Clapperton, J.L., Kelly, M.E., Banks J.M. and Rook, J.A.F. (1980) The production of milk rich in protein and low in fat, the fat having a high polyunsaturated fatty acid content. *Journal of the Science of Food and Agriculture*, **31**, 1295-1302

Corl, B.A., Chouinard, P.Y., Bauman, D.E., Dwyer, D.A., Griinari, J.M. and Nurmela, K.V. (1998) Conjugated linoleic acid in milk fat of dairy cows originates in part by endogenous synthesis from trans-11 octadedenoic acid. *Journal of Dairy Science,* **81** (Suppl.) 1, 233

Davis, C.L. and Brown, R.E. (1970) Low-fat milk syndrome. In *Physiology of Digestion and Metabolism in the Ruminant* pp. 545-565 Edited by A.T. Phillipson. Oriel Press, Newcastle, UK

Demeyer, D. and Doreau, M. (1999) Targets and means for altering meat and milk lipids. *Proceedings of the Nutrition Society,* in press

Doreau, M., Batisse, V. and Bauchart, D. (1989) Appréciation de l'hydrogénation des acides gras alimentaires dans le rumen de la vache : étude méthodologique préliminaire. *Annales de Zootechnie,* **38**, 139-144

Doreau, M. and Ferlay, A. (1994) Digestion and utilisation of fatty acids by ruminants. *Animal Feed Science and Technology*, **45**, 379-396

Doreau, M. and Chilliard, Y. (1997a) Digestion and metabolism of dietary fat in farm animals. *British Journal of Nutrition*, **78** (Suppl.) 1, S15-S35

Doreau, M. and Chilliard, Y. (1997b) Effects of ruminal or postruminal fish oil supplementation on intake and digestion in dairy cows. *Reproduction Nutrition Development*, **37**, 113-124

Doreau, M., Demeyer, D.I. and Van Nevel, C.J. (1997) Transformations and effects of unsaturated fatty acids in the rumen. Consequences on milk fat secretion. In *Milk Composition, Production and Biotechnology* pp 73-92. Edited by R.A.S. Welch, D.J.W. Burns, S.R. Davis, A.I. Popay and C.G. Prosser. CAB International, Oxon, UK

Drackley, J.K., Klusmeyer, T.H., Trusk, A.M. and Clark, J.H. (1992) Infusion of long-chain fatty acids varying in saturation and chain length into the abomasum of lactating dairy cows. *Journal of Dairy Science,* **75**, 1517-1526

Drackley, J.K. and Elliott, J.P. (1993) Milk composition, ruminal characteristics, and nutrient utilization in dairy cows fed partially hydrogenated tallow. *Journal of Dairy Science,* **76**, 183-196

Enjalbert, F., Nicot, M.C., Bayourthe, C., Vernay, M. and Moncoulon, R. (1997) Effects of dietary calcium soaps of unsaturated fatty acids on digestion, milk composition and physical properties of butter. *Journal of Dairy Research*, **64**, 181-195

Enjalbert, F., Nicot, M.C., Bayourthe, C. and Moncoulon, R. (1998) Duodenal infusions of palmitic, stearic or oleic acids differently affect mammary gland metabolism of fatty acids in lactating dairy cows. *Journal of Nutrition*, **128**, 1525-1532

Fellner, V., Sauer, F.D. and Kramer, J.K.G. (1995) Steady-state rates of linoleic acid biohydrogenation by ruminal bacteria in continuous culture. *Journal of Dairy Science,* **78**, 1815-1823

Fellner, V., Sauer, F.D. and Kramer, J.K.G. (1997) Effect of nigericin, monensin and tetronasin on biohydrogenation in continuous flow-through ruminal fermenters. *Journal of Dairy Science*, **80**, 921-928

Ferlay, A., Chilliard, Y. and Doreau, M. (1992) Effects of calcium salts differing in fatty acid composition on duodenal and milk fatty acid profiles in dairy cows. *Journal of the Science of Food and Agriculture*, **60**, 31-37

Focant, M., Mignolet, E., Marique, M., Clabots, F., Breyne, T., Dalemans, D. and Larondelle, Y. (1998) The effect of vitamin E supplementation of cow diets containing rapeseed and linseed on the prevention of milk fat oxidation. *Journal of Dairy Science*, **81**, 1095-1101.

France, J. and Siddons, R.C. (1993) Volatile fatty acid production. In *Quantitative Aspects of Ruminant Digestion and Metabolism*, pp. 107-121. Edited by M. Forbes and J. France. CAB International, Oxon, UK

Franklin, S.T., Schingoethe, D.J. and Baer, R.J. (1998) Production and feed intake of cows fed diets high in omega-3 fatty acids from unprotected and ruminally protected algae. *Journal of Dairy Science*, **81** (Suppl.) 1, 353

Fredeen, A.H. (1996) Considerations in the nutritional modification of milk composition. *Animal Feed Science and Technology*, **59**, 185-197

Gagliostro, G., Chilliard, Y. and Davicco, M.J. (1991) Duodenal rapeseed oil infusion in early and midlactation cows. 3. Plasma hormones and mammary apparent uptake of metabolites. *Journal of Dairy Science,* **74**, 1893-1903

Gaynor, P.J., Erdman, R.A., Teter, B.B., Sampugna, J., Capuco, A.V., Waldo, D.R. and Hamosh, M. (1994) Milk fat yield and composition during abomasal infusion of *cis* or *trans* octadecenoates in Holstein cows. *Journal of Dairy Science,* **77**, 157-165

Gaynor, P.J., Waldo, D.R., Capuco, A.V., Erdman, R.A., Douglass, L.W. and Teter, B.B. (1995) Milk fat depression, the glucogenic theory and *trans*-$C_{18:1}$ fatty acids. *Journal of Dairy Science*, **78**, 2008-2015

Gerson, T., John, A. and King, A.S.D. (1985) The effects of dietary starch and fibre on the *in vitro* rates of lipolysis and hydrogenation by sheep rumen digest. *Journal of Agricultural Science, Cambridge*, **105**, 27-30

Givens, D.I., Cottrill, B.R., Davies, M., Lee, P., Mansbridge R. and Moss A.R. (1997) Sources of n-3 polyunsaturated fatty acids additional to fish oil for livestock diets. *Report, MAFF project OC9514*, Ministry of Agriculture, Fisheries and Food, London, UK

Griinari, J.M., Bauman, D.E. and Jones, L.R. (1995) Low milk fat in New York Holstein herds. *Proceedings Cornell Nutrition Conference for Feed Manufacturers*, pp. 96-105, Cornell Univ., Ithaca, NY, USA

Griinari, J.M., Chouinard, P.Y. and Bauman, D.E. (1997a) *Trans* fatty acid hypothesis of milk fat depression revised. *Proceedings Cornell Nutrition Conference for Feed Manufacturers,* pp. 208-216, Cornell Univ. Ithaca, NY, USA

Griinari, J.M., McGuire, M.A., Dwyer, D.A., Bauman, D.E. and Palmquist, D.L. (1997b) Role of insulin in the regulation of milk fat synthesis in dairy cows. *Journal of Dairy Science*, **80**, 1076-1084

Grummer, R.R. (1991) Effect of feed on the composition of milk fat. *Journal of Dairy Science*, **74**, 3244-3257

Gulati, S.K., Scott, T.W. and Ashes, J.R. (1997) In-vitro assessment of fat supplements for ruminants. *Animal Feed Science and Technology*, **64**, 127-132

Hagemeister, H., Precht, D. and Barth, C.A. (1988) Zum Transfer von Omega-3-Fettsäuren in das Milchfett bei Kühen. *Milchwissenschaft*, **43**, 153-158

Hansen, H.O. and Knudsen, J. (1987) Effect of exogenous long-chain fatty acids on lipid biosynthesis in dispersed ruminant mammary gland epithelial cells: esterification of long-chain exogenous fatty acids. *Journal of Dairy Science*, **70**, 1344-1349

Harfoot, C.G. and Hazlewood, G.P. (1997). Lipid metabolism in the rumen. In: *The Rumen Microbial Ecosystem* pp. 382-426. Edited by P.N. Hobson and C.S. Stewart. Blackie Acad. Prof., London, UK

Havel, R.J. (1997) Milk fat consumption and human health: Recent NIH and other American governmental recommendations. In: *Milk Composition, Production and Biotechnology* pp 13-22. Edited by R.A.S. Welch, D.J.W. Burns, S.R. Davis, A.I. Popay and C.G. Prosser. CAB International, Oxon, UK

Hurtaud, C., Vérité, R. and Rulquin, H. (1991) Effet du niveau et de la nature des nutriments énergétiques sur la composition du lait et ses aptitudes technologiques : effet du monopropylène glycol. *Annales de Zootechnie*, **40**, 259-269

Hurtaud, C., Rulquin, H. and Vérité, R. (1993) Effect of infused volatile fatty acids and caseinate on milk composition and coagulation in dairy cows. *Journal of Dairy Science,* **76**, 3011-3020

Hurtaud, C., Rulquin, H. and Vérité, R. (1998a) Effects of level and type of energy source (volatile fatty acids or glucose) on milk yield, composition and coagulating properties in dairy cows. *Reproduction Nutrition Development*, **38**, 315-330

Hurtaud, C., Rulquin, H. and Vérité, R. (1998b) Effects of graded duodenal infusions of glucose on yield and composition of milk from dairy cows. 1. Diets based on corn silage. *Journal of Dairy Science*, **81**, 3239-3247

Hussein, H.S. and Jordan, R.M. (1991) Fish meal as a protein supplement in ruminant diets : a review. *Journal of Dairy Science*, **69**, 2147-2156

Jarrige, R. and Journet, M. (1959) Influence des facteurs alimentaires et climatiques sur la teneur en matières grasses du lait. *Annales de Nutrition et d'Alimentation*, **13**, A233-A277

Jenkins, T.C. (1995) Butylsoyamide protects soybean oil from ruminal biohydrogenation: effects of butylsoyamide on plasma fatty acids and nutrient digestion in sheep. *Journal of Animal Science,* **73**, 818-823

Jenkins, TC, Bateman, H.G. and Block, S.M. (1996) Butylsoyamide increases unsaturation of fatty acids in plasma and milk of lactating dairy cows. *Journal of Dairy Science*, **79**, 585-590

Jenkins, T.C. (1998) Fatty acid composition of milk from Holstein cows fed oleamide or canola oil. *Journal of Dairy Science*, **81**, 794-800

Jenny, B.F., Polan, C.E. and Thye, F.W. (1974) Effects of high grain feeding and stage of lactation on serum insulin, glucose and milk fat percentage in lactating cows. *Journal of Nutrition*, **104**, 379-385

Jiang, J., Bjoerck, L., Fondén, R. and Emanuelson, M. (1996) Occurrence of conjugated *cis*-9, *trans*-11-octadecadienoic acid in bovine milk : effects of feed and dietary regimen. *Journal of Dairy Science*, **79**, 438-445

Joblin, K.N. and Hudson, J.A. (1997) Management of milk flavour through the manipulation of rumen microorganisms. In : *Milk Composition, Production and Biotechnology* pp. 455-463. Edited by R.A.S. Welch, D.J.W. Burns, S.R. Davis, A.I. Popay and C.G. Prosser. CAB International, Oxon, UK

Jouany, J.P., Michalet-Doreau, B. and Doreau, M. (1998) Manipulation of the rumen ecosystem to support high-performance beef cattle. *Proceedings, Symposium Series 2, 8th World Conf. Anim. Prod, Seoul, Korea*, 109-130

Journet, M. and Chilliard, Y. (1985) Influence de l'alimentation sur la composition du lait. 1. Taux butyreux : facteurs généraux. *Bulletin Technique CRZV Theix, INRA*, **60**, 13-23

Kalscheur, K.F., Teter, B.B., Piperova, L.S. and Erdman, R.A. (1997a) Effect of dietary forage concentration and buffer addition on duodenal flow of trans-C18:1 fatty acids and milk fat production in dairy cows. *Journal of Dairy Science,* **80**, 2104-2114

Kalscheur, K.F., Teter, B.B., Piperova, L.S. and Erdman, R.A. (1997b) Effect of fat source on duodenal flow of *trans*-$C_{18:1}$, fatty acids and milk fat production in dairy cows. *Journal of Dairy Science*, **80**, 2115-2126

Kankare, V., Antila, V., Väätäinen, H., Miettinen, H. and Setälä, J. (1989) The effect of calcium salts of fatty acids added to the feed of dairy cows on the fatty acid composition of milk fat. *Finnish Journal of Animal Science,* **1**, 1-9

Kelly, M.L. and Bauman, D.E. (1996) Conjugated linoleic acid : a potent anticarcinogen found in milk fat. *Proceedings Cornell Nutrition Conference for Feed Manufacturers,* pp 68-74. Cornell Univ. Ithaca NY, USA

Kemp, P., Lander, D.J. and Gunstone, F.D. (1984) The hydrogenation of some *cis*- and *trans*-octadecenoic acids to stearic acid by a rumen *Fusocillus* sp. *British Journal of Nutrition,* **52**, 165-170

Kennelly, J.J. (1996a) The fatty acid composition of milk fat as influenced by feeding oilseeds. *Animal Feed Science and Technology*, **60**, 137-152

Kennelly, J.J. (1996b) Producing milk with 2.5 % fat - the biology and health implications for dairy cows. *Animal Feed Science and Technology*, **60**, 161-180

Kinsella, J., Lokesh, B. and Stone, R.A. (1990) Dietary n-3 polyunsaturated fatty acids and amelioration of cardiovascular disease : possible mechanisms. *American Journal of Clinical Nutrition*, **52**, 1-28

Kobayashi, Y., Wakita, M. and Hoshino, S. (1992) Effects of ionophore salinomycin on nitrogen and long-chain fatty acid profiles of digesta in the rumen and the duodenum of sheep. *Animal Feed Science and Technology*, **36**, 67-76

Kowalski, Z.M., Pisulewski, P.M. and Spanghero, M. (1999) Effects of calcium soaps of rape seed fatty acids and protected methionine on milk yield and composition in dairy cows. *Journal of Dairy Research*, in press

Lacasse, P., Kennelly, J.J. and Ahnadi, C.E. (1998) Feeding protected and unprotected fish oil to dairy cows : II Effect on milk fat composition. *Journal of Animal Science*, **76** (Suppl.1), 231

Latham, M.J., Storry, J.E. and Sharpe, M.E. (1972) Effect of low-roughage diets on the microflora and lipid metabolism in the rumen. *Applied Microbiology*, **24**, 871-877

Léger, C., Sauvant, D., Hervieu, J. and Ternois, F. (1994) Influence of duodenal infusions of EPA and DHA on the lipidic milk secretion of the dairy goat. *Annales de Zootechnie*, **43**, 297

Lemosquet, S., Rideau, N., Rulquin, H., Faverdin, P., Simon, J. and Vérité, R. (1997) Effects of a duodenal glucose infusion on the relationship between plasma concentrations of glucose and insulin in dairy cows. *Journal of Dairy Science*, **80**, 2854-2865

Lin, H., Boylston, T.D., Chang, M.J., Luedecke, L.O. and Shultz, T.D. (1995) Survey of the conjugated linoleic acid contents of dairy products. *Journal of Dairy Science,* **78**, 2258-2365

Maczulak, A.E., Dehority, B.A. and Palmquist, D.L. (1981) Effect of long-chain fatty acids on growth of rumen bacteria. *Applied and Environmental Microbiology*, **42**, 856-862

Mansbridge, R.J. and Blake J.S. (1997) The effect of feeding fish oil on the fatty acid composition of bovine milk. *Proceedings of the British Society of Animal Science*, p. 22, BSAS, Penicuik, UK

Massart-Leën, A.M., Roets, E., Peeters, G. and Verbeke, R. (1983) Propionate for fatty acid synthesis by the mammary gland of the lactating goat. *Journal of Dairy Science*, **66**, 1445-1554

McClymont, G.L. and Vallance, S. (1962) Depression of blood glyderides and milk-fat synthesis by glucose infusion. *Proceedings of the Nutrition Society,* **21**, XLI-XLII

McDonald, I.W. and Scott, T.W. (1977) Foods of ruminant origin with elevated content of polyunsaturated fatty acids. *World Review of Nutrition and Dietetics*, **26**, 144

McGuire, M.A., Griinari, J.M., Dwyer, D.A. and Bauman, D.E. (1995) Role of insulin in the regulation of mammary synthesis of fat and protein. *Journal of Dairy Science,* **78**, 816-824

Nagajara, T.G., Newbold, C.J., Van Nevel, C. and Demeyer, D.I. (1997) Manipulation of ruminal fermentation. In : *The rumen microbial ecosystem,* pp. 523-632. Edited by P.N. Hobson and C.S. Stewart, Blackie Acad. Prof., London

Nevens, W.B., Alleman, M.B. and Peck, L.T. (1926) The effect of fat in the ration upon the percentage fat content of the milk. *Journal of Dairy Science*, **9**, 307-345

Nicholson, J.W.G. and Sutton, J.D. (1971) Some effects of unsaturated oils given to dairy cows with rations of different roughage content. *Journal of Dairy Research*, **38**, 363-372

Noble, R.C., Moore, J.H. and Harfoot, C.G. (1974) Observations on the pattern on biohydrogenation of esterified and unesterified linoleic acid in the rumen. *British Journal of Nutrition*, **31**, 99-108

Nozière, P., Martin, C., Rémond, D., Kristensen, N.B., Bernard, R. and Doreau, M. (to be published) Effect of composition of ruminally-infused short-chain fatty acids on net fluxes of nutrients across portal-drained viscera in underfed ewes. *British Journal of Nutrition*.

Opstvedt, J. (1984) Fish fats. In: *Fats in Animal Nutrition* pp 53-83. Edited by J. Wiseman, Butterworths, London, UK

Palmquist, D.L. and Jenkins, T.C. (1980) Fat in lactation rations for dairy: a review. *Journal of Dairy Science*, **63**, 1-14

Palmquist, D.L. and Moser, E.A. (1981) Dietary fat effects on blood insulin, glucose utilization, and milk protein content of lactating cows. *Journal of Dairy Science*, **64**, 1664-1670

Palmquist, D.L. (1984) Calcium soaps of fatty acids with varying unsaturation as fat supplements for lactating cows. *Canadian Journal of Animal Science*, **64** (Suppl.) 240-241

Palmquist, D.L., Beaulieu, A.D. and Barbano, D.M. (1993) Feed and animal factors influencing milk fat composition. *Journal of Dairy Science*, **76**, 1753-1771

Pennington, J.A. and Davis, L. (1975) Effects of intraruminal and intra-abomasal additions of cod-liver oil on milk fat production in the cow. *Journal of Dairy Science* **58**, 49-55

Pike, I.H., Miller, E.L. and Short, K. (1994) The role of fish meal in dairy cow feeding. IFOMA *Tech. Bull.*, no 27, 26 pp., IFOMA, St Albans, UK

Prins, R.A., Van Nevel, C.J. and Demeyer, D.I. (1972) Pure culture studies of inhibitors for methanogenic bacteria. *Antonie van Leeuwenhoek*, **38**, 281-287

Rémond, B. and Journet, M. (1971) Alimentation des vaches laitières avec des rations à forte proportion d'aliments concentrés. I. Quantités ingérées et production laitière. *Annales de Zootechnie*, **20**, 165-184

Romo, G.A., Casper, D.P., Erdman, R.A. and Teter, B.B. (1996) Abomasal infusion of *cis* or *trans* fatty acid isomers and energy metabolism of lactating dairy cows. *Journal of Dairy Science*, **79**, 2005-2015

Rulquin, H. (1997) Régulation de la synthèse et de la sécrétion des constituants du lait chez les ruminants. *Rencontres Recherches Ruminants*, **4**, 327-338

Selner, D.R. and Schultz, L.H. (1980) Effects of feeding oleic acid or hydrogenated vegetable oils to lactating cows. *Journal of Dairy Science*, **63**, 1235-1241

Singh, S. and Hawke, J.C. (1979) The in vitro lipolysis and biohydrogenation of monogalactosyldiglyceride by whole rumen contents and its fractions. *Journal of the Science of Food and Agriculture,* **30**, 603-612

Sprott, L.R., Goehring, T.B., Beverly, J.R. and Corah, L.R. (1988) Effects of ionophores on cow herd production : a review. *Journal of Animal Science*, **66**, 1340-1346

Steele, W., Noble, R.C. and Moore, J.H. (1971) The effects of 2 methods of incorporating soybean oil into the diet on milk yield and composition in the cow. *Journal of Dairy Science,* **38**, 43-48

Storry, J.E., Brumby, P.E., Hall, A.J. and Tuckley, B. (1974) Effects of free and protected forms of codliver oil on milk fat secretion in the dairy Cow. *Journal of Dairy Science*, **57**, 1046-1049

Sundstøl, F. (1974) Hydrogenated marine fat as feed supplement. IV. Hydrogenated marine fat as feed supplement. *Meldinger fra Norges Landbrukshogskole*, **162**, 50 pp

Sutton, J.D., Oldham, J.D. and Hart, I.C. (1980) Products of digestion, hormones and energy utilization in milking cows given concentrates containing various proportions of barley or maize. In *Energy Metabolism*, pp. 303-306. Edited by L.E. Mount. Butterworths, London, UK

Sutton, J.D., Hart, I.C., Broster, W.H., Elliott, R.J. and Schuller, E. (1986) Feeding frequency for lactating cows : effects on rumen fermentation and blood metabolites and hormones. *British Journal of Nutrition*, **56**, 181-192

Sutton, J.D., Bines, J.A., Morant, S.V., Napper, D.J. and Givens, D.I. (1987) A comparison of starchy and fibrous concentrates for milk production, energy utilization and hay intake by Friesian cows. *Journal of Agricultural Science, Cambridge*, **109**, 375-386

Sutton, J.D. (1989) Altering milk composition by feeding. *Journal of Dairy Science*, **72,** 2801-2814

Thivierge, M.C., Chouinard, P.Y., Lévesque, J., Girard, V., Seoane, J.R. and Brisson, G.J. (1998) Effects of buffers on milk fatty acids and mammary arteriovenous differences in dairy cows fed Ca salts of fatty acids. *Journal of Dairy Science*, **81**, 2001-2010

Thompson, G.E. and Christie, W.W. (1991) Extraction of plasma triacylglycerols by the mammary gland of the lactating cow. *Journal of Dairy Research*, **58**, 251-255

Tice, E.M., Eastridge, M.L. and Firkins, J.L. (1994) Raw soybeans and roasted soybeans of different particle sizes. 2. Fatty acid utilization by lactating cows. *Journal of Dairy Science*, **77**, 166-180

Urbach, G. (1990) Effect of feed on flavor in dairy foods. *Journal of Dairy Science*, **73**, 3639-3650

Van Amburgh, M.E. (1997) Effect of ionophores on growth and lactation in cattle. *Proceedings Cornell Nutrition Conference for Feed Manufacturers* pp. 93-103, Cornell Univ., Ithaca, NY, USA

Van der Werf, J.H.J., Jonker, L.J. and Oldenbroek, J.K. (1998) Effect of monensin on milk production by Holstein and Jersey cows. *Journal of Dairy Science*, **81**, 427-433

Van Nevel, C.J. and Demeyer, D.I. (1995) Lipolysis and biohydrogenation of soybean oil in the rumen *in vitro*: inhibition by antimicrobials. *Journal of Dairy Science*, **78**, 2797-2806

Van Nevel, C.J. and Demeyer, D.I. (1996a) Influence of pH on lipolysis and biohydrogenation of soybean oil by rumen contents *in vitro*. *Reproduction Nutrition Development*, **36**, 53-63

Van Nevel, C.J. and Demeyer, D.I. (1996b) Effect of pH on biohydrogenation of polyunsaturated fatty acids and their Ca-salts by microorganisms *in vitro*. *Archives of Animal Nutrition,* **49**, 151-158

Van Nevel, C.J., Fievez, V. and Demeyer, D.I. (1999) Lipolysis and biohydrogenation of PUFA's from fish oil during in vitro incubations with rumen contents. *Proceedings of the Nutrition Society*, in press

Viallon, C., Verdier-Metz, I., Denoyer, C., Pradel, P., Coulon, J.B. and Berdagué, J.L. (1999) Desorbed terpenes and sesquiterpenes from forages and cheeses. *Journal of Dairy Research*, **66**, 319-326

Willett, W.C., Stampfer, M.J., Manson, J.E., Colditz, G.A., Speizer, F.E., Rosner, B.A., Sampson, L.A. and Hennekens, C.H. (1993) Intake of *trans* fatty acids and risk of coronary heart disease among women. *Lancet,* **341**, 581-585.

Wolff, R.L. (1994) Les isomères 18:1 trans dans l'alimentation des Européens. Evaluations quantitative et qualitative. *Oléagineux Corps Gras Lipides*, **1**, 209-218

Wonsil, B.J., Herbein, J.H. and Watkins, B.A. (1994) Dietary and ruminally derived *trans*-18:1 fatty acids alter bovine milk lipids. *Journal of Nutrition,* **124**, 556-565

Zinn, R.A. (1988) Comparative feeding value of supplemental fat in finishing diets for feedlot steers supplemented with and without monensin. *Journal of Animal Science*, **66**, 213-227

6

EVALUATION OF PHYSICAL STRUCTURE IN DAIRY CATTLE NUTRITION

D.L. DE BRABANDER, J.L. DE BOEVER, J. M. VANACKER, CH.V. BOUCQUÉ and S.M. BOTTERMAN

Department Animal Nutrition and Husbandry, Agricultural Research Centre - Ghent Ministry of Small Enterprises, Traders and Agriculture
Scheldeweg 68, B-9090 Melle-Gontrode, Belgium

Introduction

Optimum feeding of dairy cattle requires the maintenance of good rumen function. A disturbed rumen fermentation is the main outcome of a shortage of physical structure. This results in depressed feed intake, reduced digestion, decreased production efficiency, health problems like rumen acidosis and parakeratosis, and lowered fat milk content. Although physical structure is difficult to define, it could be considered as an expression of "the extent to which a feedstuff, through its content and properties of the carbohydrates, contributes to an optimum and stable rumen function". Feed containing physical structure stimulates chewing activity and this in its turn increases saliva secretion. Since saliva buffers the rumen contents, it reduces acidosis and helps to achieve an optimum pH and ratio of volatile fatty acids. Roughage creates a fibrous layer in the rumen, which is important for frequent and strong rumen contractions. It is generally accepted that physical structure deficiency depresses milk fat content. Most likely, this is mainly caused by a lower acetic acid:propionic acid ratio in the rumen.

Due to breeding, improved nutrition and better management, the milk production level of dairy cattle has increased considerably and will no doubt increase still further in the future. Although feed intake capacity increases with increasing milk production potential, the higher feed intake of certain rations is insufficient to meet the increased energy requirements. Consequently, the quality of the diet has to improve. This can be achieved by raising the proportion of concentrates in the ration, and/or by supplementing with energy-rich byproducts, and/or by using better roughages. All these measures reduce the structure value of the ration. Hence, problems with physical structure occur more frequently now than in the past and will increase still further in the future. Therefore the need for an adequate physical structure system has become more and more important.

Different methods have been proposed to optimise dairy cattle diets with regard to rumen activity. Some authors suggested minimum roughage:concentrate ratios e.g. 40:60 (Flatt, Moe, Munson and Cooper, 1969) or 45% long roughage (Sutton, 1984). Due to the great variation in chemical and physical characteristics of roughages, such standards are useless. Other authors suggested crude fibre (CF) content as a standard. The big variation among the proposed CF levels, i.e. 100-160 g/kg (Hoffmann, 1983), 130-140 g/kg (Kesler and Spahr, 1964), 170 g/kg (NRC, 1978), 180 g/kg (Kaufmann, 1976) and 180-220 g/kg in the German technical literature (Guth, 1995), as well as the wide ranges, support the conclusion that CF is not appropriate as a general standard. Moreover, CF is not a nutritional, chemical or physically uniform entity (Van Soest, Robertson and Lewis, 1991). Nevertheless, within a type of roughage with well-defined physical characteristics, CF content could be a suitable criterion. In the USA, a system based on NDF is used (NRC, 1988). However, it was recognised that NDF as such could not be used as a general parameter for physical structure (Weiss, 1993). Therefore, the system was modified by attributing an effectivity coefficient to the NDF of feedstuffs, and the unit became "effective NDF" (Sniffen, O'Connor, Van Soest, Fox and Russell, 1992; Mertens, 1997). Besides eNDF, which is aimed at providing a normal milk fat content, peNDF (physical effective NDF) was also introduced (Armentano and Pereira, 1997; Mertens, 1997). The latter takes the physical properties as well as particle size into account, and is related to chewing activity. However, the standards are vague and a modification was suggested. Because most factors influencing rumen activity are related to chewing activity, the latter became an important research topic (Balch, 1971; Sudweeks, Ely, Mertens and Sisk, 1981; De Boever, 1991; Sauvant, 1992; Dulphy, Rouel, Jailler et Sauvant, 1993; Nørgaard, 1985, 1993).

Conscious of the need for a scientifically-based system for evaluation of physical structure for dairy cattle, a research programme was started some years ago at our Department to determine the physical structure value of feedstuffs in current use on the one hand and to derive physical structure standards for dairy cattle on the other. It is an important step towards an innovative system for evaluation of physical structure for dairy cattle. Obviously, the system may need further adjustments with the evolution of knowledge in this field.

Materials and methods

CHEWING ACTIVITY

Several arguments support a relationship between physical structure and chewing activity. Therefore, eating and ruminating time were measured for a large number of feeds according to a standardised experimental design. Chewing activity was measured continuously during 4 days, according to the method described by De Boever, De Smet, De Brabander and Boucqué (1993). An adaptation period of 10 days between batches of the same forage type was considered sufficient, whereas this period was 17 days for transition to another type of forage. The recordings enabled a clear distinction between eating and ruminating activity. In the following, eating, ruminating and chewing time, expressed per kg dry matter(DM)intake, are indicated as the indices EI, RI and CI, respectively.

In each trial, 8 low-yielding Holstein cows were used. Because of the moderate milk production level, the amount of concentrates could be restricted without risks to metabolism. First lactation cows were avoided, except when they had attained the age of 2nd lactation cows, because of the less efficient chewing behaviour of primiparous cows (Dehareng and Godeau, 1991; Dado and Allen, 1994; Beauchemin and Rode, 1994, 1997).

The roughage under investigation was fed *ad libitum*. To meet requirements for minerals and vitamins, and to assure a good rumen function, the basal ration was supplemented with 3 kg balanced concentrates or, in case of protein shortage, with 2 kg soya bean meal and 200 g of a mineral-vitamin mixture. Byproducts or other feeds, that could not be fed as single feedstuffs, were combined with maize or grass silage and chewing indices were derived by difference.

Eighty-one chewing trials, carried out according to the standardised experimental design, produced 644 cow observations from 95 different cows.

CRITICAL ROUGHAGE PART

The physical structure requirements were derived from trials (standard trials), in which the roughage (R) part of the diet of Holstein cows was decreased weekly until symptoms of physical structure deficiency appeared (decreased milk fat content, decreased milk yield, off feed). The roughage part just before problems occur, is called "the critical roughage part (R_{crit})". Thus, a lower R_{crit} corresponds with a higher structure value. These data in combination with the chewing indices were also used to derive structure values.

During the first two weeks of the standard trials, the roughage part of the diet was always ± 600 g/kg on DM basis. Then, the proportion was changed weekly to a lower roughage part: 500, 450 or 400 g/kg and further in steps of 50 g/kg. After the problematic ratio, most cows received the ration with 600 g R/kg again for one week. Cows were fed approximately *ad libitum*, though without appreciable refusals.

Milk yield was measured at each milking and the last 4 milkings of each week were sampled for determination of milk fat and protein content. A decrease in milk fat content was the main indicator to terminate the experiment. The concentrate (C) was always the same, except in trials concerning the nature of the concentrate. The main ingredients were sugar beet pulp (300 g/kg), wheat (180 g/kg), soya bean meal (140 g/kg), malt sprouts (100 g/kg), maize gluten feed (100 g/kg) and sugar beet molasses (70 g/kg). The ingredients were ground through a sieve of 6 mm, whereas the concentrates were pelleted (6 mm). Concentrates were supplied twice daily, except in trials where the effect of more frequent feeding was studied.

Because the R:C-ratios were changed in steps of 50 g/kg, the R_{crit} could not be determined precisely. Statistically, it is likely that the exact mean R_{crit} of a ration is 20 g/kg lower than the observed value. Therefore, all values of R_{crit} were diminished by 20 g/kg. The values of R_{crit} were also corrected for the effects of milk yield, milk fat content and age, so that they are valid for a standard cow producing 25 kg milk with 44 g fat/kg in the 1st, 2nd or 3rd lactation.

The R_{crit} was determined for 56 rations, resulting in 510 cow observations. The average milk yield, when the diets contained 600 g roughage/kg was 25.1 kg. The mean fat and protein concentrations amounted to 43.8 and 33.0 g/kg, respectively. Total DM intake averaged 18.1 kg/d.

General results

CHEWING ACTIVITY

The average eating, ruminating and chewing time for all observations was 287, 487 and 774 min/day with a SD of 64, 85 and 104 min/day. Ruminating generally lasts longer than eating, and usually takes between 7 and 9 hours per day (Welch, 1982; Sniffen, Hooper, Welch, Randy and Thomas, 1986; Beauchemin, Farr, Rode and Schaalje, 1994; Dado and Allen, 1994). The observed EI, RI and CI of the basal rations (C not included) averaged (SD) 25.2 (11.6), 41.4 (12.2) and 66.6 (21.7) min/kg DM. When animal species or live-weight classes are compared, chewing time is usually expressed in minutes per gram DM per kg $LW^{0.75}$. On average for the 644 observations, the chewing time was 8.4 min/kg DM/kg $LW^{0.75}$.

The main parameters representing the eating and ruminating behaviour for rations with preserved grassland products, fresh grass and maize silage as the sole roughage are presented in Table 6.1.

The chewing indices are corrected to a live weight of 600 kg based on metabolic weight. The EI, RI and CI averaged 26.2, 47.4 and 73.6 min/kg DM for 30 preserved grassland products, 33.5, 37.3 and 70.8 min/kg DM for 13 samples of fresh grass and 19.9, 39.3 and 59.3 min/kg DM for 23 maize silages. Variation was usually at least as high between cows as between samples of the same kind of roughage. Mean number of meals per day was 9.2, 8.6 and 8.5, respectively, with an average duration of 34.8, 43.5 and 34.6 minutes. Feeding is followed by a long eating time. For preserved grassland products and maize silage the first meal after the 2 feeding times represented 51% and

Table 6.1 Eating and ruminating behaviour for rations based on grassland products and maize silage

Roughage	*Preserved grassland products*	*Fresh grass*	*Maize silage*
Number of samples	30	13	23
NDF content (g/kg DM)	471	432	386
CF content (g/kg DM)	261	230	202
Roughage intake (kg DM/day)	11.8	11.0	14.1
Chewing time[1] (min/day)			
Eating	301	356	273
Ruminating	526	391	522
Chewing	827	746	795
Chewing indices[1] (min/kg DM)			
EI mean	26.2	33.5	19.9
SD_f - SD_c	4.9–4.5	4.4–6.6	2.2–3.1
RI mean	47.4	37.3	39.3
SD_f - SD_c	6.4–6.0	3.9–4.5	4.2–5.1
CI	73.6	70.8	59.3
SD_f - SD_c	10.1–8.6	7.7–9.5	5.9-6.7
Meals			
Number/day	9.2	8.6	8.5
SD_f - SD_c	1.8–1.9	1.1–1.5	1.5–1.9
Duration/meal (min)	34.8	43.5	34.6
SD_f - SD_c	7.5–7.5	4.6–7.7	7.1–8.1
Duration 2 main meals (min)	154	-	146
Ruminating periods			
Number/day	14.8	14.1	15.0
SD_f - SD_c	0.8–1.9	1.6–1.8	0.8–1.8
Duration/period (min)	36.6	28.7	35.6
SD_f - SD_c	2.8–7.0	2.8–5.5	3.1–5.0

[1] Chewing time and chewing indices refer to the roughage
SD_f: standard deviation of forages
SD_C: standard deviation of cow data within each trial

53% of the total eating time. The mean number of ruminating periods was 14.8, 14.1 and 15.0 per day, and lasted on average 36.6, 28.7 and 35.6 minutes, respectively. Other experiments confirm the higher frequency of ruminating compared with eating (Nørgaard, 1989; Dado and Allen, 1994). The longer eating time after feeding is generally found, whereas ruminating is more regularly spread over the day (Rémond and Journet, 1972; Dado and Allen, 1994).

CRITICAL ROUGHAGE PART

With an increasing proportion of concentrates, milk yield increased until the R:C-ratio just before the R_{crit}. When physical structure appeared to be deficient, daily milk yield decreased by an average of 0.4 kg in one week. At that moment, the decrease in milk fat content averaged 5.8 g/kg. In the subsequent (recovery) week with 600 g roughage/kg, fat content rose by 5.6 g/kg, whereas milk yield was 2.6 kg lower due to the depressed feed intake. Consequently, it can be assumed that the rumen function of most cows is normalised within one week.

For the 510 cow observations, average R_{crit} amounted to 299 g/kg, with a SD of 115 g/kg. The wide variation illustrates that a "fixed roughage proportion" in the ration cannot be a useful standard in a structure evaluation system. The R_{crit} seems to depend to a great extent on the cow as well as on the type of ration. The values for R_{crit} of the rations with a grassland product or maize silage as the sole roughage are presented in Table 6.2. The R_{crit} was appreciably higher for maize silage than for grassland products. The difference is greater than would be expected based on the difference in chewing indices.

Table 6.2 Characteristics and R_{crit} of grassland products and maize silage

Roughage	*Grassland products*		*Maize silage*	
Number of samples	*17*		*14*	
	Mean	$SD_f^{(1)}$	*Mean*	$SD_f^{(1)}$
Characteristics				
NDF (g/kg DM)	495	56	397	31
CF (g/kg DM)	259	24	208	14
Chewing index (min/kg DM)	74.9	8.8	59.3	7.1
R_{crit} (g/kg)	205	39	348	56
$SD_c^{(1)}$	52	21	65	18

(1) SD_f : standard deviation on samples of forages
SD_c : mean of standard deviations on cow data within each trial

Variation between samples of forages was relatively higher for R_{crit} than for chewing indices. The standard deviation of the individual cow data within each trial amounted to 52 and 65 g/kg for grassland products and maize silage, respectively. This relatively large individual variation necessitates a large safety margin in the structure evaluation system.

Animal-linked factors

CHEWING ACTIVITY

To make the results (n=644) of the different trials useful for studying the effect of animal characteristics, all chewing indices, as well as potential influences (live weight, milk yield, roughage intake), were expressed as a percentage of the mean for each trial. Relationships were studied by means of regression analysis.

The EI, RI and CI, were negatively correlated (P<0.001) with live weight(LW). This is in agreement with the results of Bae, Welch and Gilman (1983). To eliminate the effect of LW, all chewing indices were corrected to a LW of 600 kg, based on metabolic LW, (x $LW^{0.75}/600^{0.75}$). However, CI appeared to be overcorrected since a positive relationship was found between the corrected CI and LW. A log-transformation of both variables led to the choice of 0.3 as exponent rather than 0.75. The corrected index CI x $LW^{0.3}/600^{0.3}$ was independent of LW. The other animal-linked factors were studied for chewing indices corrected for $LW^{0.3}$.

In the present experimental circumstances, where young primiparous cows were not involved in the trials, age had no effect on chewing activity.

The results demonstrate a weak but significant negative relationship between milk yield and chewing index. When milk yield was expressed as the absolute difference (x) from the mean milk yield in each trial, the following regression equation was obtained:

$$y = 100.0 - 0.32 * x \text{ , where y is the CI (in \%)} \qquad \text{(Eq. 6.1)}$$

Since feed intake capacity is related to LW, the correction of CI for LW probably implies some correction for differences in feed intake capacity. Nevertheless, a highly significant negative effect of feed intake on the corrected ($LW^{0.3}$) chewing indices was obtained. EI, RI and CI were depressed by 0.5, 0.4 and 0.5% per one per cent higher feed intake. Consequently, total chewing time per day increased 0.5% for each increase of feed intake by one per cent. The negative relationship between chewing indices and feed intake is confirmed by Coulon, Doreau, Rémond and Journet (1987), Deswysen, Ellis and Pond (1987), Deswysen and Ellis (1990) and Dado and Allen (1994).

From the results of cows (n=76), which were involved in several chewing experiments, it appeared that 0.31, 0.50 and 0.46 of the total variance in EI, RI and CI was due to differences between cows.

CRITICAL ROUGHAGE PART

To study the animal linked factors, R_{crit} and the independent variables were also expressed as a percentage of the mean of each trial. First the independent variables were considered within each of five age groups (1st, 2nd, 3rd, 4th and 5th+ lactation) with respectively 188, 126, 97, 53 and 46 cow observations.

For each age group, relative milk yield was positively correlated with the relative R_{crit}, whereas milk fat content was negatively correlated with R_{crit} in four of the five age groups. The effect of LW was not consistent and feed intake capacity seemed to have no effect.

Subsequently, the effects of milk yield, fat content, live weight and age were studied using a regression model with categorical (dummy) variables which enables detection of the fixed effects of variables (age). From this analysis it was concluded that live weight has no effect on physical structure requirements. Regression equations with milk yield and/or milk fat content were highly significant ($P<0.001$). This was the case irrespective of whether these variables were expressed as a percentage of, or as absolute differences from the average of each trial. As milk yield and fat content are negatively correlated, their regression coefficients were lower when both were considered in a multiple regression than when each was separately included in a single regression. When both parameters were taken into account, the R_{crit} was 1.2% higher per kg higher milk yield, and 0.7% lower per g/kg higher milk fat content. Considered separately, these effects were 1.5 and 1.1%, respectively.

After correcting for the effect of milk yield and milk fat content, the relative R_{crit} in the first 6 lactations were 103, 105, 99, 94, 83 and 79%. Differences were not significant between cows in the first three lactations, nor between the cows in the 5th and 6th lactations, whereas these two groups differed significantly from each other. Therefore, the first three lactations were considered as a reference group, cows in the 5th lactation and older as another group and the 4th lactation cows as a separate group. Compared with the reference group, R_{crit} in the 4th and 5th+ lactation was 9 and 20% lower, respectively. Considering the 5th lactation cows separately, the difference was 19%.

Three-hundred and forty-six R_{crit} values, from 67 cows, were used to study the effect of individual cows on physical structure requirements. 0.21 of the total variance in R_{crit} could be explained by differences between cows. This highly significant effect was lower than for chewing activity.

Feed-linked factors

GRASSLAND PRODUCTS

Growth stage

Three trials were carried out with prewilted grass silage, originating from grass that was cut at two different growth stages. Mean NDF content was 437 and 515 g/kg DM respectively, for the young and older growth stage. Consequently, EI, RI and CI (corrected to a LW of 600kg based on metabolic weight) were significantly higher for the latter. These indices averaged 23.0, 44.0 and 67.0 min/kg DM for the young growth stage vs. 26.9, 52.7 and 79.6 min/kg DM for the more advanced growth stage.

The relationships between chewing index and characteristics depending on growth stage, such as CF, NDF, ADF, OM digestibility and roughage intake, were studied using the results of 13 prewilted grass silages that were chopped to a nominal length of 24 mm. All these parameters were significantly ($P < 0.05$) related to the chewing index; CF content was the best predictor.

Regression analysis to estimate R_{crit} from parameters reflecting the growth stage, was carried out using the results of 12 rations, of which 10 used prewilted and 2 used direct-cut grass silage as the sole roughage. Here too, CF content was the best predictor, whereas the relationship between R_{crit} and CI was not significant ($P = 0.16$). The main regression equations to estimate CI and R_{crit} are given in Table 6.3. Figure 6.1 represents the relationship between CF content and R_{crit}.

Table 6.3 Regression equations to estimate chewing index and critical roughage part of grass silage

Regression equation	*P*	*R^2*	*RSD*
CI (min/kg DM) n = 13			
- 15.1 + 0.326 * CF [(1)]	0.000	0.86	3.6
- 18.7 + 0.188 * NDF [(1)]	0.001	0.64	5.7
9.7 + 0.197 * ADF [(1)]	0.014	0.44	7.1
230.1 - 2.13 * Dig.OM [(2)]	0.000	0.77	4.5
135.8 - 5.422 * Rint. [(3)]	0.001	0.68	5.4
R_{crit} (g/kg) n = 12			
406 - 0.79 * CF	0.03	0.40	20
342 - 0.29 * NDF	0.05	0.34	21
- 475 + 3.24 * Dig.OM	0.07	0.29	21
301 - 1.40 * CI	0.16	0.19	23

[(1)] CF, NDF, ADF in g/kg DM
[(2)] Dig.OM = in vitro organic matter digestibility in % (Tilley and Terry, 1963)
[(3)] Rint. = roughage intake in kg dry matter/day

The present results, as well as data from the literature (Castle, Gill and Watson, 1981; Dulphy and Michalet-Doreau, 1983; Nørgaard, 1985; Kaiser and Combs, 1989; Beauchemin, 1991; Kamatali, 1991; Sauvant, 1992; Dulphy *et al.*, 1993) clearly demonstrate an increased CI when grass is cut at a later growth stage.

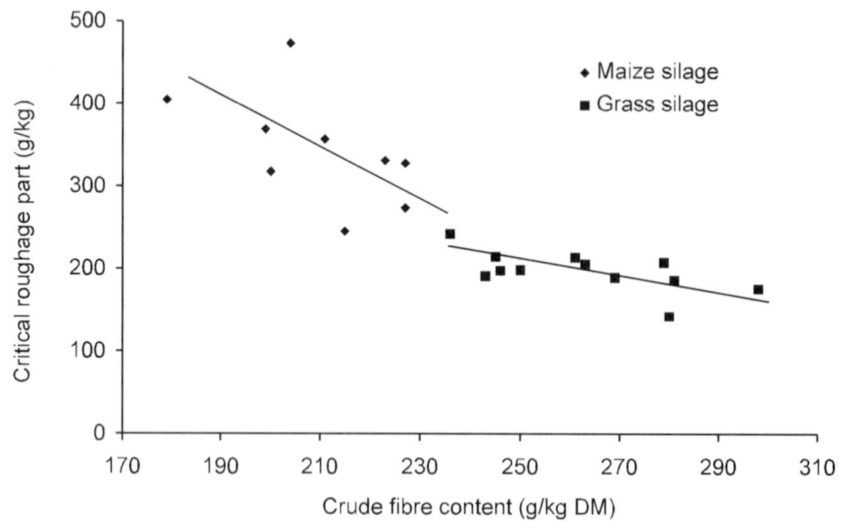

Figure 6.1 Relationship between crude fibre content and critical roughage part

Particle length

The effect of chopping grass at harvest versus long silage on chewing activity was investigated in five trials. In four trials the grass was either chopped at a nominal length of 24 mm or harvested with a self-loading wagon equipped with 16 knives, whereas in one trial, three lengths were compared, i.e. 3.5 mm, 24 mm and long material.

The effect of particle length on chewing index is ambiguous. In three of the five trials CI significantly ($P < 0.05$) increased with increasing particle length, but in most trials fibre content tended to be higher in the longer silage. When chewing indices were corrected to equal CF contents, the differences became negligible, except for one trial. Corrected to the CF content of the long silages, EI, RI and CI averaged, for the five trials, 23.9 and 24.7, 47.3 and 49.6, and 71.1 and 74.3 min/kg DM, respectively for the 24 mm and the long silage. Compared with the large difference in measured particle length between treatments, the effect on CI was very small. Chewing time, expressed per kg NDF, increased by an average of only 0.1 min per mm increase in particle length.

In two trials the R_{crit} was compared for rations with either chopped or long prewilted grass silage from the same origin as the sole roughage. The nominal chopping length was 24 mm, whereas a self-loading wagon with 16 knives was used for the long material. Although CI was lower for the chopped silage in both trials, the R_{crit} was by no means higher for that silage. A third long silage was used in this research, whereas all other silages were chopped. The R_{crit} of all three long silages was situated around the regression line which represents the relationship between CF content and R_{crit}.

This again indicates no effect of particle length of grass silage on structure value. One has to keep in mind that for grass silage, particle length is several centimetres.

Fine chopping (< 20 mm), which is rarely used in practice for grass silage , mostly resulted in a depressed EI and CI (Voskuil and Metz, 1973; Piatkowski, Nagel und Bergner, 1977; Castle, Retter and Watson, 1979; Deswysen, 1980; Dulphy, Michalet-Doreau and Demarquilly, 1984). However, it appears that beyond a certain limit, increasing particle length does not further stimulate chewing activity. According to the review of Beauchemin *et al.* (1994), that upper limit should be situated between 6.4 and 10 mm. Clark and Armentano (1997) tend to conclude from their literature study that the critical particle length with regard to chewing activity, rumen pH and acetic acid:propionic acid ratio is situated between 4 and 8 mm. Lammers, Heinrichs and Buckmaster (1996) suggest a length of ± 2 cm as critical. Therefore it may be concluded that the structure value of normal grass silages in practice (> 20 mm) is independent of particle length.

Preservation method

In five trials, chewing indices of direct-cut and prewilted grass silages were compared. As expected, DM intake of the prewilted grass silages exceeded intake of the direct-cut silages by an average of 1.1 kg. The direct-cut silages were mainly characterised by a higher EI. On average for the five comparisons of direct-cut and prewilted grass silage, EI amounted to 31.1 and 23.6, RI to 49.0 and 46.2 and CI to 80.1 and 69.8 min/kg DM, respectively. The higher EI for direct-cut silage is supported by other researchers (Teller, Vanbelle, Kamatali and Wavreille, 1989; Teller, Vanbelle, Kamatali, Collignon, Page and Matatu, 1990). This is probably due to the higher volume of feed that has to be eaten, as well as to reduced palatability, which is a result of the stronger silage fermentation.

Two trials were carried out to compare direct-cut with prewilted grass silage; with regard to the R_{crit}, EI and CI were considerably higher for the direct-cut silages. The R_{crit} values for the diets with direct-cut silage and prewilted silage amounted to 214 and 191 g/kg in the first trial, and to 215 and 198 g/kg in the second one. Contrary to the EI and CI, R_{crit} does not demonstrate a higher structure value for direct-cut grass silage. Meyer, Bartley, Morrill and Stewart (1964) showed that during eating of fresh grass, saliva secretion per kg DM intake decreases with decreasing DM content; the same may be true for silage. Moreover, direct-cut silage has a lower pH and a higher acid content. Therefore direct-cut silage has a higher base buffer capacity than prewilted silage (Playne and McDonald, 1966). The reduced saliva secretion and the higher base buffer capacity could probably compensate for the higher CI.

Chewing activity was compared between grass hay and prewilted silage in two trials. CF content was higher for the hay, whereas the difference in NDF content seems to depend on DM content of silage. It appears that drying the grass to a high DM content or hay making considerably increases NDF content. EI was significantly higher for hay, whereas RI did not differ. After correction to the same CF content, CI was on average approximately 1.06 times higher for hay. Brouk and Belyea (1993) obtained both a higher RI and EI for alfalfa hay than for prewilted silage. Moreover, Meyer *et al.* (1964) observed a 1.2 times higher saliva secretion per kg DM during eating of alfalfa hay compared with alfalfa silage with 200 g/kg DM. In the first trial R_{crit} was considerably lower (157 vs. 186 g/kg) for the ration including hay instead of prewilted silage, whereas no difference was noted in the second trial. Obviously, a higher structure value can be attributed to hay than to silage.

Fresh grass

Thirteen samples of fresh grass, containing on average 432 g NDF and 230 g CF per kg DM, had a mean EI, RI and CI of 33.5, 37.3 and 70.8 min/kg DM. Eight and five chewing trials with respectively spring and autumn grass, each originating from two different years, were carried out. Although fibre content did not differ, chewing indices were markedly lower for the spring grass, i.e. EI: 31.1 vs. 37.4 , RI: 36.0 vs. 39.3, and CI: 67.1 vs. 76.7 min/kg DM. The differences could perhaps be explained by the higher digestibility, the lower milling resistance and the higher intake level of the spring grass.

R_{crit} was determined in two concurrent trials with spring grass and in one trial with autumn grass. Spring grass was supplemented with either Concentrate 1 or Concentrate 3 (Table 6.6), whereas autumn grass was supplemented only with Concentrate 1. The R_{crit} values were 310, 380 and 212 g/kg, respectively. The higher R_{crit} of the spring grass diets, although they contained more fibre, clearly indicate a lower structure value of spring grass. The difference between the two first trials illustrates the effect of concentrate composition (Table 6.7).

MAIZE SILAGE

Stage of maturity

The effect of the stage of maturity of whole crop maize on chewing activities was studied in three trials. Three stages were compared in the first trial, and two in each of the other two. As fibre content decreased with advancing stage of maturity, CI significantly decreased. On average for the three trials, maize silage harvested in the milk dough and the hard dough dent stage had a DM content of 271 and 353 g/kg and a NDF content of 395 and 372 g/kg DM, respectively. The mean chewing indices were 61.6 and 57.4 min/kg DM, respectively.

Regression analysis with 17 maize silages, chopped at a length of 8 mm, confirms the dependence of chewing activity on stage of maturity. The main relationships between CI and parameters reflecting stage of maturity are presented in Table 6.4. CI was best related to roughage intake level. Among the chemical parameters, CF seems to be the best predictor.

Relationships between R_{crit} and other parameters, were investigated using nine maize silages (Table 6.4, Figure 6.1). Here NDF seems to be a somewhat better predictor than CF, whereas CI was not significantly related to R_{crit}.

Table 6.4 Regression equations to estimate chewing index and critical roughage part for maize silage

Regression equation		*P*	R^2	*RSD*
CI (min/kg DM)	n = 17			
4.7 + 0.272 * CF [1]		0.001	0.50	4.5
- 10.5 + 0.182 * NDF [1]		0.014	0.34	5.1
18.6 + 0.176 * ADF [1]		0.013	0.35	5.1
219.9 - 2.16 * Dig.OM [2]		0.001	0.57	4.3
127.4 - 4.789 * Rint. [3]		0.000	0.70	3.5
$\mathbf{R_{crit}}$ (g/kg)	n = 9			
844 - 2.38 * CF		0.11	0.32	58
783 - 1.09 * NDF		0.07	0.39	55
607 - 4.37 * CI		0.19	0.23	62

[1] CF, NDF, ADF in g/kg DM
[2] Dig.OM = in vitro organic matter digestibility in % (Tilley and Terry, 1963)
[3] Rint. = roughage intake in kg dry matter/day

Chopping length

Five chewing trials were carried out to study the effect of chopping length of maize silage. Four trials were carried out to compare a nominal length of 4 and 8 mm; in one of these trials a 16 mm treatment was also used. In a fifth trial, 6.5 mm was compared with 8 mm. Chopping length had no effect on EI, except for the 16 mm treatment. In three of the four trials, chopping at 8 mm significantly increased RI compared with 4 mm. Mean CI for the 4 and 8 mm comparisons were 54.2 and 58.2 min/kg DM, respectively. Expressed per mm increase in particle length, chewing time per kg NDF was approximately 8 minutes longer, compared with only 0.1 minutes for prewilted grass silage. Chopping at 16 mm length also significantly increased CI compared with 8 mm. The comparison of 6.5 with 8 mm resulted in chewing indices of 63.9 and 66.7 min/kg DM, which were not significantly different. The observed effect of the chopping length of maize on chewing activity agrees with other experiments (Rohr, Honig and Daenicke, 1983; Kuehn, Linn and Jung, 1997).

The maize silages of the trial with the 6.5 and 8 mm treatments were also used to compare R_{crit} values. The latter were 397 and 368 g/kg, respectively, indicating a higher structure value for the more coarsely chopped maize silage. Demarquilly (1994) also concluded that fine chopping of maize can reduce milk fat content. In contrast to grass silage, chopping length of maize affects its structure value, probably because particle size is smaller.

SUPPLEMENTS

The physical structure of fodder beets, raw potatoes, ensiled pressed sugar beet pulp and ensiled brewers grains was evaluated by determining the chewing indices and the critical roughage part in the ration. Two samples of brewers grains were tested – conventional brewers grains (A) and pressed brewers grains (B). The latter had a higher DM content (280 vs. 236 g/kg) and originated from ground malt, resulting in a smaller particle size. In the chewing trials, beet pulp was used as a supplement to either grass silage or maize silage A, and beets were supplemented to the same grass silage and maize silage B, each time in two ratios of 20:80 and 35:65 on DM basis. Potatoes and brewers grains A and B were used as supplements to maize silage C, C and D respectively in a fixed amount of 5 to 6 kg DM. In addition, cows received a restricted amount of concentrates. The chewing indices of the test feeds were derived by difference.

The critical roughage part was determined for rations including plus concentrates either the control roughage alone or the roughage and one of the supplements. Supplements were given in constant amounts of 4.9, 4.2, 4.0, 4.3 and 4.6 kg DM respectively for beet pulp, fodder beets, potatoes and brewers grains A and B.

Chewing indices derived for the experimental feeds are presented in Table 6.5. Data obtained by difference are mostly characterised by a great variation. The CIs of beet pulp and beets were hardly affected by the nature of the roughage or by the inclusion ratio, and amounted on average to 32.3 and 34.3 min/kg DM. The CI of potatoes, brewers grains A and B were 23.7, 56.6 and 40.7 min/kg DM, respectively.

The critical grass silage part was 206 g/kg and decreased to 102 and 115 g/kg when fed in combination with beet pulp and beets, respectively. Also the critical maize silage part was markedly lower when these two supplements were added. In agreement with the chewing activity, the R_{crit} values also decreased by adding potatoes or brewers grains to the diet. Although the R_{crit} indicates a similar structure value for potatoes and brewers grains A, chewing indices were very different. In agreement with the findings of Vérité and Journet (1973), Rohr, Daenicke, Honig and Lebzien (1986), Dulphy *et al.* (1993) and Swain and Armentano (1994), these results clearly demonstrate that the supplements investigated here contain physical structure.

Table 6.5 Chewing indices (min/kg DM) of the supplements – R_{crit} (g/kg)

	Chewing trials			*Critical roughage part*		
Diet[1]	*R:S*[2]	***CI***	*(SD)*	*Diet*	$\boldsymbol{R_{crit}}$	*(SD)*
GS+PBP	80:20	**35.0**	(18.9)	GS	**206**	(44)
	65:35	**33.3**	(9.3)	GS+PBP	**102**	(40)
				GS+FB	**115**	(43)
MS A+PBP	80:20	**30.1**	(12.1)			
	65:35	**30.6**	(8.4)	MS A	**369**	(63)
				MS A+PBP	**229**	(75)
GS+FB	80:20	**32.4**	(15.9)	MS A+FB	**276**	(96)
	65:35	**40.0**	(11.9)			
MS B+FB	80:20	**29.8**	(40.2)			
	65:35	**35.0**	(22.4)	MS C	**473**	(45)
				MS C+PO	**385**	(77)
MS C+PO	64:36	**23.7**	(7.5)	MS C+BG A	**380**	(38)
MS C+BG A	65:35	**56.6**	(17.5)			
				MS D	**405**	(78)
MS D+BG B	64:36	**40.7**	(14.6)	MS D+BG B	**321**	(85)

[1] GS = grass silage, MS = maize silage, PBP = ensiled pressed beet pulp, FB = fodder beets, PO = potatoes, BG = ensiled brewers grains,
[2] R:S = roughage:supplement

STRUCTURE CORRECTORS

Straw

Because straw is often used as a structure corrector in dairy rations, three chewing experiments were carried out, two with wheat and one with barley straw. The NDF and CF contents were 813 and 469 g per kg DM on average. Mean eating, ruminating and chewing indices were 72.5, 85.8 and 158.3 min/kg DM. Barley straw had higher NDF and CF contents as well as a higher CI. In one standard trial with wheat straw, a R_{crit} of 155 g/kg was obtained.

Dehydrated chopped alfalfa

The alfalfa used was chopped to a mean particle length of 11.8 mm and was pressed in big bales. NDF and CF contents were 524 and 367 g/kg DM. EI, RI and CI were 21.6, 29.4 and 51.0 min/kg DM, whereas a R_{crit} of 249 g/kg was found.

CONCENTRATES

Acidotic effect

Concentrates could, on the one hand, contribute to rumen buffering by the chewing activity they induce and by their intrinsic buffering capacity. On the other hand, rapid degradation of the carbohydrates can cause rumen acidosis. The structure value of concentrates will mainly depend on these two characteristics.

In a chewing trial with maize silage and dried sugar beet pulp, an EI, RI and CI of 7.7, 14.0 and 21.6 min/kg DM were derived for beet pulp. This illustrates how concentrate ingredients could even be ruminated. According to Mertens (1997), chewing index is mainly related to NDF content, particle size and physical form. CI of pelleted concentrates could be estimated by CI = 150 * 0.01 * 0.3 * NDF(%/DM) (Mertens, 1997). This equation assumes a chewing time of 150 minutes per kg NDF and a physical effectivity of 0.3 for pelleted concentrates.

The acidotic effect of three concentrates and nine ingredients was studied by *in sacco* and *in vitro* incubations (De Smet, De Boever, De Brabander, Vanacker and Boucqué, 1995). DM degradation was determined *in sacco* and pH decrease *in vitro*. Even though pH decrease is also influenced by the intrinsic buffering capacity, a close relationship between both parameters could still be expected. The ingredients were used to formulate three concentrates with different composition and rates of carbohydrates degradation (Table 6.6).

Two lactating Holstein cows, fitted with rumen cannulae, were used to determine the *in sacco* DM degradation of the feedstuffs. Sets of bags were incubated in the rumen for 0, 3, 6, 12, 24 and 48 hours. The *in vitro* pH decrease was determined using a rumen fluid / buffer solution (50/50 v/v). The buffer solution was prepared according to the method described by Tilley and Terry (1963).The rumen fluid was taken from the two previously mentioned fistulated cows. Figures 6.2 and 6.3 show the DM degradation and pH decrease, respectively. Based on the degradation after 3h incubation, ingredients could be placed in order of decreasing DM degradation: manioc, wheat, barley, maize gluten feed, beet pulp, soya bean meal, maize, sorghum, soya bean hulls. After 5h *in vitro* incubation, ingredients were placed in order of declining pH decrease: manioc, wheat, beet pulp, maize gluten feed, barley, maize, soya bean meal, sorghum, soya bean hulls. The best correlation was obtained between DM degradation after 3 or 6h and *in vitro* pH decrease after 5h incubation. Between 3h degradation and 5h pH decrease (Figure 6.4) a correlation coefficient of 0.91 was found.

Critical roughage part – type of concentrate

The effect of the composition (starch + sugar: S+S) and the degradation rate of the compound feed on R_{crit} was investigated in five experiments with three different concentrates (Table 6.6). The R_{crit} values are shown in Table 6.7.

Table 6.6 Composition of the concentrates

Concentrate	*1*	*2*	*3*
Starch + sugar content	*Low*	*High*	*High*
DM Degradation rate	*Low*	*Low*	*High*
Main ingredients (g/kg)			
Beet pulp	357	-	-
Soya bean hulls	120	-	-
Maize	-	230	-
Sorghum	-	230	-
Barley	-	-	200
Wheat	-	-	170
Manioc	-	-	140
Soya bean meal	205	185	207
Maize gluten feed	140	207	120
Chemical composition (g/kg DM)			
Crude protein	194	203	192
NDF	291	137	143
Crude fibre	143	41	54
Starch	56	377	342
Sugars	100	68	78
DM degraded 3h (g/kg)	490	480	670

Table 6.7 Effect of concentrate composition on R_{crit}

Concentrate	*1*	*2*	*3*	
Starch + sugar content	*Low*	*High*	*High*	*SD*
Degradation rate	*Low*	*Low*	*High*	
Trial 1 - Maize silage	262 g DM/kg, 468 g NDF/kg DM, CI: 72.8 min/kg DM			
R_{crit} (g/kg)	**305**[a]	**331**[ab]	**391**[b]	68
Trial 2 - Prewilted grass silage	317 g DM/kg, 479 g NDF/kg DM, CI: 61.0 min/kg DM			
R_{crit} (g/kg)	**204**[a]	**262**[ab]	**329**[b]	64
Trial 3 - Maize silage 2 concentrate meals	334 g DM/kg, 406 g NDF/kg DM, CI: 59.0 min/kg DM			
R_{crit} (g/kg)	**299**		**375***	58
Trial 4 - Maize silage 6 concentrate meals	see Trial 3			
R_{crit} (g/kg)	**240**		**344***	92
Trial 5 - Fresh grass	198 g DM/kg, 460 g NDF/kg DM, CI: 70.8 min/kg DM			
R_{crit} (g/kg)	**310**		**380**[ns]	79

[a,b]: means on a same line with the same superscript letter are not significantly different (P>0.05)

* : significantly different ($P \leq 0.05$)

[ns] : not significantly different (P>0.05)

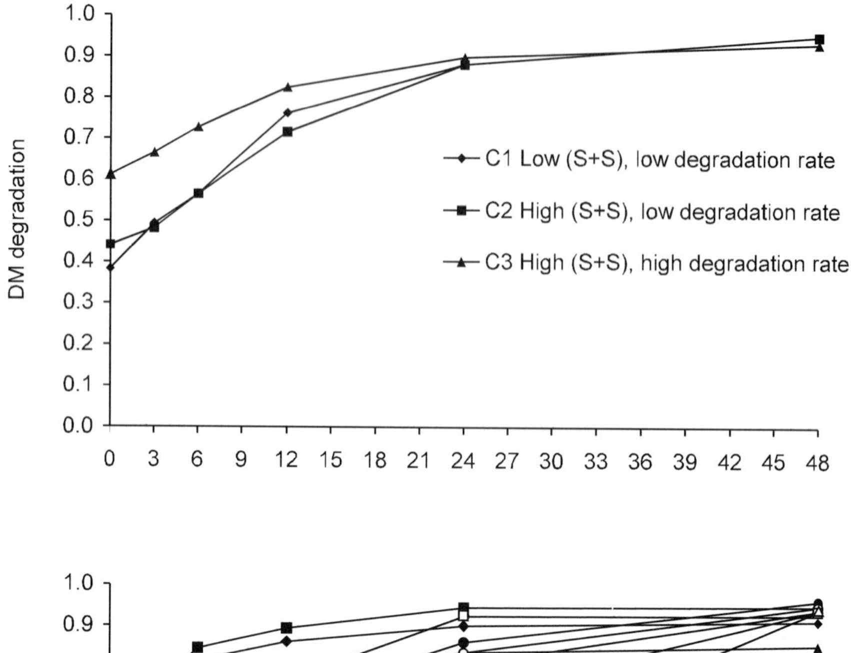

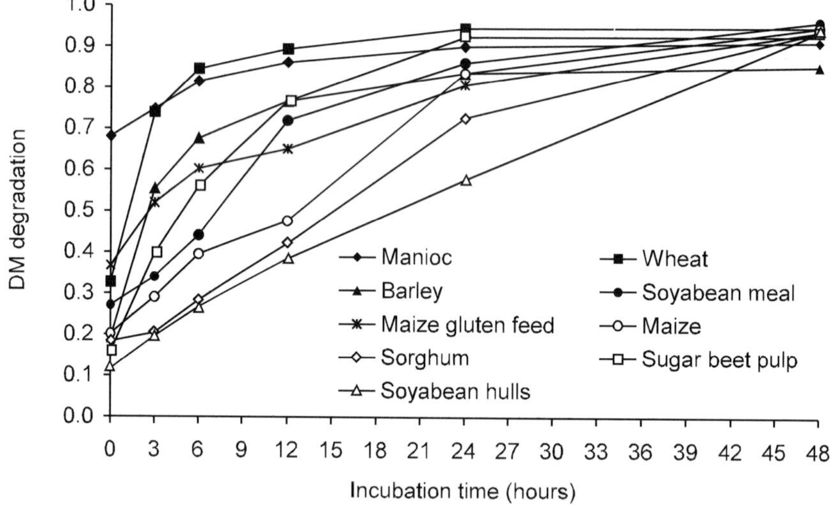

Figure 6.2 *In sacco* dry matter degradation of concentrates and ingredients

The trials were carried out with an average of eight cows per treatment. Concentrates were given twice daily, except in Trial 4. The R_{crit} was consistently lower when concentrates with a low (S+S)-content were used. However, due to high individual variation, differences were not always significant. The mean difference between the two extreme concentrates was 92 g/kg, corresponding to 29% of the average R_{crit}. Taking into account the fact that increasing the proportion of concentrates in a

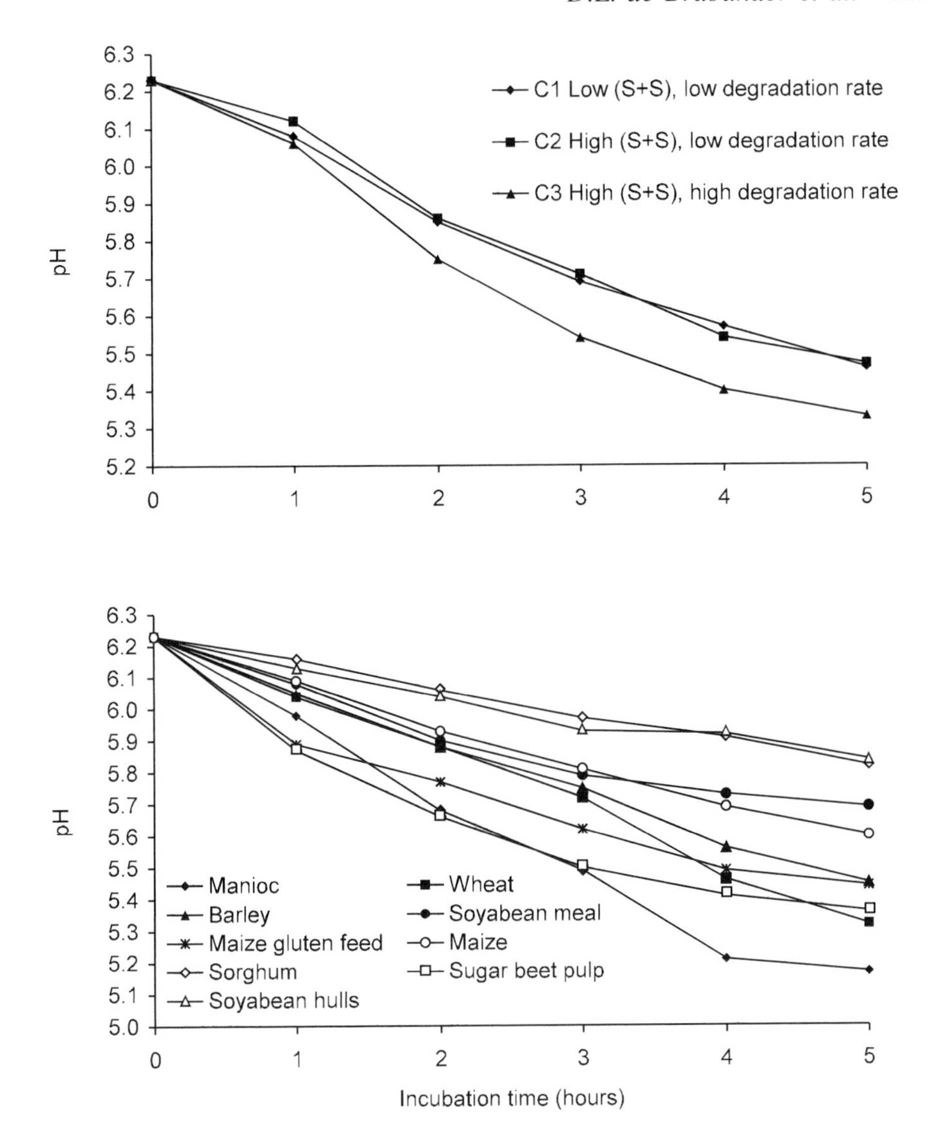

Figure 6.3 pH of *in vitro* incubated concentrates and ingredients

diet increases the ingestibility of the total diet, markedly more concentrate can be incorporated in the diet when the concentrate has a low (S+S)-content. In the present trials 3.9 kg more concentrate was given in the critical phase when C1 was given instead of C3. These results, as well as results from similar trials with fewer cows not discussed here, indicate that *in sacco* DM degradation after 3h and *in vitro* pH decrease after 5h incubation are closely related to the acidotic effect of concentrates.

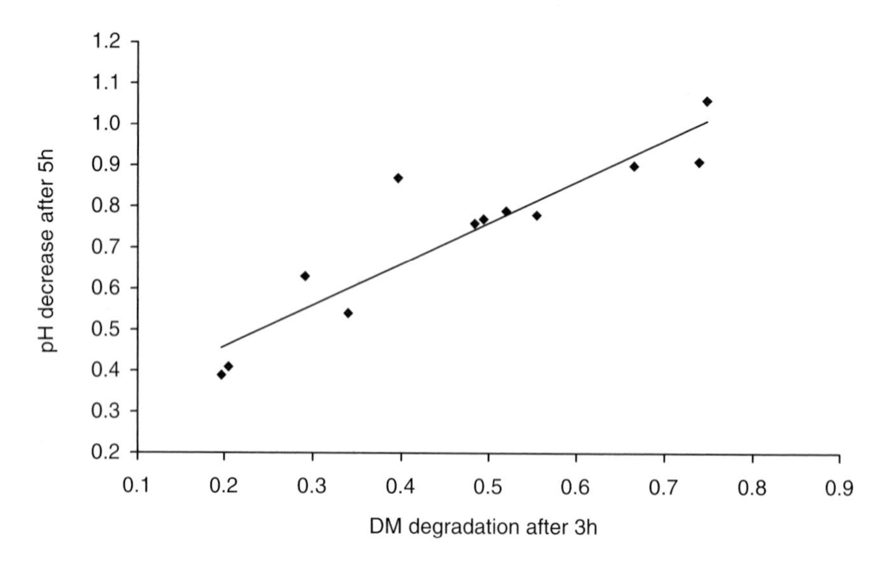

Figure 6.4 Relationship between *in sacco* DM degradation and *in vitro* pH decrease of concentrates and ingredients

Critical roughage part – Number of concentrate meals

It is well-documented that more frequent feeding of concentrates results in a more stable pH and fatty acid concentration in the rumen and consequently reduces the risk of critical values (Satter and Baumgardt, 1962; Bath and Rook, 1963; Kaufmann, 1976; Malestein, van 't Klooster, Counotte and Prins, 1981; Sutton, Hart, Broster, Elliott and Schuller, 1986; Robinson, 1989). The effect of 6 versus 2 concentrate meals per day on R_{crit} was investigated in three trials. Two trials are mentioned in Table 6.7 (Trials 3 and 4). A third trial was linked to Trial 2 in Table 6.7, where concentrate 3 was given in 2 and 6 meals per day, resulting in R_{crit} values of 329 and 299 g/kg, respectively. On average for the 3 trials the R_{crit} was 40 g/kg lower when the concentrates were given in 6 meals, corresponding to a higher concentrate intake of 1.8 kg in the critical phase.

Formation of a structure evaluation system

WHICH STANDARD?

The following parameters, related to physical structure, were investigated as potential standards in the structure system:

1. chewing time, ruminating time + half the eating time
2. NDF, effective NDF, crude fibre
3. derived unit

For a standard to be perfect, its value in the critical phase of the present trials has to be the same independent of the ration type. This critical value, including a safety margin to allow for individual variation, could be considered as the minimum structure requirement. However, some variation in critical values is acceptable, as the R_{crit} values could not be determined precisely. The values of the parameters concerned in the critical phase were calculated for the 510 observations. The chewing indices of the concentrates were estimated according to the formula proposed by Mertens (1997), i.e. CI = 150 * 0.01 * peNDF with NDF expressed as % of DM and assuming a pe-factor of 0.3 for pellets. Besides CI, RI + 0.5 * EI was also considered, because there are reasons for reducing the weighting of eating index. A summary of the results is given in Table 6.8. The average critical chewing time for all the rations concerned was 26.5 min/kg DM with a standard deviation of 4.0 min.

For rations with a grassland product or maize silage, the critical chewing time was 24.1 and 27.4 min/kg DM, respectively. This parameter also varied considerably between rations with brewers grains, potatoes or maize cobs + husks. For the ration with straw, the critical chewing time was 35.1 min/kg DM. When only half the eating time was taken into account, similar results were found. The generally high SD, the difference between grassland products and maize silage and the difference between other types of diets, lead to the conclusion that chewing time is not an ideal standard for physical structure. This is supported by some data from the literature and other considerations. Beauchemin, Farr and Rode (1991) found structure deficiency when a ration with a CI of 36.8 min/kg DM was fed, whereas this was not the case with similar rations with a lower CI. The structure-deficient diets contained more barley. The rapidly degradable barley could have depressed microbial digestion in the rumen, which is probably compensated by a higher chewing activity. Grummer, Jacob and Woodford (1987), De Boever (1991) and Dulphy (1995) also obtained a higher CI as a result of compensation due to a high proportion of concentrates in the ration. We also found that certain animal-linked factors, such as live weight and feed intake level, influence chewing indices, but not R_{crit}. Moreover we could not demonstrate an individual relationship between chewing activity and physical structure requirements. In addition, the chewing index can only be considered as a measure of rumen buffering through saliva but does not take the acidotic effect and the intrinsic buffering capacity of the feed into account.

The R_{crit} of grass and maize silage was more closely related to NDF content than to CI (Tables 6.3 and 6.4). The critical NDF content amounted on average to 302 g/kg DM with a SD of 32 g, which is relatively low compared with the SD for critical CI. This is not surprising as the standard compound feed already contained 254 g NDF

Table 6.8 Value of some parameters in the R_{crit} phase

		n[1]	*CT*[2]	*RT + 0.5 * ET*	*NDF*	*eNDF*	*CF*
			(min/kg DM)			*(g/kg DM)*	
All rations	Mean	56	26.5	21.7	302	161	136
	SD		4.0	3.2	32	13	18
Grassland products[3]	Mean	21	24.1	19.6	295	160	130
	SD		2.7	1.9	218	13	17
Maize silage[4]	Mean	21	27.4	22.7	300	160	137
	SD		2.7	2.3	30	12	18
Individual trials							
Grass silage			22.3	18.4	301	154	130
Grass silage + PBP[5]			22.0	18.2	329	148	137
Grass silage + FB			22.4	18.0	249	129	107
Maize silage A			25.5	21.1	304	168	134
Maize silage A + PBP			25.8	21.4	325	159	145
Maize silage A + FB			26.7	21.6	261	153	116
Maize silage C			31.8	25.9	304	172	147
Maize silage C + PO			30.7	24.2	261	157	127
Maize silage C + BG A			37.4	30.8	354	169	155
Maize silage C + MCH A			34.6	28.2	295	171	141
Maize silage D			27.2	22.8	325	173	144
Maize silage D + BG B			30.7	25.2	372	168	144
Maize silage D + grass silage			25.9	21.5	33.7	176	140
Maize silage D + grass silage + MCH B			34.0	28.1	307	184	139
Straw			35.1	26.9	340	194	153

[1] n : number of rations
[2] : CT, RT, ET: chewing, ruminating, eating time
[3] : results of rations with a grassland product as the sole roughage
[4] : results of rations with maize silage as the sole roughage
[5] : PBP = pressed beet pulp, FB = fodder beets, PO = potatoes, BG = brewers grains, MCH = maize cobs + husks

per kg DM. Therefore, this parameter is not sensitive enough to detect changes in the roughage:concentrate ratio. Moreover the minimum requirement should mainly depend on the NDF content of the concentrate. In the trials with three concentrates (Table 6.7), critical NDF contents amounted to an average of 311, 237 and 256 min/kg DM, respectively.

Both the chewing trials and the standard trials attribute a similar structure value to pressed beet pulp as to fodder beets. However, since beet pulp contains markedly more NDF, critical NDF content was also a lot higher for the rations with beet pulp. A similar observation was made for the rations containing brewers grains. On the other hand, the structure value of beets and potatoes based on NDF content would be considerably underestimated. The high variation in critical NDF content between different rations, as well as the dependence of it on the NDF content of the concentrate lead to the conclusion that NDF is also not a suitable parameter for physical structure. Data from the literature support this conclusion. This is not surprising because NDF content does not take particle size and physical form of feedstuffs into account, nor the acidotic effect of ingredients. The different chewing times per kg NDF also make NDF doubtful as a suitable structure index (Kaiser and Combs, 1989; Beauchemin and Buchanan-Smith, 1990; Beauchemin, 1991; Beauchemin *et al.*, 1991; Beauchemin and Iwaasa, 1993; Okine, Khorasani and Kennelly, 1994). Due to other influences on structure value, NDF has recently been replaced by effective NDF. This necessitates the attribution of an effectivity coefficient to the NDF of feedstuffs (Sniffen *et al.*, 1992; Armentano and Pereira, 1997; Mertens, 1997). For a lot of feedstuffs these coefficients are missing or are estimated through extrapolation. Table 8 again demonstrates great differences in critical eNDF content between certain types of rations. Although an effectivity coefficient of 1.0 was attributed to fodder beet and potatoes, the critical eNDF contents of the ration with those feeds were clearly lower than the mean value. Furthermore, the two experimental compound feeds with a high starch + sugar content clearly affect R_{crit} in a different way, despite similar eNDF content (49 and 47 g/kg DM). Hence there are few reasons to use eNDF as a standard in a structure evaluation system.

From the critical crude fibre contents, it can also be concluded that CF is not suitable as a general standard for physical structure. This explains why in the past, strongly diverging standards, ranging from 100 to 220 g CF/kg DM were proposed.

Considering the above, a structure evaluation system with a derived unit was chosen. Primarily, this system is based on the obtained R_{crit} results, but chewing indices, NDF and eNDF contents are also taken into account for several feedstuffs.

PRINCIPLE FOR DERIVING THE STRUCTURE VALUES (SV)

With a derived unit, it does not matter what the unit is. The feeding values simply have to correspond to the requirements. For practical reasons, the minimum structure requirements of a standard cow is assumed equal to 1. A standard cow is presumed to be a cow in 1st, 2nd or 3rd lactation, producing 25 kg milk daily with a fat content of 44 g/kg and receiving concentrates in two meals a day.

It was postulated that in the critical phase of the experiments the minimum requirement was met precisely. Then, roughage, concentrates and eventually supplements provided just enough structure.

Principle: $(R_{crit}/1000 * SV_R) + (C_{crit}/1000 * SV_C) = 1$ R_{crit} in g/kg

When a supplement was also fed, a third term was added i.e. $S_{crit} * SV_S$, with S_{crit} being the proportion of the supplement in the critical phase and SV_S the structure value of the supplement. The critical R-, C- and S-proportions were determined individually in the experiments. If for the rations with only a roughage and a concentrate, the SV of the concentrate is known, the SV of the roughage can be derived from the equation. For that purpose, the concentrate poorest in structure, the one with the high (S+S) content and rapidly degradable, Concentrate 3 (Table 6.6), was attributed a SV of 0.05 per kg DM. Once the SV of the roughage was known, the SVs of the other two experimental concentrates, which were given in combination with the same roughage (Table 6.7), could be deduced. Following this procedure, the SV of Concentrate 1 was derived five times and the SV of Concentrate 2 twice. If a higher or lower SV was attributed to Concentrate 3, only a small effect on SV of the concentrates and roughages was noticed. Consequently, the SV of 0.05 for concentrate 3 was maintained. From the SV of the 3 concentrates, the SV of the ingredients were estimated. This was based on the buffering effect of the concentrates and ingredients on the one hand and on their acidotic effect on the other hand. The rumen buffering from saliva secretion during mastication was estimated from NDF content according to the system proposed by Mertens (1997). The acidotic effect was mainly based on the determined *in vitro* pH decrease after 5h incubation. Once the SVs of the ingredients had been estimated, the SV of the standard concentrate, which was used in most trials, could be calculated. Finally, the SVs of the roughages and supplements were derived by means of the equation mentioned above.

SAFETY MARGIN

Because of the great individual variation in structure requirements, and the variation in structure value among samples of the same forage type, a large safety margin has to be built in to minimise the risks for cows fed at the limit (according to the system). However, a larger safety margin lowers the tolerable concentrate proportion in the ration, which could jeopardise the energy supply of high yielding cows. As a compromise, a risk of 5% was chosen. In principle, this implies that, when the system is applied strictly, 5% of the cows fed at the limit would show physical structure deficiency.

Based on individual variation and variation among samples of forages, the safety margin could be based on a standard deviation of 25% of the R_{crit}. At a risk of 5%, the R_{crit} have to be increased by 1.65 * 25% = 41%. Statistically, 5% of the R_{crit} exceed the value given by the average + 1.65*SD. This safety margin would increase the physical structure requirements by 0.3. Because a requirement for a standard cow equal to 1 was preferred, the SVs were divided by 1.3.

STRUCTURE VALUES

Roughages and supplements

The derived structure values, shown in Table 6.9, are the so-called "safe" structure values. The SV of grassland products and maize silage are mainly based on the R_{crit} values, although the chewing indices were taken into account to adjust the intercept and regression coefficients somewhat, when desirable. CF and NDF contents seem to be the best predictors of SV. The SV of maize silage can also be estimated with the same precision from starch content. Chewing index was a poor predictor of SV. From the R_{crit} values, no difference appeared between direct-cut and prewilted grass silage, nor between normal chopped and longer grass silage. The same regression equations can be used for hay as for silage. If the CF formula is used, the SV obtained has to be increased by 0.06. This correction is not necessary when the SV of hay is calculated from NDF, because NDF content increases during the hay making process. At present, the number of trials with fresh grass is too small to propose structure values.

The regression equations to estimate the SV of maize silage are valid for a theoretical chopping length of 6 mm. A correction of 2% per mm deviation in chopping length is proposed.

From the R_{crit}, chewing index and eNDF content of wheat straw, a SV of 4.30 per kg DM was derived. Although the barley straw used in the trials had a higher NDF content, a higher milling resistance and a higher chewing index than wheat straw, it is not justified to make a distinction between barley and wheat straw, since only one sample of barley straw was used. Fodder beets and ensiled pressed sugar beet pulp were studied extensively. Although NDF contents differed markedly, both chewing indices and R_{crit} values demonstrate similar SVs for the two supplements. For pressed brewers grains, with its smaller particle size than conventional brewers grains, a SV of 0.85 could be derived, compared with 1.00 for conventional brewers grains. The R_{crit} as well as chewing index of rations containing raw potatoes demonstrated a SV of approximately 0.70 per kg DM.

Table 6.9 Structure values (per kg DM) of currently used feedstuffs

1. Grassland products	
Grass silage	
$SV = +0.15 + 0.0060 * NDF$[1]	($R^2 = 0.26$; RSD = 0.40)
$SV = -0.20 + 0.0125 * CF$	($R^2 = 0.40$; RSD = 0.35)
Direct cut grass silage = prewilted grass silage	
Chopped silage = longer silage	
Hay	
NDF equation for grass silage without correction	
CF equation for grass silage * 1.06	
2. Maize silage chopping length 6 mm	
$SV = -0.57 + 0.0060 * NDF$	($R^2 = 0.33$; RSD = 0.25)
$SV = -0.10 + 0.0090 * CF$	($R^2 = 0.26$; RSD = 0.26)
Correction for deviating chopping length: +(-) 2% / +(-) 1 mm length	
3. Straw	
SV = 4.30	
4. Supplements	SV
Ensiled pressed beet pulp	1.05
Fodder beets	1.05
Ensiled brewers grains	1.00
Ensiled pressed brewers grains	0.85
Raw potatoes	0.70
5. Concentrates and ingredients	
Ground ingredients incorporated in pelleted concentrates	
$SV = 0.175 + 0.00082 * NDF + 0.00047 * USt.$[2] $- 0.00100 * (SU + a * DSt.)$	($R^2 = 0.97$; RSD = 0.04)
$SV = 0.321 + 0.00098 * CF + 0.00025 * USt. - 0.00112 * (SU + a * DSt.)$	($R^2 = 0.91$; RSD = 0.07)
$a = 0.90 - 1.3 *$ starch resistance	

[1]Contents in g/kg DM
[2]USt. = undegradable starch, DSt. = degradable starch, SU = sugars

Concentrates and ingredients

According to the principle already described, the SV of the concentrates used and their ingredients were derived. Subsequently, a method was sought to estimate the SV of these feedstuffs on the basis of known or currently determined parameters. The latter were chosen for logical reasons. They imply characteristics which are related to the rumen buffering capacity on the one hand and to the rumen acidotic effect on the other hand. NDF or CF content, as well as undegradable starch content, were considered as rumen stabilising factors, whereas sugars and degradable starch were considered as acidifying factors in the rumen. As sugar content was low and less variable, sugar and degradable starch content were considered as one term. Since the acidifying effect of degradable starch is lower compared with sugars, the first was only partially taken into account. This fraction "a" can be considered as the sugar equivalent of the degradable starch in terms of affecting rumen acidosis. It is logical to assume that this acidifying effect "a" is not a fixed value but depends on the degradation rate, which is related to the solubility of the degradable starch. From solubility results published by Tamminga, van Vuuren, van der Koelen, Ketelaar and van der Togt (1990), Nocek and Tamminga (1991) and Sauvant, Chapoutot and Archimède (1994), we derived the solubility of the degradable starch fraction. For manioc, wheat, barley, maize gluten feed, sorghum and maize, mean solubilities were 0.75, 0.73, 0.62, 0.65, 0.42 and 0.40, respectively. These solubility coefficients could be considered as the sugar equivalent (a) of the degradable starch. However, this parameter is not reported in current feed tables. Expecting an inverse relationship between the solubility of degradable starch and starch resistance to rumen degradation, a relationship between both parameters was sought. Based on the data from Tamminga *et al.* (1990), Nocek and Tamminga (1991), Sauvant *et al.* (1994) and the Dutch feed tables (Centraal Veevoederbureau, 1998), as well as on a database of 56 concentrates from this Institute, the following relationship, valid for pelleted concentrates, could be derived: a = 0.90 – 1.3 * starch resistance. Regression analysis using the calculated SV of 13 ingredients and concentrates from the present experiments, resulted in the equations shown in Table 6.9. The formulae are valid for ground ingredients (6 mm) which are incorporated into pelleted concentrates.

Example: maize grain NDF: 139 g, St.: 676 g, SU: 12 g, starch resistance: 0.42

USt. = 676 g * 0.42 = 284 g, DSt. = 676 g * 0.58 = 392 g

a = 0.90 – 1.3 * 0.42 = 0.35

SV = 0.175 + 0.00082 * 139 + 0.00047 * 284 – 0.00100 * (12 + 0.35 * 392)

= 0.27 /kg DM

The relationship between estimated and the derived SV of the ingredients and concentrates is presented in Figure 6.5.

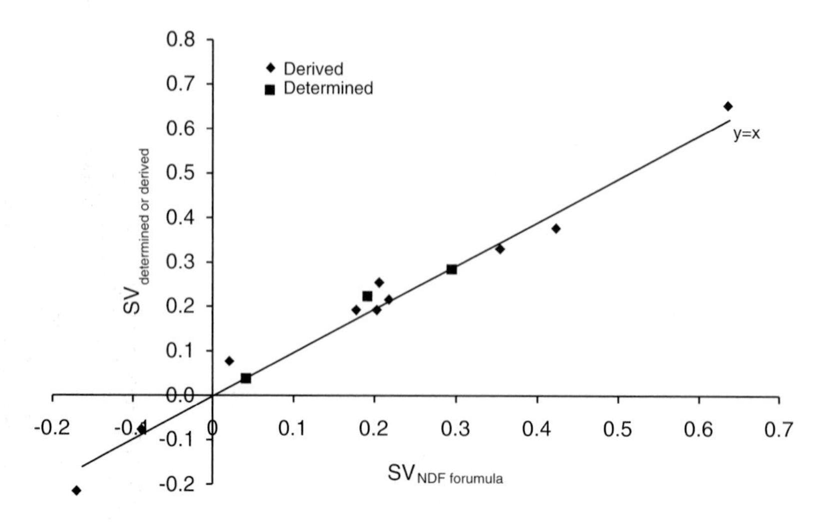

Figure 6.5 Relationship between estimated (NDF formula) and determined or derived structure value of concentrates and ingredients

STRUCTURE REQUIREMENTS

In the system it is assumed that the structure value of the diet has to be at least 1 per kg DM for a standard cow (25 kg milk, 44 g/kg fat, 1st, 2nd, 3rd lactation), when concentrates are provided in two meals (Table 6.10). For other situations, the following corrections have to be applied. Under circumstances similar to the present experiments, a correction to the R_{crit} of 1%, modified the structure requirements by 0.8%.

Table 6.10 Structure requirements

SV_{ration} ≥1 per kg DM

Cow 25 kg milk; 44 g/kg fat; 1st, 2nd, 3rd lactation; two concentrate meals

Corrections for

1) - Milk yield and fat content
   +(-) 0.01 / kg milk higher (lower) than 25 kg
   +(-) 0.005 / g fat lower (higher) than 44 g/kg
   - Only milk yield
   +(-) 0.012 / kg milk higher (lower) than 25 kg

2) Age
   4th lactation: -0.07
   ≥5th lactation: -0.15

3) Frequent concentrate feeding (at least 6 meals): -0.10

Milk yield and fat content

Cows with a higher daily milk yield and/or a lower fat content need a ration with a higher structure value. The effect of both parameters on R_{crit} resulted in an increase (decrease) of the SV requirement by 0.01 per kg higher (lower) milk yield, and by 0.005 per g/kg lower (higher) fat content. If a correction is made for milk yield alone, the requirement has to be increased (decreased) by 0.012 per kg higher (lower) milk yield.

Age (number of lactations)

From the experimental data it could be concluded that the R_{crit} for cows in their 4th or 5th lactation is 9 or 19% lower respectively than for cows in the 1st, 2nd or 3rd lactation. This corresponds with a decrease in structure requirement of 0.07 and 0.15.

Frequent concentrate feeding

Three trials were carried out to study the effect of frequent feeding of concentrates on structure requirements. Six concentrate meals per day resulted in structure requirements of 0.91, 0.92 and 0.88 compared with 1 when the concentrates were given in 2 meals. Hence, the structure requirements with at least 6 meals are assumed to be 0.10 lower.

APPLICATION OF THE SYSTEM

The structure evaluation system allows calculation of the minimum roughage portion in the ration necessary to maintain normal rumen function.

Example: a cow in the 4th lactation producing 35 kg milk with 38 g fat/kg receives concentrates with a SV of 0.20/kg DM in two meals, and roughage consisting of 30% prewilted grass silage and 70% maize silage (6 mm chopping length) containing respectively 460 and 380 g NDF per kg DM.

Structure requirement: $1 - 0.07 + (35 - 25) * 0.01 - (38 - 44) * 0.05 = 1.06$

$SV_{prewilted\ grass\ silage}$: $+0.15 + 0.0060 * 460 = 2.91$

$SV_{maize\ silage}$: $-0.57 + 0.0060 * 380 = 1.71$

$SV_{roughage}$: $(2.91 * 0.30) + (1.71 * 0.70) = 2.07$

$(R_{crit}/1000 * SV_R) + (C_{crit}/1000 * SV_C) = 1.06$

$R_{crit}/1000 = x$ $C_{crit}/1000 = 1\text{-}x$

$x * 2.07 + (1\text{-}x) * 0.20 = 1.06$

x * (2.07 – 0.20) = 1.06 – 0.20
x = 0.86/1.87 = 0.46

The roughage proportion in the diet has to be at least 0.46.

If the concentrates are fed in a complete diet (i.e. concentrates are eaten spread throughout the day):

Structure requirements: 1.06 – 0.10 = 0.96
x * 2.07 + (1 – x) * 0.20 = 0.96
x * (2.07 – 0.20) = 0.96 – 0.20
x = 0.76/1.87 = 0.41

When concentrates are spread, a roughage proportion of 0.41 is sufficient.

Acknowledgements

The authors wish to thank A. Opstal and all the personnel of the Department for their excellent technical assistance. The Institute for the Encouragement of Scientific Research in Industry and Agriculture is gratefully acknowledged for funding this research.

References

Armentano, L. and Pereira, M. (1997) Measuring the effectiveness of fiber by animal response trials. *Journal of Dairy Science*, **80**, 1416-1425.

Bae, D.H., Welch, J.G. and Gilman, B.E. (1983) Mastication and rumination in relation to body size of cattle. *Journal of Dairy Science*, **66**, 2137-2141.

Balch, C.C. (1971) Proposal to use time spent chewing as an index to the extent to which diets for ruminants possess the physical property of fibrousness characteristic of roughages. *British Journal of Nutrition*, **26**, 383-392.

Bath, I.H. and Rook, J.A.F. (1963) The evaluation of cattle foods and diets in terms of the ruminal concentration of volatile fatty acids. *Journal of Agricultural Science*, **61**, 341-348.

Beauchemin, K.A. (1991) Effects of dietary neutral detergent fiber concentration and alfalfa hay quality on chewing, rumen function, and milk production of dairy cows. *Journal of Dairy Science*, **74**, 3140-3151.

Beauchemin, K.A. and Buchanan-Smith, J.G. (1990) Effects of fiber source and method of feeding on chewing activities, digestive function, and productivity of dairy cows. *Journal of Dairy Science,* **73**, 749-762.

Beauchemin, K.A., Farr, B.I. and Rode, L.M. (1991) Enhancement of the effective fiber content of barley-based concentrates fed to dairy cows. *Journal of Dairy Science*, **74**, 3128-3139.

Beauchemin, K.A., Farr, B.I., Rode, L.M. and Schaalje, G.B. (1994) Effects of alfalfa silage chop length and supplementary long hay on chewing and milk production of dairy cows. *Journal of Dairy Science*, **77**, 1326-1339.

Beauchemin, K.A. and Iwaasa, A.D. (1993) Eating and ruminating activities of cattle fed alfalfa or orchard-grass harvested at two stages of maturity. *Canadian Journal of Animal Science*, **73**, 79-88.

Beauchemin, K.A. and Rode, L.M. (1994) Compressed baled alfalfa hay for primiparous and multiparous dairy cows. *Journal of Dairy Science*, **77**, 1003-1012.

Beauchemin, K.A. and Rode, L.M. (1997) Minimum versus optimum concentrations of fiber in dairy cow diets based on barley silage and concentrates of barley or corn. *Journal of Dairy Science*, **80**, 1629-1639.

Brouk, M. and Belyea, R. (1993) Chewing activity and digestive responses of cows fed alfalfa forages. *Journal of Dairy Science*, **76**, 175-182.

Castle, M.E., Gill, M.S. and Watson, J.N. (1981) Silage and milk production: a comparison between grass silages of different chop lengths and digestibilities. *Grass and Forage Science*, **36**, 31-37.

Castle, M.E., Retter, W.C. and Watson, J.N. (1979) Silage and milk production: comparisons between grass silage of three different chop lengths. *Grass and Forage Science*, **34**, 293-301.

Centraal Veevoederbureau - Nederland (1998) *Veevoedertabel 1998 - Chemische samenstelling, verteerbaarheid en voederwaarde van voedermiddelen.* Editor Centraal Veevoederbureau, 8203 AD Lelystad, The Netherlands.

Clark, P.W. and Armentano, L.E. (1997) Influence of particle size on the effectiveness of beet pulp fiber. *Journal of Dairy Science*, **80**, 898-904.

Coulon, J.B., Doreau, M., Rémond, B. andJournet, M. (1987) Evolution des activités alimentaires des vaches laitières en début de lactation et liaison avec les quantités d'aliments ingérées. *Reproduction, Nutrition et Développement*, **27**, 67-75.

Dado, R.G. and Allen, M.S. (1994) Variation in and relationship among feeding, chewing, and drinking variables for lactating dairy cows. *Journal of Dairy Science*, **77**, 132-144.

De Boever, J. (1991) Roughage evaluation of maize and grass silage based on chewing activity measurements with cows. Ph. D. Diss., Univ. Ghent, Belgium, 171 pp.

De Boever, J.L., De Smet, A., De Brabander, D.L. and Boucqué, Ch.V. (1993) Evaluation of physical structure. 1. Grass silage. *Journal of Dairy Science*, **76**, 140-153.

Dehareng, D. and Godeau, J.M. (1991) The durations of masticating activities and the feed energetic utilisation of Friesian lactating cows on maize silage-based rations. *Journal of Animal Physiology and Animal Nutrition*, **65**, 194-205.

Demarquilly, C. (1994) Facteurs de variation de la valeur nutritive du maïs ensilage. *INRA Productions Animales*, **7**, 177-189.

De Smet, A.M., De Boever, J.L., De Brabander, D.L., Vanacker, J.M. and Boucqué, Ch.V. (1995) Investigation of dry matter degradation and acidotic effect of some feedstuffs by means of in sacco and in vitro incubations. *Animal Feed Science and Technology*, **51**, 297-315.

Deswysen, A. (1980) Influence de la longueur des brins et de la concentration en acides organiques des silages sur l'ingestion chez les ovins et bovins. Ph. D. Dissertation, Univ. Louvain-la-Neuve, Belgique, 254 pp.

Deswysen, A.G. and Ellis, W.C. (1990) Fragmentation and ruminal escape of particles as related to variations in voluntary intake, chewing behavior and extent of digestion of potentially digestible NDF in heifers. *Journal of Animal Science*, **68**, 3861-3879.

Deswysen, A.G., Ellis, W.C. and Pond, K.R. (1987) Interrelationships among voluntary intake, eating and ruminating behavior and ruminal motility of heifers fed corn silage. *Journal of Animal Science*, **64**, 835-841.

Dulphy, J.P. (1995) L'indice de fibrosité des aliments des ruminants. Intérêt et utilisation. Personal communication.

Dulphy, J.P. and Michalet-Doreau, B. (1983) Comportement alimentaire et mérycique d'ovins et de bovins recevant des fourrages verts. *Annales de Zootechnie*, **32**, 465-474.

Dulphy, J.P., Michalet-Doreau, B. and Demarquilly, C. (1984) Etude comparée des quantités ingérées et du comportement alimentaire et mérycique d'ovins et de bovins recevant des ensilages d'herbe réalisés selon différentes techniques. *Annales de Zootechnie*, **33**, 291-320.

Dulphy, J.P., Rouel, J., Jailler, M. and Sauvant, D. (1993) Données complémentaires sur les durées de mastication chez des vaches laitières recevant des rations riches en fourrage: influence de la nature du fourrage et du niveau d'apport d'aliment concentré. *INRA Productions Animales*, **6**, 297-302.

Grummer, R.R., Jacob, A.L. and Woodford, J.A. (1987) Factors associated with variation in milk fat depression resulting from high grain diets fed to dairy cows. *Journal of Dairy Science*, **70**, 613-619.

Flatt, W.P., Moe, P.W., Munson, A.W. and Cooper, T. (1969) Energy utilization by dairy Holstein cows. In Energy metabolism of farm animals. Publication n° 12 Symposium European Association of Animal Production - sept. 1969, Warsaw, 235-251. Edited by K.L. Blaxter, J. Kielanowski and G. Thorbek.

Guth, N. (1995) Unterschiedliche Häckselgutstruktur von Halmfutter: Einfluß auf Futteraufnahme, Leistung und Kauverhalten von Rindern, Silagequalität und

Häckselleistungsbedarf sowie bildanalytische Vermessung der Futterstruktur. Dissertation, Editor Rosa Fischer - Löw Verlag, Gieben, 305 pp.

Hoffmann, M. (1983) Tierfütterung. *VEB Deutscher Landwirtschaftsverlag*, DDR-Berlin.

Kaiser, R.M. and Combs, D.K. (1989) Utilization of three maturities of alfalfa by dairy cows fed rations that contain similar concentrations of fiber. *Journal of Dairy Science*, **72**, 2301-2307.

Kamatali, P. (1991) L'ingestion volontaire d'ensilages d'herbe, l'efficience digestive et le comportement alimentaire et mérycique chez les bovins. Doctoral thesis, Louvain-la-Neuve, Belgium, 297 pp.

Kaufmann, W. (1976) Influence of the composition of the ration and the feeding frequency on pH-regulation in the rumen and the feed intake in ruminants. *Livestock Production Science*, **3**, 103-114.

Kesler, E.M. and Spahr, S.L. (1964) Physiological effects of high level concentrate feeding. *Journal of Dairy Science*, **47**, 1122-1128.

Kuehn, C.S., Linn, J.G. and Jung, H.G. (1997) Effect of corn silage chop length on intake, milk production, and milk composition of lactating dairy cows. *Journal of Dairy Science*, **80**, (Suppl. 1), 219 (Abstr.).

Lammers, B., Heinrichs, J. and Buckmaster, D. (1996) Method helps in determination of forage, TMR particle size requirements for cattle. *Feedstuffs*, September 30, 14-16.

Malestein, A., van't Klooster, A.Th., Counotte, G.H.M. and Prins, R.A. (1981) Concentrate feeding and ruminal fermentation. 1. Influence of the frequency of feeding concentrates on rumen acid composition, feed intake and milk production. *Netherlands Journal of Agricultural Science*, **29**, 239-248.

Mertens, D.R. (1997) Creating a system for meeting the fiber requirements of dairy cows. *Journal of Dairy Science,* **80**, 1463-1481.

Meyer, R.M., Bartley, E.E., Morrill, J.L. and Stewart, W.E. (1964) Salivation in cattle. I. Feed and animal factors affecting salivation and its relation to bloat. *Journal of Dairy Science*, **47**, 1339-1345.

Nocek, J.E. and Tamminga, S. (1991) Site of digestion of starch in the gastrointestinal tract of dairy cows and its effect on milk yield and composition. *Journal of Dairy Science*, **74**, 3598-3629.

Nørgaard, P. (1985) Physical structure of feeds for dairy cows. (A new system for evaluation of the physical structure in feedstuffs and rations for dairy cows). *CEC-workshop. New developments and future perspectives in research on rumen function*, 25-27 June, Ørum Sønderlyng, Denmark.

Nørgaard, P. (1989) The influence of physical form of ration on chewing activity and rumen motility in lactating cows. *Acta Agriculturae Scandinavica*, **39**, 187-202.

Nørgaard, P. (1993) The effect of carbohydrate composition and physical form of feed on chewing activity and rumen motility in dairy cows. *44th Annual Meeting of the European Association for Animal Production*, Aarhus, Denmark 16-19 August, CN 2.2.

NRC (1978) *Nutrient requirements of dairy cattle*. Fifth Revised Edition. National Academy of Sciences, Washington, 76 pp.

NRC (1988) *Nutrient requirements of dairy cattle*. Sixth Revised Edition. National Academy Press, Washington, 158 pp.

Okine, E.K., Khorasani, G.R. and Kennelly, J.J. (1994) Effects of cereal grain silages versus alfalfa silage on chewing activity and reticular motility in early lactation cows. *Journal of Dairy Science*, **77**, 1315-1325.

Piatkowski, B., Nagel, S. and Bergner, E. (1977) Das Wiederkauverhalten von Kühen bei unterschiedlicher Trockensubstanzaufnahme und verschiedener physikalischer Form von Grasheu. *Archiv für Tierernährung*, **27**, 563-569.

Playne, M.J. and McDonald, P. (1966) The buffering constituents of herbage and of silage. *Journal of the Science of Food and Agriculture*, **17**, 264-268.

Rémond, B. and Journet, M. (1972) Alimentation des vaches laitières avec des rations à forte proportion d'aliments concentrés. *Annales de Zootechnie*, **21**, 191-205.

Robinson, P.H. (1989) Dynamic aspects of feeding management for dairy cows. *Journal of Dairy Science*, **72**, 1197-1209.

Rohr, K., Honig, H. and Daenicke, R. (1983) Zur Bedeutung des Zerkleinerungs-grades von Silomaïs. 2. Mitteilung: Einflub des Zerkleinerungsgrades auf Wieder-kauaktivität, Pansenfermentation und Verdaulichkeit der Rohnährstoffe. *Das Wirtschaftseigene Futter*, **29**, 73-86.

Rohr, K., Daenicke, R., Honig, H. and Lebzien, P. (1986) Zum Einsatz von Prebschnitzelsilage in der Milchviehfütterung. *Landbauforschung Völkenrode*, **36**, 50-55.

Satter, L.D. and Baumgardt, B.R. (1962) Changes in digestive physiology of the bovine associated with various feeding frequences. *Journal of Animal Science*, **21**, 897-900.

Sauvant, D. (1992) Compléments sur la fibrosité des rations des ruminants. *Journées CAAA-AFTAA*, Tours, 26-27/02/1992.

Sauvant, D., Chapoutot, P. and Archimède, H. (1994) La digestion des amidons par les ruminants et ses conséquences. *INRA Productions Animales*, **7**, 115-124.

Sniffen, C.J., Hooper, A.P., Welch, J.G., Randy, H.A. and Thomas, E.V. (1986) Effect of hay particle size on chewing behavior and rumen mat consistency in steers. *Journal of Dairy Science*, **69**, 135 (Suppl. 1).

Sniffen, C.J., O'Connor, J.D., Van Soest, P.J., Fox, D.G. and Russell, J.B. 1992. A net carbohydrate and protein system for evaluating cattle diets: II. Carbohydrate and protein availability. *Journal of Animal Science*, **70**, 3562-3577.

Sudweeks, E.M., Ely, L.O., Mertens, D.R. and Sisk, L.R. (1981) Assessing minimum amounts and form of roughages in ruminant diets: roughage value index system. *Journal of Animal Science*, **53**, 1406-1411.

Sutton, J.D. (1984) Feeding and milk fat production. In Milk compositional quality and its importance in future markets. Occasional publication No.9 of the *British Society of Animal Production* - 1984, pp. 43-52. Edited by M.E. Castle and R.G. Gunn.

Sutton, J.D., Hart, I.C., Broster, W.H., Elliott, R.J. and Schuller, E. (1986) Feeding frequency for lactating cows: effects on rumen fermentation and blood metabolites and hormones. *British Journal of Nutrition*, **56**, 181-192.

Swain, S.M. and Armentano, L.E. (1994) Quantitative evaluation of fiber from nonforage sources used to replace alfalfa silage. *Journal of Dairy Science*, **77**, 2318-2331.

Tamminga, S., van Vuuren, A.M., van der Koelen, C.J., Ketelaar, R.S. and van der Togt, P.L. (1990) Ruminal behaviour of structural carbohydrates, non-structural carbohydrates and crude protein from concentrate ingredients in dairy cows. *Netherlands Journal of Agricultural Science*, **38**, 513-526.

Tilley, J.M.A. and Terry, R.A. (1963) A two-stage technique for the in vitro digestion of forage crops. *Journal of the British Grassland Society*, **18**, 104-111.

Teller, E., Vanbelle, M., Kamatali, P., Collignon, G., Page, B. and Matatu, B. (1990) Effects of chewing behavior and ruminal digestion processes on voluntary intake of grass silages by lactating dairy cows. *Journal of Animal Science*, **68**, 3897-3904.

Teller, E., Vanbelle, M., Kamatali, P. and Wavreille, J. (1989) Intake of direct cut or wilted grass silage as related to chewing behavior, ruminal characteristics and site and extent of digestion by heifers. *Journal of Animal Science*, **67**, 2802-2809.

Van Soest, P.J., Robertson, J.B. and Lewis, B.A. (1991) Methods for dietary fiber, neutral detergent fiber, and nonstarch polysaccharides in relation to animal nutrition. *Journal of Dairy Science*, **74**, 3583-3597.

Vérité, R. and Journet, M. (1973) Utilisation de quantités élevées de betteraves par les vaches laitières: étude de l'ingestion, de la digestion et des effets sur la production. *Annales de Zootechnie*, **22**, 219-235.

Voskuil, G.C.J. and Metz, J.H.M. (1973) The effect of chopped hay on feed intake, rate of eating and rumination of dairy cows. *Netherlands Journal of Agricultural Science*, **21**, 256-262.

Weiss, W.P. (1993) Fiber requirements of dairy cattle: emphasis NDF. *54th Minnesota Nutrition Conference & National Renderers Technical Symposium*, September 20-22, Bloomington, Minnesota.

Welch, J.G. (1982) Rumination, particle size and passage from the rumen. *Journal of Animal Science*, **54**, 885-894.

7

PHYTASE IN POULTRY NUTRITION

S.P. TOUCHBURN, S. SEBASTIAN and E.R. CHAVEZ
Macdonald Campus of McGill University, Ste-Anne-de-Bellevue, Quebec, Canada H9X 3V9

Phosphorus

Phosphorus (P) is an important component in the animal body although it represents only 6 - 7.5 g/kg adult weight. 0.80 of body P is present in the skeletal structure. The remaining 0.20 is critical to homeostasis: in proteins such as enzymes, in lipids such as the phospholipids of membranes, in nucleic acids such as in DNA and adenosine triphosphate (ATP), the currency for energy exchange in the body. Because of its many interactions with calcium (Ca), the two elements are normally considered together. A dietary deficiency of P results in decreased growth and production or bone lesions. An early sign of deficiency is depressed appetite which, of course, leads to decreased growth but not necessarily lower efficiency of feed utilization. The digestibility of both Ca and P is depressed by an excess of either one but most commonly occurs because of an excess of Ca, the least expensive ingredient in feed.

Sources of P

Sources of dietary P include the rock phosphates which are relatively unavailable to animals but the process of heating to eliminate fluorine alters their crystal structure and yields tricalcium-phosphate which is as digestible as animal product sources such as steamed bone meal (Scott, Nesheim and Young 1982). Along with the synthetic mono- and dicalcium phosphates, these products are generally considered to have an availability approaching 1.00. In contrast, plant sources of P are generally calculated to be only about 0.30 available to simple-stomached animals which normally consume mostly grains, oilseed meals and their by-products in which most of the P is in the form of phytic acid complexes.

Two recent developments have led to the expanding practice of adding phytase enzymes to the diets of pigs and poultry; first, the need to reduce P pollution of ground and surface waters from manure - a problem most critical in regions of high animal density and in some countries which are subject to strict regulation; second, the advent of genetically engineered micro-organisms capable of yielding acid phytase enzyme in sufficient concentration and at reasonable cost (Cantor, 1995). The high cost of P supplements of course, makes the use of these microbial phytase products even more attractive.

Phytin

Phytin, phytic acid, phytate or myo-inositol hexaphosphate is the major storage form of phosphorus in plants, comprising 0.60 to 0.80 of the P in grains and their by-products (Ravindran, Bryden and Kornegay 1995) and in oil seeds but only 0.15 to 0.30 in green plants (National Research Council, 1994). This storage form provides reactive anions capable of binding or chelating divalent or trivalent cations, forming insoluble salts. It also complexes with and binds proteins, rendering them resistant to proteolytic enzymes. It may also bind endogenous digestive enzymes, reducing their activity.

Phytate can be considered a unique compound since it bears six phosphate groups on one 6-carbon molecule of low molecular weight (Figure 7.1). At neutral pH, each of the phosphate groups has either one or two negatively charged oxygen atoms; hence various cations could strongly chelate between two phosphate groups or weakly with a single phosphate group (Figure 7.2). Thus phytate is considered a nutrient because it contains P but it can also be considered a toxin or anti-nutrient because it binds various essential mineral elements and reduces their availability for absorption from the diet (Reddy, Sathe and Salunkhe, 1982). The concentration of phytate in feedstuffs depends on the part of the plant from which they are derived. Oilseed meals contain large amounts compared to cereal by-products and grain legumes (Ravindran et al., 1995). Furthermore, the phytate is deposited in different parts of the seed as shown in Table 7.1 (Reddy *et al.*, 1982). In maize it is almost all in the germ; in wheat it is in the aleurone and the germ; in rice in the germ and the pericarp. In oilseeds the phytate is dispersed throughout the seed (Erdman, 1979).

Phytase

ENDOGENOUS PHYTASE IN POULTRY

Simple-stomached animals, like poultry and pigs, elaborate very little phytase, hence dietary phytates are largely undigested and end up in the manure. However, on diets low in non-phytin P (nP), that is lacking in added mineral phosphates or animal products, chick intestinal phytase has been shown to increase more than three-fold over that in controls fed P-supplemented diets (Davies and Motzok 1972). Most recently, Maenz and Classen (1998) studied the kinetics of phytate hydrolysis by chicken small intestinal

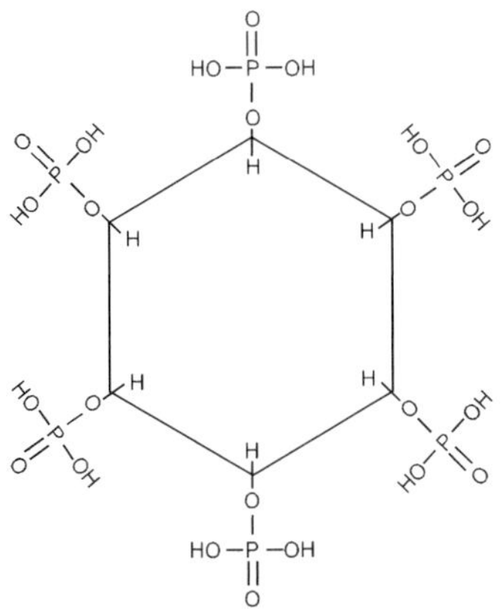

Figure 7.1 Structure of phytic acid proposed by Anderson (1914). (After illustration by Reddy *et al.*, 1982).

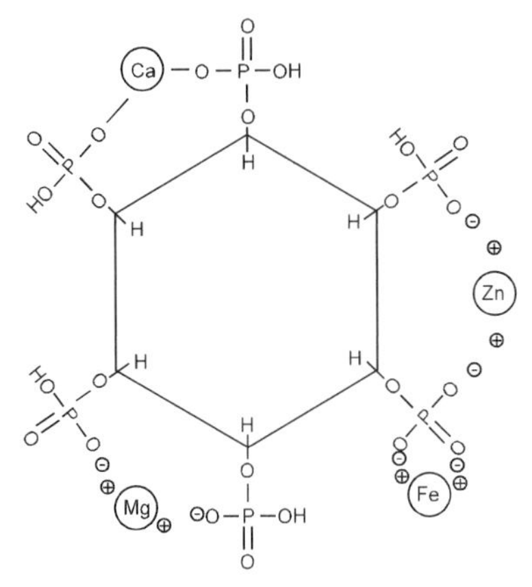

Figure 7.2 Phytic acid chelate at neutral pH. (After Erdman, 1979)

brush boarder membrane vesicles. Effective pH of the enzyme ranged from 5 to 6.5 with a maximum at 6, the pH of the intestinal unstirred water layer. Activity was highest in the duodenum and declined in the jejunum and still further in the ileum. Comparing 4-wk-old broilers and laying hens, they found the same specific activity in each category but the hens had 35% higher total activity. They suggested that the expression of intestinal phytase activity is subject to regulatory mechanisms and is important in maintaining P status. This raises the possibility of selecting birds genetically for increased ability to secrete phytase or employing bioengineering successfully in introducing the phytase gene in poultry.

Genetic differences in poultry

In early studies, Edwards (1983) reported that White Leghorn cockerels were able to utilise sub-optimal levels of P in a diet more efficiently than did commercial broilers as

Table 7.1 Phytic acid and phytic acid P concentration in morphological components of cereals

Cereal	*Sample*	*Phytic acid (g/kg)*	*Phytic acid P (g/kg)*
Maize	Commercial hybrid	8.9	2.5
	Endosperm	0.4	0.1
	Germ	63.9	18.0
	Hull	0.7	0.2
Maize	High lysine	9.6	2.7
	Endosperm	0.4	0.1
	Germ	57.2	16.1
	Hull	2.5	0.7
Wheat	Soft	11.4	3.2
	Endosperm	0.04	0.01
	Germ	39.1	11.0
	Hull	0.0	0.0
	Aleurone	41.2	11.6
Rice	Brown	8.9	2.5
	Endosperm	0.1	0.04
	Germ	34.8	9.8
	Pericarp	33.7	9.5

(After Reddy *et al.*, 1982)

measured by growth, liveability and bone calcification. Differences were exaggerated by higher levels of Ca in the diet. The Athens-Canadian Randombred Control broiler population were intermediate in their response. Carlos and Edwards (1997) compared strains and crosses of young broiler chicks and found large individual differences in their ability to utilise phytate-P. Body weight, bone ash, toe ash, and plasma minerals were measured and it was concluded that, on a P-deficient diet, 21-day body weight might be a satisfactory criterion for selection studies. Zhang, McDaniel and Roland (1998) compared broiler lines for their ability to utilise phytate-P. An unselected control line was compared with lines selected over eleven generations for a high- or low incidence of tibial dyschondroplasia and it was found the controls were superior in liveability, mortality and growth performance. However, variability was greater in the selected lines, which led to the conclusion that selection could be effective for improved utilization of phytate-P.

ENDOGENOUS PHYTASE IN FEEDSTUFFS

Endogenous plant phytases may contribute to phytate hydrolysis. They are present in considerable amounts in the cereals, rye and wheat but their activity is highly variable. Maize, Sorghum and the oilseed meals are very low in phytase content (Eeckhout and de Paepe 1994). It is relevant to note that the optimum pH for plant phytase activity is in the range of 4.0-6.0 (Table 7.2) though some activity may be retained at pH 3.0. The activity will vary depending on cultivar, climate and handling conditions. Optimum temperatures range mostly from 40 to 55°C and activity of the phytase will be reduced if the diet or ingredients containing the enzyme are subjected to temperatures much above 60 to 65°C during processing.

Table 7.2 Optimum pH and temperature of phytase from cereals and legumes

Phytase source	*Optimum pH*	*Temperature (°C)*	*Reference*
Triticale	5.4	45	Singh and Sedeh (19790
Corn	5.6	50	Chang (1967)
Wheat flour	5.15	55	Peers (1953)
Wheat bran	5.0	-	Nagai and Funahashi (1962)
Rice aleurone particles	4.0-5.0	45	Yoshida *et al.* (1975)
Navy bean	5.3	50	Lolas and Markakis (1977)
Dwarf french bean	5.2	40	Gibbins and Norris (1963)
California small white bean	5.2	60	Chang (1975)
Mung bean (germinating)	7.5	57	Mandal and Biswas (1970)

Microbial phytase as poultry feed supplement

Rojas and Scott (1969) reported that phytase from the mould, *Aspergillus ficuum* improved the metabolizable energy value of cottonseed meal. It effectively rendered the P from phytate available to the birds and also freed some protein from the phytate complex. Nelson, Shieh, Wodzinski and Ware (1971) also observed the improvement in phytin-P utilization from supplementation of the diet with this mould phytase. Subsequently, Simons, Versteegh, Jongbloed, Kemme, Slump, Bas, Wolters, Buedekers and Vershoor (1990) carefully prepared a batch of a crude phytase dry powder from *Aspergillus ficuum*. *In-vitro* studies showed the phytase to be active over a wide range in pH from 2.5 to 5.5. The microbial phytase supplement in the diets of broilers to 24 and 28 d of age overcame the poor growth performance of the birds fed the low-P basal diet. At a phytase dose level of 800 U/kg of diet, maximum apparent availability of P was obtained. The availability of the P rose to over 0.60 and the amount in the faeces decreased by 50%. It was also shown that the enzyme preparation had good stability at pelleting temperatures of 80°C with 0.46 activity remaining after pellet temperatures had reached 87°C (Table 7.3). Since that time, many studies have demonstrated the value of microbial phytase in broiler diets during the first 3 to 4 wk of life. More details are contained in the comprehensive review by Ravindran *et al.* (1995).

Table 7.3 influence of microbial phytase on growth performance and retention of P in broiler chickens fed maize-soya bean meal diets

Source	*Age (d)*	*Phytase units/kg*	*Non phytate-P in diet*	*% improvement*		
				BW gain	*FCR*	*P retention*
Simons *et al.* (1990)	28	750	1.6	38	0.63	20
Brox *et al.* (1994)	22	500	-	13	3.4	15
Aoyagi and Baker (1995)	20	600	-	77	3.7	-
Sebastian *et al.* (1996a)	21	600	3	13.2	0.67	24
Kornegay *et al.* (1996)	21	600	2	36	2.6	5.3
Yi *et al* (1996)	20	600	4.5	11	3.53	-

With the advent of genetically engineered microorganisms capable of producing large amounts of phytase at much more reasonable cost, there has been a flurry of research in this area. Campbell and Bedford (1992), reviewing research on enzyme applications for non-ruminant feeds, commented that the renewed interest in dietary phytase comes with the realisation that phytase provides a cost-effective alternative to inorganic P supplementation and that it becomes of even greater importance where excessive P in animal waste is a national concern.

Table 7.3 lists a few of the many papers indicating that microbial phytase supplementation increases the availability of phytate P in young broiler chickens (Simons *et al.*, 1990; Mohammed, Gibney and Taylor, 1991; Broz, Oldale, Perrin-Voltz, Rychen, Schulze and Simoes Nunes, 1994; Denbow, Ravindran, Kornegay, Yi and Hulet, 1995; Yi, Kornegay, Ravindran and Denbow, 1996; Kornegay, Denbow, Yi and Ravindran, 1996; Sebastian, Touchburn, Chavez and Laguë, 1996a,b) and in young turkeys (Yi, Kornegay and Denbow 1996a, Ravindran, Kornegay, Denbow, Yi and Hulet, 1995; Qian, Kornegay and Denbow 1996). The improvements in P retention by phytase supplementation ranged from 5.3% to 24% . Almost all of the trials involved chicks and turkey poults to 3- and some to 4-wk of age. Kornegay, Denbow and Zhang, (1997) concentrated on broilers in the 3- to 7-wk age period, supplementing a low-P maize-soyabean meal diet with 0, 200 or 400 U of phytase/kg of diet. It was observed that ileal digestibility and total retention of P (as a proportion of P intake) increased linearly with increasing phytase level. P excretion was reduced 35.4% by 200 U of phytase and 37.7% by 400 U of phytase per kg of diet. Calculated equivalency values averaged 290 U of phytase for 1 g of P based on improvements in toe- and tibia ash, and 231 U of phytase based on averages of body weight and gain. Thus, supplemental phytase proved to be very effective in broiler finisher diets.

The researchers at Virginia Polytechnic Institute and State University have also applied these equivalency calculations to young broilers and young turkeys. Yi *et al.* (1996) calculated an equivalency of 785 U of phytase for 1 g of P in young broilers fed a corn-soybean meal diet. Qian, Kornegay and Denbow (1996) determined these values for young male turkeys fed a maize-soyabean meal diet and found 963 U of phytase equivalency for 1 g of P when the nP level of the diet was 3.6 g/kg and 652 U of phytase with a dietary content of 2.7 g of nP/kg. In this trial it was observed that supplemental phytase yielded linear improvements in body weight gain, feed intake, feed-to-gain ratio, toe ash and apparent retention of Ca and P at each Ca:tP ratio and nP level. Widening the Ca:tP ratio had a detrimental effect at each phytase- and P level and was greatest at the lower levels of phytase and P. Many of these observations are in agreement with work in our laboratory with maize-soyabean meal broiler diets containing 3.5 g nP/kg (Sebastian 1996 a,b, Sebastian, Touchburn, Chavez and Laguë, 1997). Supplemental microbial phytase yielded increases in feed intake, growth, bone mineralization and retention of P, Ca, Cu, Zn and N. In the first trial (Sebastian *et al.*, 1996a) the best results occurred with dietary Ca at 6.0 or 10 g/kg of diet versus 12.5 g/kg. In a subsequent similar trial (Sebastian *et al.*, 1996b), feed intake and gain were depressed when the diet contained 3.5 g nP and 10 g Ca/kg but performance was restored to normal when the Ca level was reduced to 6 g/kg, i.e., a lower nP:Ca ratio. This negative response to a widened nP:Ca ratio has been apparent in other studies (Qian *et al.*, 1996). Since microbial phytase is active over a wide range of pH, the optimum range being 2.5 to 5.5 (Simons *et al.*, 1990), it can be active in the crop (pH 5-6) and in the proventriculus and gizzard (pH 2-4). Excess Ca may raise the pH to the point of affecting phytase function. The low solubility of phytate chelates and complexes formed would then further inhibit absorption of Ca, P and other nutrients such as mineral cations and protein.

Effect of microbial phytase on availability of cations

Theoretically, when phytic acid is hydrolysed by microbial phytase, all minerals bound to it will be released. There is ample evidence to indicate that microbial phytase supplementation improves the availability of Ca (Table 7.4) in broiler chickens (Schoner, Hoppe and Schwartz, 1991; Broz *et al.*, 1994; Kornegay *et al.*, 1996; Sebastian *et al.*, 1996a) and turkeys (Qian *et al.*, 1996). In a broiler study designed to measure the effect of phytase on Ca availability, Schoner *et al.* (1994) reported that 500 U of microbial phytase were equivalent to 0.35 g Ca, as measured by body weight gain and 0.56 g Ca as measured by phalanx ash. Qian *et al.* (1996) reported that both P and Ca retention were sensitive to the addition of phytase at varying nP levels and Ca:total P ratios in turkeys. Calcium retention increased linearly as the amount of supplemented phytase increased, and decreased as the Ca:total P ratio became wider and as the level of nonphytate P increased (Qian *et al.*, 1996). In a study with turkeys, Kornegay *et al.* (1996) estimated the value of phytase based on body weight gain, gain:feed and digested Ca within a Ca range of 5.3 to 7.4 g/kg of diet and a phytase level up to 500 U /kg of diet. On this basis they suggested that 500 U of phytase are equivalent to approximately 0.87 g of Ca.

Zinc availability has been shown to be affected by phytate in chicks, pigs, rats and other species (Prasad, 1966). Zinc is the most vulnerable to phytate complexation

Table 7.4 The effect of supplemental microbial phytase on availability of minerals in broiler chickens

Source	*Age (d)*	*Phytase units/kg*	*Mineral*	*Improvement (% units)*
Yi *et al.* (1996)	20	600	ZN	6.9
Sebastion *et al.* (1996a)	21	600	Zn	62.3
Aoyagi and Baker (1995)	20	600	Cu	-21
Sebastion *et al.* (1996a)	21	600	Cu	19.2
Simons *et al.* (1990)	28	750	Ca	13.1
Sebastian *et al.* (1996a)	21	600	Ca	12.2

based on *in vitro* studies. Thus, zinc deficiency is the most likely result from the feeding of a diet with phytate-containing plant seed protein. Oberleas and Harland (1996) reported that results clearly showed phytate was a significant factor in the genesis of zinc deficiency. Studies on the influence of phytase on other phytate-bound minerals are limited. Pallauf, Hohler and Rimbach (1992) found that phytase addition increased the apparent absorption of Mg, Zn, Cu and Fe by up to 13, 13, 7, and 9%, respectively, in pigs. The addition of 800 U of phytase /kg to a diet containing 27 mg of Zn/ kg increased the retention of Zn and decreased Zn excretion of chicks (Thiel and Weigand, 1992). Thiel *et al.* (1993) reported that the femoral Zn content of chicks fed a diet containing 30 mg Zn /kg plus 700 U of phytase /kg diet was equal to that of chicks fed a diet containing 39 mg of Zn/ kg of diet without phytase. Roberson and Edwards (1994) reported that phytase supplementation had no effect on Zn retention in broiler chicks, but improved its retention when given along with vitamin D_3. In a study conducted in our laboratory (Sebastian *et al.*, 1996a), it has been shown that microbial phytase supplementation increased the relative retention of Zn by 62.3% compared with control in broiler chickens fed a low P maize-soyabean meal diet (Table 7.4). Yi *et al.* (1996a) reported that Zn retention in broiler chickens was linearly increased by adding microbial phytase to a low-Zn basal diet. It was further indicated that 0.9 mg of Zn was released per 100 U of phytase over a range of 150 to 600 U of phytase. Based on Biehl *et al.*, (1995) estimates using tibia Zn, the Zn equivalency would be 0.63 mg Zn per 100 U of phytase when 600 U of phytase were added, but 0.46 mg Zn per 100 U of phytase when 1200 U of phytase were added.

Aoyagi and Baker (1995) studied the effect of microbial phytase supplementation on dietary Cu utilization in chickens; it was found that phytase supplementation to soyabean meal had no effect on Cu availability and in fact it reduced the Cu bioavailability by 21 % (Table 7.4). Sebastian *et al.* (1996a) showed that microbial phytase supplementation increased the relative retention of Cu by 19.3% in broiler chickens fed a low P maize-soyabean meal diet (Table 7.4). Studies on the influence of microbial phytase on other minerals are very limited.

Effects of microbial phytase on availability of protein and amino acids

Theoretically, phytase supplementation must also be able to release the phytate-bound protein for utilization, but published data on this aspect are scanty. Officer and Batterham (1992) observed improvements of 7-12 % in the ileal digestibility of protein and essential amino acids in pigs fed a diet with a supplemental microbial phytase. Several studies have reported that phytase supplementation improved nitrogen digestibility in pigs (Mroz, Jongbloed, Kemme, 1994; Yi, Kornegay, Lindemenn and Ravindran, 1994a) and nitrogen retention in broiler chickens (Farrell, Martin, Preez, Bongarts, Sudamen and Thomson, 1993) and in laying hens (Van der Klis, Versteegh, Simons and Kies, 1997). Yi *et al.* (1996b) determined the effect of phytase on N and amino acid digestibility in female turkey poults; they found that, with a dietary level of 4.5 g of nonphytate P/kg, adding phytase to diets containing 225g crude protein (CP)/kg tended to improve the apparent and true ileal digestibility of N and amino acids with the exception of cystine and methionine (Table 7.5). In a recent study with male and female broiler chickens fed a maize-soyabean meal diet to 28-d, Sebastian *et al.* (1996b) showed that, in the males, phytase supplementation increased the apparent ileal digestibility (AID) of CP but had no influence on AID of any amino acid except methionine and phenylalanine. In contrast, in the females, adding phytase increased the AID of all amino acids except lysine, methionine, phenylalanine and proline (Table 7.5). The possible effect of phytase on protein and amino acid utilization is of immense practical interest and needs to be confirmed. The factors contributing to the variability of the response must also be identified.

Supplemental phytase in diets for laying hens

Until recently, there has been very little interest in the effect of phytase in diets for laying hens. The reasons are obvious: first is the knowledge that young birds produce very low levels of the enzyme and have hence been shown to respond favourably to dietary supplementation. Secondly, the literature clearly documents the depressed response to supplemental phytase caused by elevated Ca levels in young broiler and turkey diets and it is known that higher Ca levels are usual for laying hen diets. Summers (1995) fed laying hens a diet in which the nP level was reduced from 4 to 3 g/kg and found no depression of egg production but a 20% reduction of the P content of the excreta. The 20% lower dietary P was achieved simply by reducing the amount of supplemental P from calcium phosphate and represented a considerable saving in feed cost. Van der Klis *et al.* (1997) fed White Leghorn hens practical diets providing Ca at 30 g/kg and nP at 2.7 g/kg with 250 or 500 U of phytase/kg of diet. Ileal absorption of Ca and P measured at 24 wk of age appeared to be near maximum with 250 U of phytase/kg and were not further improved by 500 U of phytase/kg of diet. Feeding a similar diet with 40 g Ca/kg of diet resulted in a drop of 12% in absorption of P. The

Table 7.5 The effect of microbial phytase on improving availability of N and amino acids in poultry[1]

Species	*CP (g/kg)*	*Phytase (units/kg)*	*Improvement in the availability of N and amino acids (percentage units of control)*											*Source*
			N	*Essential amino acids*										
				Met	*Cys*	*Lys*	*Thr*	*Arg*	*His*	*Ile*	*Leu*	*Phe*	*Val*	
Turkey	225	750	1.0	0.6	2.1	0.7	1.0	0.6	0.8	0.7	0.9	0.8	0.8	Yi *et al*. (1996b)
	280	750	1.5	0.5	3.1	1.3	1.2	1.1	1.3	1.9	1.5	1.6	1.8	Yi *et al*. (1996b)
Broiler	220	600	1.3	0.3	-	0.6	4.3	1.8	0.3	3.1	1.9	1.1	2.5	Sebastian *et al* (1996b)
					Non-essential amina acids									
					Pro	*Asp*	*Ser*	*Glu*	*Gly*	*Ala*	*Try*			
Turkey	225	750			0.8	0.9	1.2	0.8	1.1	0.8	1.3			Yi *et al*. (1996b)
	280	750			2.4	1.7	0.7	1.3	1.6	1.3	1.2			Yi *et al*. (1996b)
Broiler	220	600			0.2	3.7	4.1	2.1	0.8	1.9	2.8			Sebastian *et al* (1996b)

[1] Female

250 U of phytase/kg in the diet containing only 30 g Ca/kg released phytate-P equivalent to 1.3 g P from monocalcium phosphate (MCP). A second trial at 36 wk of age showed ileal hydrolysis of phytate-P such that 250 U released the equivalent of 0.89 g of P from MCP. This drop was attributed to the increased age of the latter hens (36 vs. 24 wk). In a long-term trial (18 to 68 wk age) a control diet providing 28 g of Ca and 2.6 g of nP/kg of diet proved inadequate to support normal levels of egg production and egg weight. It also resulted in decreased tibia weight and bone ash content. Supplementation of the diet with either 100 U of phytase- or 0.33 g of MCP-P/kg was sufficient to maintain normal performance. Thus, 2.9 g of nP/kg proved adequate in a diet containing 28 g of Ca/kg. Similarly, Gordon and Roland (1997) fed laying hens a diet containing 40 g of Ca/kg with an nP level of 1 g/kg and found it to reduce the rate of egg production. Addition of 300 U of phytase/kg of diet restored the egg production to normal levels. In a subsequent study (Gordon and Roland, 1998), the same low level of nP, 1 g/kg of diet, was fed in a diet containing only 25 g of Ca/kg of diet. This low nP- low Ca diet resulted in poor performance (egg production, shell quality, feed consumption) all of which were normal in birds fed the same diet supplemented with 300 U of phytase/kg of diet. Nakashon, Nakau and Mirosh (1994) also observed inferior production performance and poor retention in a diet providing 2.5 g nP/kg and a Ca level of 35 g/kg of diet. Normal performance was recorded in birds fed this diet supplemented with a product of condensed cane molasses plus a lactobacillus fermentation probiotic. The lactobacillus product supplied active phytase but it also lowered the pH of the crop and gastro-intestinal tract which, the authors suggested, could increase the solubility of various chelates and thus explain the observed enhanced absorption of P and some cations.

In a trial with growing pullets to 18 weeks of age, Punna and Roland (1997) observed a depression in feed intake, body weight and bone quality on a diet containing only 1.3 g P/kg and Ca at 8.6 g/kg. Supplementation of this diet with 300 U of phytase/kg restored these values to levels matching those on diets containing P at 2.3, 3.3 and 4.3 g P/kg. These latter treatments showed no significant differences from each other.

Supplemental phytase and 1,25-dihydroxycholecalciferol

Interests have recently converged concerning the relationships of supplemental phytase and vitamin D_3 in their effects on utilization of dietary Ca and P. Mitchell and Edwards (1996a,b) added microbial phytase, 600 U/kg of diet, or 1,25-dihydroxycholecalciferol (5 μg/kg of diet), or a combination of the two to a basal maize-soyabean meal diet already adequate in its content of vitamin D_3. Based on improvements in body weight, bone ash weight and phytate-P retention, it was calculated that either supplement alone could replace up to 1 g of inorganic P/kg of diet whereas the combination could replace up to 2 g of inorganic P/kg of diet. The authors stated that the interaction observed between these supplements suggests that they are involved in different mechanisms of action. Qian, Kornegay, and Denbow (1997) fed young broiler chicks diets supplemented with phytase levels of 0, 300, 600 or 900 U/kg and 1,25-dihydroxycholecalciferol at 66 and 660 μg/kg alone or in combination. They observed linear improvements in weight gain, feed intake, toe ash and in P and Ca retention with increments of either supplement alone and a synergism of combined action. All of these parameters were positive at Ca:total P ratios of 1.1 to 1.4:1 but declined at the higher ratios of 1.7 and 2.0:1. Total P retention ranged from 0.509 to 0.680, which was similar to that reported by Simons et al. (1990). Baker, Biehl, and Emmert (1998) fed young broilers supplemental 1,25-dihychoxycholecalciferol levels from 0 to 37.5 μg/kg in a basal diet low in available P (1.0 g/kg) and Ca (6.3 g/kg). A linear, positive response to the added vitamin in weight gain and tibia ash content was observed. A supplemental level of 1,25-dihydroxy D_3 of 1250 μg/kg of diet yielded a further increase in bone ash without any apparent negative effects on performance. Carlos and Edwards (1998) explored the effect of supplementing the diets of laying hens during a 9-week trial period with either 5 μg of 1,25-dihydroxy D_3/kg or 600 U of microbial phytase /kg or a combination of the two. In young hens, either supplement alone increased the body weight, plasma dialyzable P and tibia ash. It also improved the P retention from the basal level of 0.43 to levels of 0.63 to 0.77. The effect was slightly greater in older hens. This level of improvement was similar to that reported by Van der Klis *et al.* (1997) in response to 250 and 500 U of phytase/kg of diet; the authors suggested that, based on bone ash values, the hens receiving the P-deficient diet would likely have developed P-deficiency cage layer fatigue whereas the others would probably not have.

Biehl, Baker, and De Luca (1998) evaluated the efficacy of two new vitamin D_3 analogues in increasing the bioavailability of phytate-bound P from maize-soybean meal diets that were more than adequate in D_3 for young chicks. The analogues, one dihydroxy compound (20-epi-19-nor-1α, 25-dihydroxycholecalciferol) and one monohydroxy compound (20-epi-19-nor-1α hydroxycholecalciferol) were compared to 1α hydroxycholecalciferol which at levels of 10 to 15 μg/kg of diet had been shown to optimise phytate-P utilization. The results showed the dihydroxy test compound to be equivalent to the 1α-hydroxy D_3 standard supplement whereas the monohydroxy test compound had only 0.45 as much phytate-P-releasing activity; it was suggested that conversion of these compounds to the fully active 1,25-dihydroxycholecalciferol occurs in the liver and that a portion is returned to the gut by biliary flow where it enhances the action of intestinal or microbial phytase.

Recombinant phytase

The phytase supplement used until now has been predominantly from the mould *Aspergillus ficuum* var. niger, now called *A. niger*. Recent developments indicate an improvement to the potency and perhaps the convenience of using the phytase enzyme preparation. Sun, Patterson, Woloshuk and Muir (1997) reported application of molecular biology to effect the transfer of the gene for the phytase, myo-inositol hexakisphosphate phosphohydrolase, EC 3.1.3.8, from *A. niger* to the yeast, *Saccharomyces cerevisiae*. By measuring the amount of inorganic P released from sodium phytate *in vitro* a 4-to-11-fold increase in the activity of the transformed yeast over that of the control yeast was estimated; it was suggested that about 20 g/kg of the recombinant yeast (dry matter basis) will be required to provide phytase supplementation in pig and poultry feeds.

Another transformed product was evaluated by Denbow, Grabau, Lacy, Kornegay, Russell and Umbeck (1998), who tested a new product of raw, transformed soyabeans expressing recombinant phytase derived from *A. niger*, comparing it to Natuphos[1] phytase at levels of 0, 400, 800 or 1200 U/kg in diets fed to broiler chicks from 7 to 21d of age. The positive linear effect on body weight, feed efficiency, feed intake, toe ash and tibia sheer force of both phytase supplements paralleled those from incremental additions of nP with the exception that the phytase treatments reduced rather than increased P excretion.

Conclusions

Phosphorus (P) is an important component of the animal body, both structurally and metabolically. In P deficiency the earliest sign is diminished appetite, which will be manifested in reduced growth rate, and later, reduced bone mineralization. Phosphorus is normally considered in conjunction with calcium (Ca) because these two elements are collectively involved in so many functions. The inorganic phosphates and animal products represent feed ingredients in which P availability approaches 1.00. In plant materials, particularly the grains, oilseeds and their by-products which represent the major portion of the diets of poultry and pigs, the P availability is only about 0.30. The plant storage form of P is phytic acid, myo-inositol hexaphosphate, a powerful chelating agent that binds di- and trivalent cations and protein, rendering the complex resistant to the digestive enzymes. It may even bind and inactivate these enzymes under certain circumstances.

Several phytases capable of hydrolysing phytates release the phosphates and the bound components making them available for absorption. Poultry intestinal mucosa secrete a phytase which is active at low concentrations of inorganic P. Its optimum pH of 2-5.5 is similar to that of microbial phytase. Genetic differences among breeds or strains of poultry suggest that the ability to elaborate this enzyme could be enhanced through selection but it is not, currently, a major factor affecting P availability.

Another phytase enzyme is present in some feedstuffs such as rye and wheat but the activity is extremely variable. Maize, sorghum and oilseeds have very little endogenous phytase. The optimum pH is considerably higher, pH 5-6, so its

[1] BASF Corporation, Mount Olive, NJ. USA

effectiveness for poultry would probably have to occur prior to consumption or primarily in the crop. Microbial phytases mainly from fungi and predominantly from *A. niger* were observed to exert an important benefit upon addition to poultry diets. They operate over a pH range of 2.5 to 5.5 and are considered to be active in the crop, proventriculus and gizzard. Furthermore, they resist feed processing temperatures up to 80°C.

Many studies with broiler chicks and some with turkey poults, mostly to 3 wk of age, have shown the increasing benefits of supplemental microbial phytase from 100 to over 1000 units of activity. Feed intake, growth rate, bone mineralization, increased absorption of P, Ca, Cu, Zn, protein and amino acids have been observed. In laying hens supplemental phytase has additionally enhanced egg production, egg size and egg shell quality. These responses have been enhanced by supplements of 1,25-dihydroxycholecalciferol, well above the established requirement level, with the action being apparently synergistic with that of phytase.

All of these responses are susceptible to the levels of Ca and nP or total P in the diet, being most effective at low levels and declining at higher levels. Reducing the dietary levels of Ca and P to match more closely the actual needs of the bird enhances the activity of the supplemental phytase and reduces the output of Ca and P in the faeces. The ultimate result is reduced cost and a more environmentally friendly operation. Finally, recombinant phytases in which the gene has been transferred to yeast or to soyabean will increase the enzyme potency and facilitate feed supplementation.

References

Anderson, R.J. (1914) A contribution to the chemistry of phytin. *Journal of Biological Chemistry*, **17:** 171-190.

Aoyagi, S. and Baker, D.H.(1995) Effect of microbial phytase and 1,25-dihydroxycholecalciferol on dietary copper utilization in chicks. *Poultry Science*, **74**, 121-126.

Baker, D.H., Biehl, R.R. and Emmert, T.L. (1998) Vitamin D_3 requirements of young chicks receiving diets varying in calcium and available phosphorus. *British Poultry Science*, **39**, 413-417.

Biehl, R.R., Baker, D.H. and De Luca, H.F. (1995) Hydroxylated cholecalciferol compounds act additively with microbial phytase to improve phosphorus, zinc and manganese in chicks fed soy-based diets. *Journal of Nutrition*, **125**, 2407-2416.

Biehl, R.R., Baker, D.H. and De Luca, H.F. (1998) Activity of various hydroxylated vitamin D_3 analogues for improving phosphorus utilization in chicks receiving diets adequate in vitamin D_3. *British Poultry Science*, **39**, 408-412.

Broz, J., Oldale, P., Perrin-Voltz, A.H., Rychen, G., Schulze, J. and Simoes Nunes, C. (1994) Effects of supplemental phytase on performance and phosphorus utilization in broiler chickens fed a low phosphorus diet without addition of inorganic phosphates. *British Poultry Science*, **35**, 273-280.

Campbell, G.L. and Bedford, M.R. (1992) Enzyme applications for monogastric feeds: A review. *Canadian Journal of Animal Science*, **72**, 449-466.

Cantor, A.H. (1995) Using enzymes to increase phosphorus availability in poultry diets. Proceedings of Alltech's Eleventh Annual Symposium, Biotechnology in the Feed Industry. 349-353.

Carlos, A.B. and Edwards, H.M., Jr. (1997) Influence of genetics on the utilization of phytate phosphorus by broilers. *Poultry Science*, **76**, Supplement 1, 142 (S158).

Carlos, A.B. and Edwards, H.M., Jr. (1998) The effects of 1,25-dihydroxycholecalciferol and phytase on the natural phytate phosphorus utilization by laying hens. *Poultry Science*, **77**, 850-858.

Davies, M.I., and Motzok, I. (1972) Properties of chick intestinal phytase. *Poultry Science*, **51**, 494-501.

Denbow, D.M., Grabau, E.A., Lacy, G.H., Kornegay, E.T., Russell, D.R. and Umbeck, P.F. (1998) Soybeans transformed with a fungal phytase gene improve phosphorus availability for broilers. *Poultry Science*, **77**, 878-881.

Denbow, D.M., Ravindran, V., Kornegay, E.T., Yi, Z., and Hulet, R.M. (1995) Improving phosphorous availability in soybean meal for broilers by supplemental phytase. *Poultry Science*, **74**, 1831-1842.

Edwards, H.M., Jr. (1983) Phosphorus I. Effect of breed and strain on utilization of suboptimal levels of phosphorus in the ration. *Poultry Science*, **62**, 77-84.

Eeckhout, W. and de Paepe, M. (1994) Total phosphorus, phytate-phosphorus and phytase activity in plant feedstuffs. *Animal Feed Science and Technology*, **47**, 19-29.

Erdman, J.W., Jr. (1979) Oilseed phytates: Nutritional implications. *Journal of the American Oil Chemists Society*, **56**, 736-741.

Farrell, D.J., Martin, E., Preez, J.J., Bongarts, M., Sudeman, A., and Thomson, E. (1993) The beneficial effects of a microbial phytase in diets of broiler chickens and ducklings. *Journal of Animal Physiology and Animal Nutrition*, **69**, 278-286.

Gordon, R.W. and Roland, D.A., Sr. (1997) Performance of commercial laying hens fed various phosphorus levels, with and without supplemental phytase. *Poultry Science*, **76**, 1172-1177.

Gordon, R.W. and Roland, D.A., Sr. (1998) Influence of supplemental phytase on calcium and phosphorus utilization in laying hens. *Poultry Science*, **77**, 290-294.

Kornegay, E.T., Denbow, D.M., Yi, Z. and Ravindran, V. (1996) Response of broilers to graded levels of Natuphos phytase added to corn-soybean meal based diets containing three levels of nonphytate phosphorus. *British Journal of Nutrition*, **75**, 839-852.

Kornegay, E.T., Denbow, D.M. and Zhang, Z. (1997) Phytase supplementation of corn-soybean meal broiler diets from three to seven weeks of age. *Poultry Science* **76:** Supplement 1, p. 6 (Abstract).

Maenz, D.D. and Classen, H.L. (1998) Phytase activity in the small intestinal brush border membrane of the chicken. *Poultry Science*, **77**, 557-563.

Mitchell, R.D. and Edwards, H.M., Jr. (1996a) Effects of phytase and 1,25-dihydroxycholecalciferol on phytate utilization and the quantitative requirement for calcium and phosphorus in young broiler chickens. *Poultry Science*, **75**, 95-100.

Mitchell, R.D. and Edwards, H.M., Jr. (1996b) Additive effects of 1,25-Dihydroxycholecalciferol and phytase on phytate-P utilization and related parameters in broiler chickens. *Poultry Science*, **75**, 111-119.

Mohammed, A., Gibney, M.J. and Taylor, T.G. (1991) The effects of dietary levels of inorganic phosphorus, calcium and cholecalciferol on the digestibility of phytate-P by the chick. *British Journal of Nutrition*, **66**, 251-259.

Mroz, Z., Jongbloed, A.W. and Kemme, P.A. (1994) Apparent digestibility and retention of nutrients bound to phytate complexes as influenced by microbial phytase and feeding regimen in pigs. *Journal of Animal Science*, **72**, 126-132.

National Research Council (1994) Nutrient Requirements of Poultry, 9th revised edition, National Academy Press, Washington, D.C.

Nakashon, S.N., Nakau, H.S. and Mirosh, L.W. (1994) Phytase activity, phosphorus and calcium retention, and performance of Single Comb White Leghorn layers fed diets containing two levels of available phosphorus and supplemented with direct-fed microbials. *Poultry Science*, **73**, 1552-1562.

Nelson, T.S., Shieh, T.R., Wodzinski, R.J. and Ware, J.H. (1971) Effect of supplemental phytase on the utilization of phytate phosphorus by chicks. *Journal of Nutrition*, **101**, 1289-1294.

Oberleas, D. and Harland, B.F. (1996) Impact of phytate on nutrient availability. In *Phytase in Animal Nutrition and Waste Management.* Edited by M.B. Coelho and E.T. Kornegay. BASF Corporation, Mount Olive, New Jersey, U.S.A.

Officer, D.I. and Batterham, E.S. (1992) Enzyme supplementation of Linola TM meal for growing pigs. Proceedings of the Australian Society of Animal Production, **9**, 288-296.

Pallauf, V.J., Hohler, D. and Rimbach, G.(1992) Effect of microbial phytase supplementation to a maize-soya-diet on the apparent absorption of Mg, Fe, Cu, Mn and Zn and parameters of Zn-status in piglets. *Journal of Animal Physiology and Animal Nutrition*, **68**, 1-8.

Prasad, A.S. (1996) Metabolism of zinc and its deficiency in human subjects. In *Zinc Metabolism* - 1996, pp 250-256. Edited by A.S. Prasad and C.C. Thomas. Springfield, Illinois, U.S.A.

Punna, S. and Roland, D.A., Sr. (1997) Effect of dietary supplementation of phytase on pullets fed varying levels of dietary phosphorus and calcium. *Poultry Science*, **76**, Supplement 1, page 6, No. 23.

Qian, H., Kornegay, E.T. and Denbow, D.M. (1996) Phosphorus equivalence of microbial phytase in turkey diets as influenced by calcium to phosphorus ratios and phosphorus levels. *Poultry Science*, **75**, 69-81.

Qian, H., Kornegay, E.T. and Denbow, D.M. (1997) Utilization of phytate phosphorus and calcium as influenced by microbial phytase, cholecalciferol, and the calcium : total phosphorus ratio in broiler diets. *Poultry Science*, **76**, 37-46.

Ravindran, V., Bryden, W.L. and Kornegay, E.T.(1995) Phytates: occurrences, bioavailability and implications in poultry nutrition. *Poultry and Avian Biology Reviews*, **6**, 125-143.

Ravindran, V., Kornegay, E.T., Denbow, D.M., Yi, Z. and Hulet, R.M. (1995) Response of turkey poults to tiered levels of Natuphos® phytase added to soybean meal-based semi-purified diets containing three levels of nonphytate phosphorus. *Poultry Science*, **74**, 1843-1854.

Reddy, N.R., Sathe, S.K. and Salunkhe, D.K. (1982) Phytates in legumes and cereals. *Advances in Food Research*, **28**, 1-91.

Roberson, K.D. and Edwards, H.M., Jr. (1994) Effects of 1,25-dihydroxycholecalciferol and phytase on zinc utilization in broiler chicks. *Poultry Science*, **73**, 1312-1326.

Rojas, S.W. and Scott, M.L. (1969) Factors affecting the nutritive value of cottonseed meal as a protein source in chick diets. *Poultry Science*, **48**, 819-835.

Schoner, F.J., Hoppe, P.P. and Schwartz, G. (1991) Comparative effects of microbial phytase and inorganic phosphorus on performance and retention of phosphorus, calcium and crude ash in broilers. *Journal of Animal Physiology and Animal Nutrition*, **66**, 248-255.

Schoner, F.J., Schwartz, G., Hoppe, P.P. and Wiesche, H. (1994) Effect of microbial phytase on Ca-availability in broilers. Third Conference of Pig and Poultry Nutrition in Halle, Germany, November 29-December 1.

Scott, M.L., Nesheim, M.C. and Young, R.J. (1982) Nutrition of the chicken, 3rd edition, M.L. Scott and Associates, Ithaca, New York, U.S.A.

Sebastian, S., Touchburn, S.P., Chavez, E.R. and Laguë, P.C. (1996a) The effects of supplemental microbial phytase on the performance and utilization of dietary calcium, phosphorus, copper and zinc in broiler chickens fed corn-soybean diets. *Poultry Science*, **75**, 729-736.

Sebastian, S., Touchburn, S.P., Chavez, E.R. and Laguë, P.C. (1996b) Efficacy of supplemental microbial phytase at different dietary calcium levels on growth performance and mineral utilization of broiler chickens. *Poultry Science*, **75**, 1516-1523.

Sebastian, S., Touchburn, S.P., Chavez, E.R. and Laguë, P.C.(1997) Apparent digestibility of protein and amino acids in broiler chickens fed a corn-soybean diet supplemented with microbial phytase. *Poultry Science*, **76**, 1760-1769.

Simons, P.C.M., Versteegh, H.A., Jongbloed, A.W., Kemme, P.A., Slump, P., Bos, K.D., Wolters, M.G.E., Buedeker, R.F. and Vershoor, G.J. (1990) Improvement of phosphorus availability by microbial phytase in broilers and pigs. *British Journal of Nutrition*, **64**, 525-540.

Summers, J.D. (1995) Reduced dietary phosphorus levels for layers. *Poultry Science*, **74**, 1977-1983.

Sun, W., Patterson, J.A., Woloshuk, C.P. and Muir, W.M. (1997) Expression of Aspergillus niger phytase in yeast Saccharomyces cerevisiae for poultry diet supplementation. *Poultry Science*, **76**, Supplement 1, No. 19, page 5.

Thiel, U. and Weigand, E. (1992) Influence of dietary zinc and microbial phytase supplementation on zinc retention and zinc excretion in broiler chickens. Proceedings of XIX World's Poultry Congress. World's Poultry Science Association, Amsterdam, The Netherlands.

Thiel, U., Weigand, E., Hoppe, P.P. and Schoner, F.J. (1993) Zinc retention of broiler chickens as affected by dietary supplements of zinc and microbial phytase. In *Trace Elements in Man and Animals* -1993, pp 658-660. Edited by M. Anke, D. Meissner and C.F. Mills.

Van der Klis, J.D., Versteegh, H.A.J., Simons, P.C. and Kies, A.K. (1997) The efficacy of phytase in corn-soybean meal-based diets for laying hens. *Poultry Science*, **76**, 1535-1542.

Yi, Z., Kornegay, E.T., Lindemann, M.D. and Ravindran, V. (1994) Effect of Natuphos phytase for improving the bioavailability of phosphorus and other nutrients on soybean meal-based semi-purified diets for young pigs. *Journal of Animal Science*, **72**, 7, Supplement 1.

Yi, Z., Kornegay, E.T. and Denbow, D.M. (1996a) Supplemental microbial phytase improves the zinc utilization in broilers. *Poultry Science*, **75**, 540-546.

Yi, Z., Kornegay, E.T. and Denbow, D.M. (1996b) Effect of microbial phytase on nitrogen and amino acid digestibility and nitrogen retention of turkey poults fed corn-soybean meal diets. *Poultry Science*, **75**, 979-990.

Yi, Z., Kornegay, E.T., Ravindran, V. and Denbow, D.M. (1996) Improving phytate phosphorus availability in corn and soybean meal for broilers using microbial phytase and calculation of phosphorus equivalency values for phytase. *Poultry Science*, **75**, 240-249.

Zhang, X., McDaniel, G.R. and Roland, D.A. (1998) Genetic variation of phytate phosphorus utilization from hatch to three weeks of age in broiler chicken lines selected for incidence of tibial dyschondroplasia. *Poultry Science*, **77**, 386-390.

8

NUTRITION OF BROILER BREEDERS[1]

C. FISHER and M.H.A. WILLEMSEN
Ross Breeders Limited, Newbridge, Midlothian, EH28 8SZ

Feeding decisions for broiler breeders involve both diet composition and feeding levels. Under controlled feeding, which is used for most of the life of the birds, dietary energy must be a limiting factor. With controlled feeding there is a natural tendency for the flock to become non-uniform. This must be countered by a range of management techniques and success in maintaining uniformity is a pre-requisite of good flock performance. Thus, in considering broiler breeder nutrition, feed composition, feeding management and general flock management are inextricably intertwined. It is convenient to view nutritional decisions in three parts. (1) control of reproductive function, (2) supplying sufficient nutrients for production and (3) transfer of nutrients to the egg for hatching and chick performance.

In a flock with satisfactory uniformity, following a body weight profile provides the essential guideline for feed allocation (dietary energy supply). The nutrients required to support production must be provided within the quantity of feed defined in this way. Breeding companies normally provides target profiles for body weight and close adherence to these is important. In the rearing phase this will ensure satisfactory tissue and organ development and prepare the birds for reproduction. In the laying stage, body weight changes provide the most sensitive, short-term indicator of over or under-supply of dietary energy.

If breeder hens are overweight at point of lay, or are overfed, then hypertrophic ovaries with multiple hierarchies of yellow follicles will develop. A number of consequences arise: double-yolked eggs, internal laying leading to peritonitis, misshapen and poorly shelled eggs. This complex has been called EODES - the Erratic Ovulation and Defective Egg Syndrome. It is essential that feeding level (dietary energy supply)

[1] Note: Due to circumstances beyond the control of the authors, editors or publishers, it was not possible to produce the full paper from this presentation. A summary is published here and it is our intention to include the full paper in the proceedings next year.

be accurately controlled at all stages to minimise the occurrence of these ovarian features. Feeding to accurately control body weight seems to provide the best way of doing this in practice.

The calculation of the nutrients required to support production appears to be straightforward. For growing birds, this can be based on results obtained with broilers. For adult birds, methods developed for laying hens seem to be applicable.

During the rearing phase, growth is severely restricted in comparison with bird potential. It is not clear whether the objective for setting nutritional 'requirements' during this phase should simply be to support this low level of growth. Nutritional status with respect to nutrient or metabolite levels in tissues may relate more closely to the well-being of the birds.

Broiler breeders are normally fed once per day early in the photoperiod. In the adult this raises questions, possibly about the utilisation of synthetic amino acids and certainly about the utilisation of calcium. There is a strong case for giving some part of the calcium required as grit, later in the day, so that provision is more closely synchronised with eggshell formation. A possible compromise is to include 28 g Ca/kg in the feed and to provide grit to maintain a daily intake of 5g Ca per day for hens.

As in many classes of stock, vitamin requirements are not well-defined. Relatively generous levels of supplementation will normally be economically justified in light of the very small improvements in hatchability or chick quality required to pay for them. The most difficult economic decision concerns vitamin E, for which a general recommendation of 100 mg/kg feed is currently made. There may be an opportunity for using vitamin C more widely than current practice.

Investigations of egg yolk lipids have not led to exact nutritional guidelines for dietary fatty acids. However, a general conclusion, that supplementary fats should only be used at low levels, seems to be justified.

A general summary of the nutrient specifications for broiler-breeder hen feeds is presented in Table 8.1.

Table 8.1 Nutrient specifications for broiler breeder hen feeds

		Starter 1 (0-21 days)		Starter 2 (22-42 days)		Grower (43 – 105 days)		Pre-breeder (105 – 154 days)		Breeder (155+ days)	
Crude Protein	g/kg	200		180–200		140–150		150–160		150–160	
Metabolisable Energy	MJ/kg	11.5		11.5		11.0		11.5		11.5	
Amino acids		*Total*	*Available*	*Total*	*Available*	*Total*	*Available*	*Total*	*Available*	*Total*	*Available*
Arginine	g/kg	11.5	10.0	9.4	8.3	6.3	5.6	7.3	6.4	7.6	6.7
iso-Leucine	g/kg	7.6	6.5	6.2	5.3	4.2	3.6	5.1	4.3	5.3	4.5
Lysine	g/kg	11.0	9.6	9.0	7.8	5.8	5.1	7.0	6.1	7.4	6.4
Methionine	g/kg	4.5	4.0	3.8	3.3	2.8	2.5	3.1	2.7	3.2	2.8
Meth. + Cyst.	g/kg	8.2	7.0	7.4	6.3	5.6	4.7	5.3	4.5	5.5	4.7
Threonine	g/kg	7.7	6.5	6.3	5.4	4.2	3.6	4.6	3.9	4.8	4.1
Tryptophan	g/kg	2.2	1.8	1.8	1.5	1.4	1.2	1.5	1.3	1.7	1.4
Minerals											
Calcium	g/kg	10.0		10.0		10.0		15.0		28.0	
Avail. Phos.	g/kg	4.5		4.5		3.5		4.0		3.5	
Sodium	g/kg	1.6		1.6		1.6		1.6-2.0		1.6-2.0	
Chloride	g/kg	1.6-2.1		1.6-2.1		1.6-2.1		1.6-2.1		1.6-2.1	
Potassium	g/kg	4.0		4.0		4.0		6.0		6.0	
Added trace minerals per kg											
Copper	mg	8		8		6		10		10	
Iodine	mg	0.50		0.50		0.50		2		2	
Iron	mg	60		60		40		60		60	

Table 8.1 (contd.)

Manganese	mg	70		70		60		60		60	
Zinc	mg	50		50		50		100		100	
Selenium	mg	0.15		0.15		0.15		0.20		0.20	
Added vitamins kg (unless shown)		*Wheat based feed*	*Maize based feed*	*Wheat based feed*	*Maize based feed*	*Wheat based feed*	*Maize based feed*	*Wheat based feed*	*Maize based feed*	*Wheat based feed*	*Maize based feed*
Vitamin A	iu/g	10	10	10	10	10	10	13	12	13	12
Vitamin D3	iu/g	3.5	3.5	3.5	3.5	3.5	3.5	3	3	3	3
Vitamin E	mg	60	60	50	50	40	40	100	100	100	100
Vitamin K	mg	2	2	2	2	2	2	5	5	5	5
Thiamin (B1)	mg	2	2	2	2	2	2	3	3	3	3
Riboflavin (B2)	mg	6	6	6	6	5	5	12	12	12	12
Nicotinic Acid	mg	25	30	25	30	20	25	50	55	50	55
Pantothenic Acid	mg	12	14	12	14	12	14	12	15	12	15
Pyridoxine (B6)	mg	3	2	3	2	3	2	6	4	6	4
Biotin	mg	0.12	0.08	0.10	0.05	0.05	0.02	0.30	0.25	0.30	0.25
Folic Acid	mg	1.5	1.5	1	1	1	1	2	2	2	2
Vitamin B12	mg	0.015	0.015	0.015	0.015	0.015	0.015	0.03	0.03	0.03	0.03
Minimum specification per kg											
Choline	mg	1300		1300		1000		1000		1000	
Linoleic Acid	g	10		10		8.5		12.0-15.0		12.0-15.0	

9

NUTRITION X ENVIRONMENTAL INTERACTIONS: PREDICTING PERFORMANCE OF YOUNG PIGS

N.S. FERGUSON
Animal Science and Poultry Science Department, University of Natal, South Africa

Introduction

Simulating animal growth provides a means of predicting animal performance to a degree of accuracy that would otherwise be impossible to accomplish. In animal production there are a number of interactions that need to be considered simultaneously in order to estimate their responses to various inputs. The interaction between nutrition and the environment is one of the most important as it is central to the prediction of voluntary food intake. Any model simulating animal performance must include an adequate description of the type of animal, the feed and the way the animal is being fed, the environment and the interactions between these components. Figure 9.1 summarizes an approach by Emmans and Oldham (1988) to modelling growth and voluntary food intake and highlights the importance of qualifying and quantifying the relationships between what is being fed and the environment to which the animal is subjected. This approach has been followed by Ferguson, Gous and Emmans (1994) in the construction of a simulation model of the growth and food intake of growing pigs.

The role of heat dissipation is central to the theory of growth and voluntary food intake in growing pigs. More specifically food intake and hence growth are often constrained by the maximum amount of heat an animal can lose to its environment. Because heat production by the animal must be balanced with the amount of heat the animal can lose, at high temperatures the desired food intake will be constrained and hence the rate of growth will be reduced to below the potential of the animal. There are a number of interactions between nutrition and the environment, responsible for the increasing discomfort felt by the animal, that are not well understood. For example when pigs are fed a protein-deficient diet they can only achieve their maximum rate of protein retention if they consume enough protein to satisfy their protein requirement. Protein intake is dependent on the ability of the animal to lose the heat produced in retaining body tissue and from the heat increment of feeding that is generated by such feed (Kyriazakis and Emmans, 1991; Kyriazakis, Stamataris, Emmans and Whittemore, 1991; Emmans, 1994; Ferguson and Gous, 1997). This would suggest that the amount of heat an animal can lose in a hot environment will determine the rate of protein deposition and this underlines the importance of knowing the maximum heat loss capability of the animal.

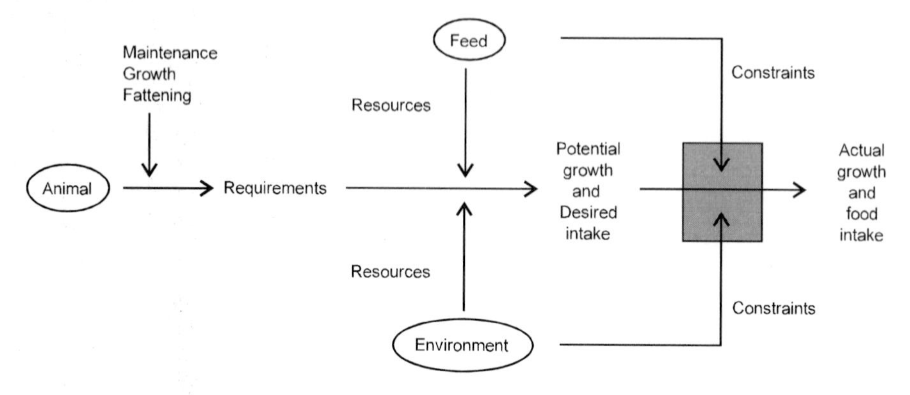

Figure 9.1 A systems approach to predicting growth and voluntary food intake in growing pigs (after Emmans and Oldham, 1988). The shaded box highlights the importance of nutrition × environment interactions in predicting food intake and growth.

It is well documented that environmental temperature affects the voluntary food intake, growth and heat production of growing pigs (Verstegen, Close, Start and Mount, 1973; Close and Mount, 1978; Dauncey, Ingram, Walters and Legge, 1983; Le Dividich and Noblet, 1986; Campbell and Taverner, 1988; Rinaldo and LeDividich, 1991). However, most studies relating the effects of temperature to the growth of pigs have been confined to the effect of low ambient temperatures on energy metabolism (Fuller and Boyne, 1971; Verstegen *et al.*, 1973, 1982; Close and Mount, 1978; Stahly, Cromwell and Aviotti, 1979; Dauncey *et al.*, 1983). Also, few authors have considered the influence of temperature on the response of pigs to decreasing dietary protein or amino acid levels (Stahly *et al.*, 1979; Schenck, Stahly and Cromwell, 1992a,b; Arnold, 1997; Ferguson and Gous, 1997).

There have been a number of attempts to measure the response of young growing pigs to feeds deficient in various amino acids but the results have been contradictory and inconsistent (Taylor, Cole and Lewis, 1981; Yen, Cole and Lewis, 1986; Batterham, Giles and Dettman, 1985; Henry, Colleaux and Seve, 1992; Adeola, 1995). The response to the deficient amino acid has not been elucidated clearly because of the diet (feed ingredients) and the feeding technique (restricted vs. *ad lib* feeding) used and because little control has been exercised over the prevailing environmental temperature. It has been recognised that the increased diversification of feed supply for pigs, and the use of fibrous and poorly digestible byproducts, has introduced additional limitations in the availability of some essential amino acids, namely threonine, tryptophan and methionine, which under certain circumstances may become limiting, thus resulting in undesirable effects on appetite and growth performance (Henry and Seve, 1993). Another issue is that the environments in which the experiments have been conducted have differed and therefore no account has been taken of the confounding effect of the growth × environment interaction. In many instances the environmental temperature was not even recorded in the publication.

The current lack of data available to assess the ability of an animal to lose heat, particularly the maximum heat loss when fed marginally deficient protein or amino acid diets, has limited our understanding of nutritional × environmental interactions and therefore the inclusion of such interactions into simulation models. The lack of information also makes it difficult to test the effect of heat production on voluntary food intake and subsequent body tissue deposition, particularly in animals fed a protein or amino acid-deficient feed. Besides the work done by Arnold (1997) and Ferguson (1997) the only other study on how dietary amino acids and environmental temperature affect the growth performance of pigs was conducted by Stahly *et al.* (1979), Schenck *et al.* (1992a,b) and Meyr, Bucklin and Fialho (1998). Unfortunately in these trials too few lysine and temperature treatments were used to determine growth responses. In order to obtain dose response curves it is essential that a wider range of dietary amino acid concentrations and temperatures be provided (Taylor *et al.*, 1981).

Although the term 'environment' is a very broad one encompassing all external factors, other than food, influencing the growth of the animal, for the purposes of this discussion environment will be defined by the ambient temperature in which the animal is kept. As lysine is usually the first limiting amino acid in diets fed to growing pigs this paper will review information about the growth responses to different dietary lysine concentrations, notwithstanding the effect of the environment and the implications for modelling pig performance.

Responses to lysine in young pigs grown in different environmental temperatures

FOOD INTAKE AND LYSINE INTAKE

It is well known that increasing the temperature above the thermoneutral zone results in a reduction in food intake, and decreasing the temperature below the lower critical temperature causes an increase in voluntary food intake (Close and Mount, 1978; Verstegen, Brandsma and Mateman, 1982; Rinaldo and Le Dividich, 1991; Nienaber, Hahn, Korthals and McDonald, 1993). According to Close (1987) and Whittemore

(1993) the lower critical temperature for pigs between 10 and 30 kg live weight is close to 26°C. At 18 °C and 22°C the environmental heat demand, associated with an ambient temperature below the thermoneutral zone, will cause the animal to increase its voluntary food intake (Close and Mount, 1978; Verstegen *et al.*, 1982; Ferguson and Gous, 1997).

From a summary of the relevant papers, as shown in Figure 9.2, it would appear that at ambient temperatures of between 18°C and 20 °C food intake will increase in response to decreasing dietary lysine content, reach a maximum and then decline. Only the data from Arnold (1997) did not follow this trend as food intake did not decline on the lowest lysine content. The dietary lysine concentration below which food intake will decline varied between 7.5 and 10.0 g/kg according to the different experiments. This is due in part to the different pig genotype, different final weights and to the slight differences in the ambient temperatures used between experiments.

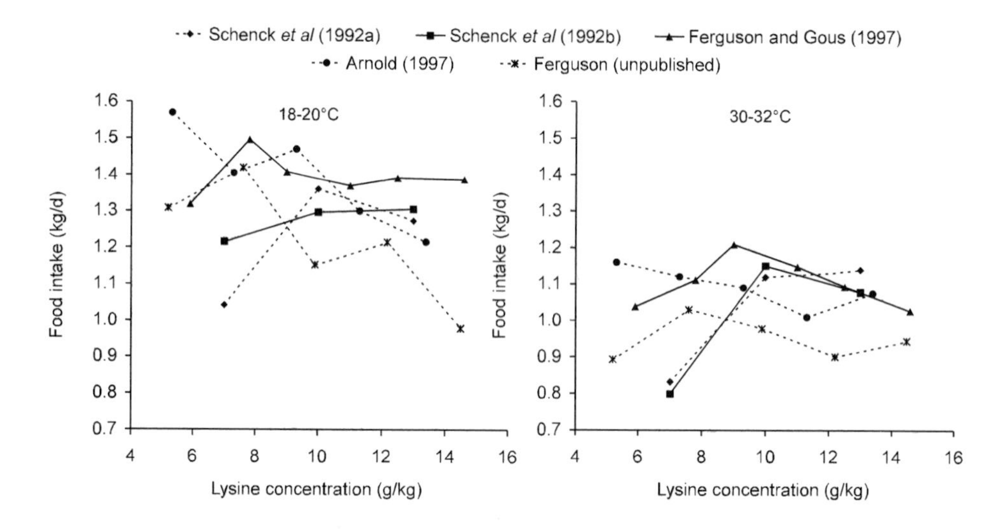

Figure 9.2 The effect of lysine concentration on food intake when young pigs between 13 and 30 kg live were kept at between 18-20°C and 30-32°C.

From Figure 9.2 it is apparent that within experiments, pigs attempt to compensate for decreasing lysine contents in the feed by increasing their food intake to satisfy their body protein requirements. Depending on the genotype of the animal, the daily requirement for lysine in pigs up to 30 kg live weight varied between 8.4 g (NRC, 1988) and 14.0 g (Ferguson, 1997). In cool environments and on high lysine diets, the limiting component is likely to be energy rather than lysine. Even with lysine contents as low as 10.0 g/kg, lysine requirements could be met as long as the conditions were cool as the animal could increase its intake sufficiently and utilize the extra heat produced to satisfy its additional requirement for cold thermogenesis (Schenck *et al.*,1992a,b;

Ferguson, 1997). However at lysine contents less than this the animals were unable to maintain an intake that would satisfy their lysine requirement and performance decreased. Similar findings were observed by Kyriazakis *et al* (1991) and Kyriazakis and Emmans (1991) in which young pigs in colder temperatures (16 °C) had substantially higher rates of food intake and compensatory protein gain than their counterparts at 22 °C after a period of feeding on protein-deficient diets.

These results are consistent with the theory that pigs attempt to consume an amount of feed that will satisfy their requirements for potential growth and maintenance. In this case they attempt to maintain a constant lysine intake as the lysine content of the feed is reduced. However, a point is reached when the animal can no longer compensate for the reduced dietary lysine concentration and food intake will subsequently decline, as observed in animals fed lysine levels < 10.0 g/kg. The reason for a decline in food intake is because of the high heat increment associated with digesting low amino acid:energy diets. What is also interesting to note is that in most cases the dietary treatment that resulted in the highest food intake (FI) was also the treatment that had the highest total heat loss (THL).

Although Ferguson (1997) and Arnold (1997) did not measure any significant interactions between the main effects of dietary lysine content and temperature on FI and ME intake they did observe a significant interaction on lysine intake. Schenck *et al.* (1992a,b) had previously observed a significant temperature x lysine content interaction for FI and lysine intake. The relationship between lysine intake and mean ambient temperature is shown in Figure 9.3. With a 0.21 g /d decrease in lysine intake per degree Celsius increase above 18 °C it would appear that the optimum lysine:ME ratio is dependent on environmental temperature, being higher as the temperature increases.

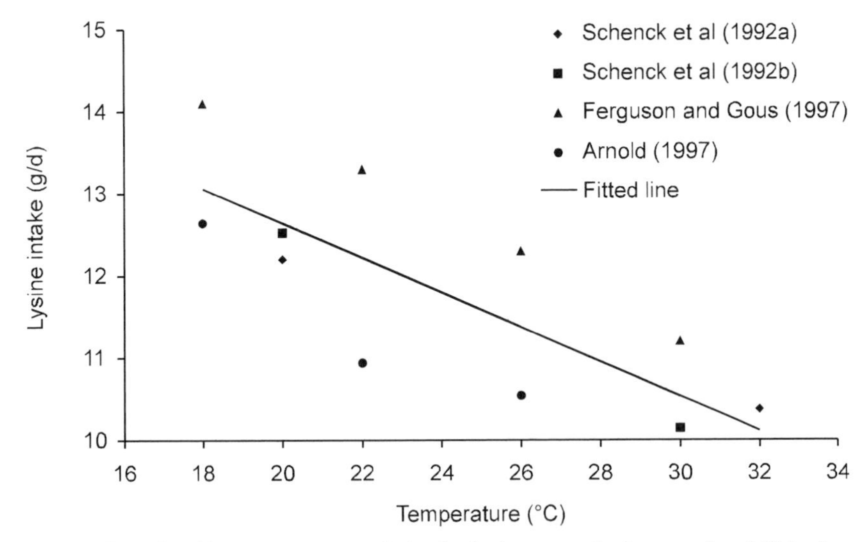

Figure 9.3 The effect of ambient temperature on lysine intake in young pigs between 8 and 30 kg live weight.

LIVE WEIGHT CHANGES AND FEED CONVERSION

In the experiments referred to in the previous section the responses of the pigs to lysine concentrations were quadratic by nature, with significant ($P<0.05$) decreases in growth rates occurring when lysine contents decreased below 0.90 of the requirement (*ca* 10 g/kg between 10 and 30 kg). However these responses were to some extent dependent on the environmental temperature (Figure 9.4).

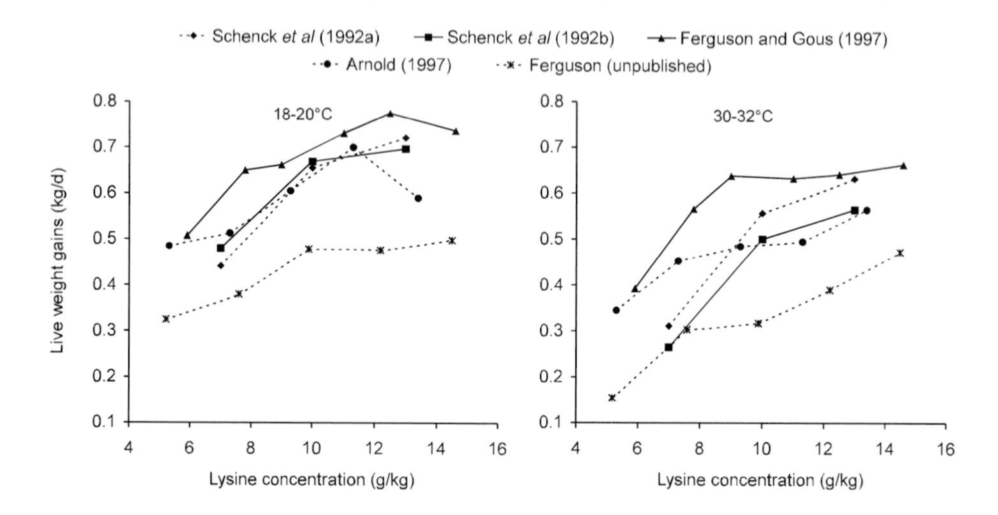

Figure 9.4 The effect of lysine concentration on live weight gain when young pigs between 13 and 30 kg live were kept between 18-20°C and 30-32°C

Animals on the lower lysine levels (between 6.0 and 7.6 g/kg) in cooler environments gained 0.30 (Schenck *et al.*, 1992a), 0.70 (Schenck *et al.*, 1992b), 0.29 (Ferguson and Gous, 1997) and 0.41 (Arnold, 1997) more weight than pigs fed similar diets in hot conditions. As the dietary lysine content increased so the difference in improved weight gains at both low and high temperatures decreased to 0.10 (Schenck *et al.*, 1992a), 0.28 (Schenck *et al.*, 1992b), 0.11 (Ferguson and Gous, 1997) and 0.04 (Arnold, 1997). The fact that there were no significant differences in energy intake confirms that the improvement in growth rate at lower temperatures was primarily due to an increase in lysine intake.

It is well accepted that an increase in the supply of lysine, when this is the first limiting in the feed, results in increased live weight gain per unit of feed consumed (FCE) (Henry *et al.*, 1992; Adeola, 1995). Feed efficiency in the experiments summarised here was with few exceptions almost linearly related to lysine concentration below 12.5 g/kg (Figure 9.5). As dietary lysine content increases less food is required per unit of growth resulting in an improvement in FCE. It has been well documented that the dietary lysine content required to maximise feed efficiency and carcass leanness are greater than those for ADG (Taylor *et al.*, 1981; Batterham *et al.*, 1985,1990; Friesen, Nelssen, Goodbrand, Tokach, Unruh, Kropf and Kerr, 1994). This indicates that animals are successful in achieving maximum growth rate at lower than required dietary amino acid contents as a result of consuming greater amounts of the marginally-limiting feed. Eventually, at non-limiting contents of amino acids, the amino acid:ME ratio is perfectly balanced and there is no need for an over consumption of energy in an attempt to consume sufficient amounts of the limiting amino acid. The over consumption of energy will lead to an increase in fat deposition and an increase in body weight gain. It is therefore expected that with decreasing lysine concentration the amount of body fat will increase and carcass leanness will decrease.

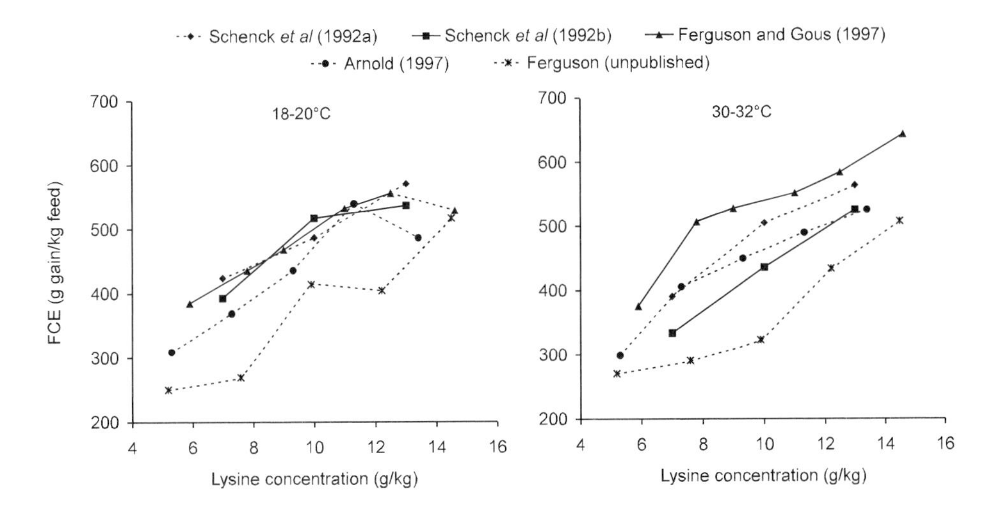

Figure 9.5 The effect of lysine concentration on feed conversion efficiency (FCE) when young pigs between 13 and 30 kg live were kept between 18-20°C and 30-32°C

The optimum dietary lysine content will differ depending on the criterion chosen: it would be highest if FCE were to be maximised, or carcass fat content minimised, lower if body weight gain were to be maximised, and somewhere in between if margin over food cost were to be maximised. A good example of how misleading ADG can be as a response criterion for determining lysine requirements is shown in Figure 9.6 using the data from Ferguson (1997). Based on ADG the requirement for these pigs would be 9.0 g/kg. This would result in an under supply of lysine and the pigs would be fatter for any given live weight and considerably less efficient with feed utilization. Although there are trends to suggest that young pigs kept in environments that are either too warm (30 °C) or too cold (< 20 °C) will be less efficient on moderate to high dietary lysine levels, the response to temperature is independent of the lysine concentration in the diet. Similar results were reported for different protein or energy and temperature experiments by Kyriazakis *et al.* (1991), Rinaldo and Le Dividich (1991) and Nienaber *et al.* (1993).

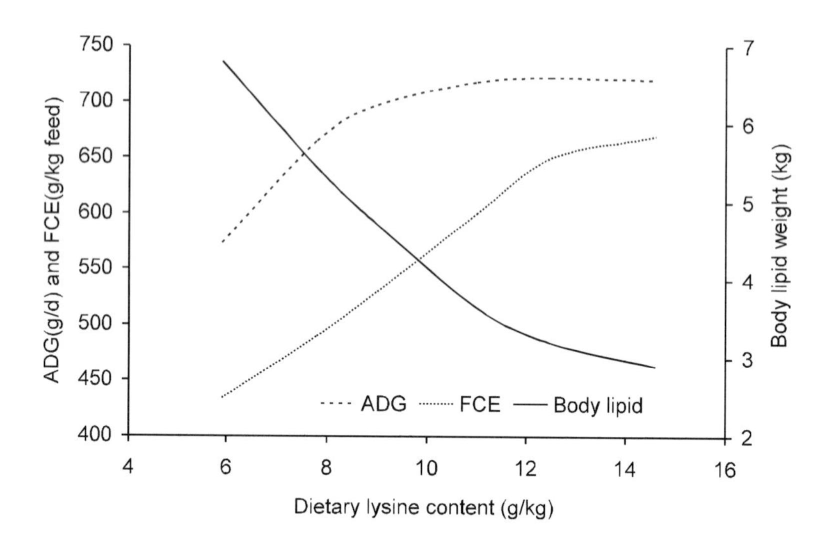

Figure 9.6 The effects of dietary lysine content on ADG, FCE and body fat weight of pigs growing between 12 and 30 kg live weight at 26°C (after Ferguson, 1997)

BODY PROTEIN AND LIPID CONTENT

The protein and lipid contents of pigs between 25 and 30 kg live weight in three of the five experiments summarised here are shown in Table 9.1. It is clear that the protein content of the body decreased as the dietary lysine concentration declined, particularly at levels below 10.0 - 11.0 g/kg. Associated with this decline was a corresponding increase in the lipid content as dietary lysine content diminished (Table 9.1). Similar to previous reports, where pigs were fed suboptimal lysine diets (Yen *et al.*, 1986; Henry *et al.*, 1992), the animals were significantly fatter and contained less protein for a given live weight. The protein content (g/kg) of carcasses was in only one case out of three significantly influenced by temperature, whereas body lipid content was influenced by temperature in two of the three cases. In only one case were both protein and fat contents significantly influenced by an interaction between lysine content and temperature.

It would appear that a diet containing less than 10.0 g lysine per kg feed will not be sufficient to allow a young pig (< 30 kg live weight) to reach its potential protein weight, unless it is grown at a temperature below the LCT (<22°C). Even if pigs were fed

Table 9.1 Protein and lipid content of pigs between 25 and 30 kg live weight given decreasing concentrations of lysine at different temperatures

Authors	*Lysine level (g/kg)*	*Protein (g/kg)* 18°C	22°C	26°C	30°C	*Lipid (g/kg)* 18°C	22°C	26°C	30°C
Ferguson (1997)	14.6	166	158	165	172	122	105	99	99
	12.5	160	161	163	173	119	121	112	125
	11.0	162	165	163	165	134	133	137	130
	9.0	166	159	150	153	151	143	162	177
	7.8	155	153	146	145	189	178	186	195
	5.9	132	137	135	137	231	207	241	241
Significance (P)									
Lysine			0.001				0.001		
Temperature			NS				0.01		
Lysine × temperature			0.001				0.001		
Arnold (1997)	13.4	165	157	158	154	93	99	77	89
	11.3	154	155	150	153	119	113	116	89
	9.3	147	146	143	144	135	141	137	137
	7.3	146	142	139	136	166	172	168	173
	5.3	132	130	130	125	207	222	217	210
Significance (P)									
Lysine			0.001				0.001		
Temperature			NS				NS		
Lysine × temperature			NS				NS		
		20°C			*32°C*	*20°C*			*32°C*
Schenck *et al* (1992a)	13.0	163			167	134			118
	10.0	166			169	122			115
	7.0	158			158	153			159
Significance (P)									
Lysine			0.10				0.10		
Temperature			0.01				0.05		
Lysine × temperature			NS				NS		

lysine levels of 10.0 g/kg they would be fatter for any given live weight than pigs fed lysine contents higher than this. From the response of food intake (Figure 9.2) it would appear that pigs do attempt to maintain their desired protein weight by eating more of the low lysine diets. However, at ±10.0 g/kg, where maximum intake occurred, the animals could not fully compensate for low lysine concentrations and therefore, body protein content decreased. Ferguson (1997) and Arnold (1997) noted a 0.19 reduction in body protein content and 1.15 to 1.34 increase in the lipid content, respectively, between pigs fed diets with the lowest and the highest lysine contents. The additional energy consumed, in an attempt to satisfy their lysine requirements, resulted in a significant ($P < 0.001$) increase in body lipid content.

PROTEIN AND LIPID RETENTION

The effects of dietary lysine and ambient temperature on the rates of body protein (PR) and lipid deposition (LR) are shown in Figures 9.7 and 9.8. A closer look at the response to dietary lysine concentration within each of the temperature treatments showed that higher dietary lysine concentrations are required to maximise PR as the environmental temperature increases. This response in PR to temperature in animals fed lysine-deficient diet suggests that animals can only achieve maximum protein growth on lysine-limiting diets if the environment is sufficiently cool to allow the extra heat increment of feeding to be dissipated. Both Arnold (1997) and Ferguson (1997) observed that within their respective experiments there were few differences in PR and LR between 18°C and 22°C which suggests that in cold conditions when voluntary energy intake is sufficient to maintain a constant energy retention (PR and LR), body tissue growth rates are independent of environmental temperature. Rinaldo and Le Dividich (1991) observed that for pigs kept at temperatures below the thermoneutral zone and fed *ad libitum*, temperature had no significant effect on protein and lipid deposition. This is in contrast to the response at high temperatures where PR decreased when pigs were housed at temperatures above 26°C. Similar findings were reported by Campbell and Taverner (1988) between 14° C and 32°C, and Rinaldo and Le Dividich (1991) between 25°C and 31.5°C. Therefore, unlike the response at low temperatures, protein and lipid deposition are more dependent on the environmental temperature at high temperatures.

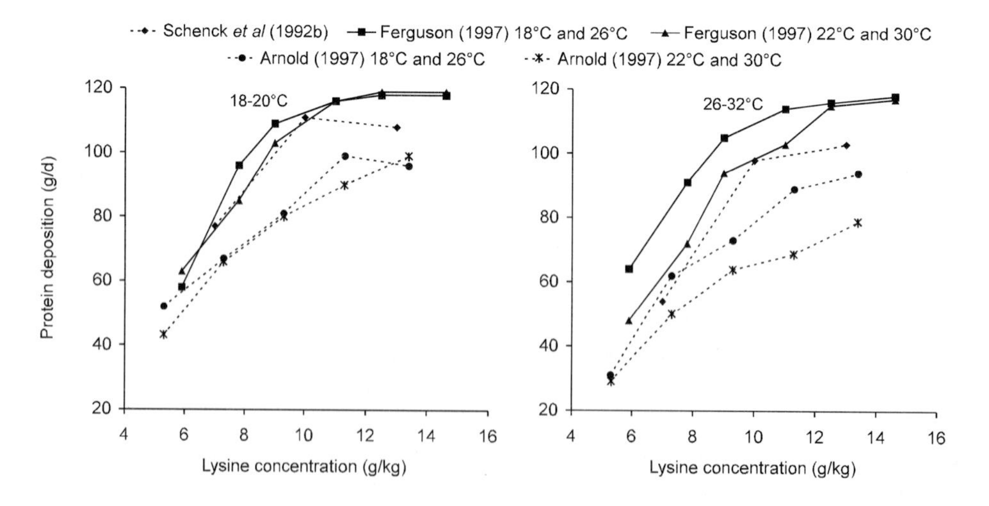

Figure 9.7 The effect of lysine concentration on the rate of protein deposition when young pigs between 13 and 30 kg live weight were kept between 18-22°C and 26-32°C

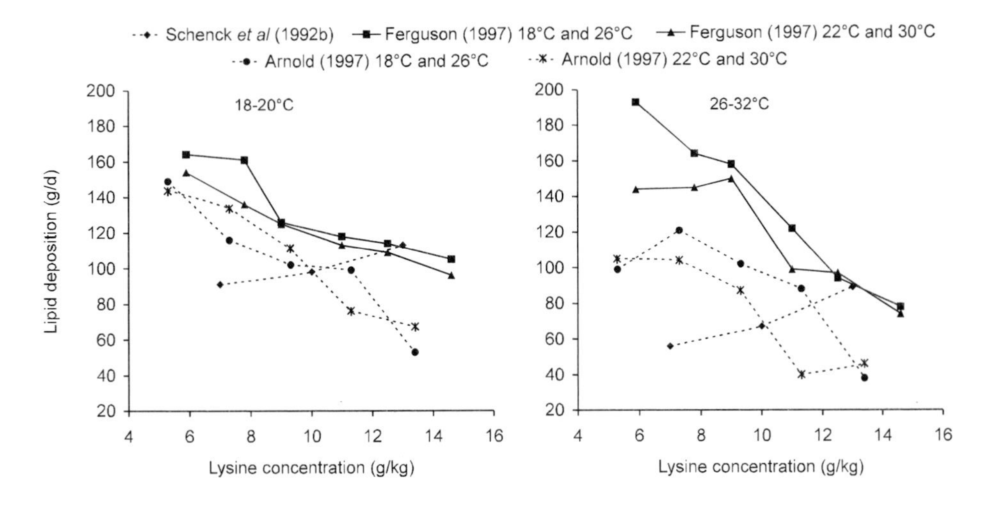

Figure 9.8 The effect of lysine concentration on the rate of lipid deposition when young pigs between 13 and 30 kg live weight were kept between 18-22°C and 26-32°C

The reduction in daily lipid retention at high temperatures is a consequence of the marked reduction in voluntary food intake (ME intake) and the resulting reduction in net energy available for tissue deposition. However, as a proportion of total body tissue deposited there was an increase in lipid retention, because insufficient lysine was available to achieve maximum protein deposition. As a consequence more energy was available for fat deposition. There is a possibility that, to reduce the internal heat burden resulting from consuming a feed with a low lysine:energy ratio, the animal may divert more dietary energy into fat deposition.

HEAT LOSS

Total heat loss (THL) can be estimated as the difference between the Metabolizable energy (ME) intake and energy retained as protein and fat. In the experiments summarised here the main effects of temperature, and to a lesser extent dietary lysine content, had a significant effect on THL. Arnold (1997) and Ferguson (1997) showed that THL across the various all treatments followed the same trend as the responses in food intake to lysine-deficient diets and to different temperatures (Figure 9.9).

Maximum THL occurred at 9.0 g lysine/kg which corresponded with the lysine treatment that resulted in pigs consuming the most food. Although there are no significant interactions between temperature and lysine content on THL, the lysine concentration responsible for maximum THL increased with temperature. For example, maximum heat output was observed in animals fed 7.8 g/kg at 18°C, 9.0 g/kg at 22°C and 26°C,

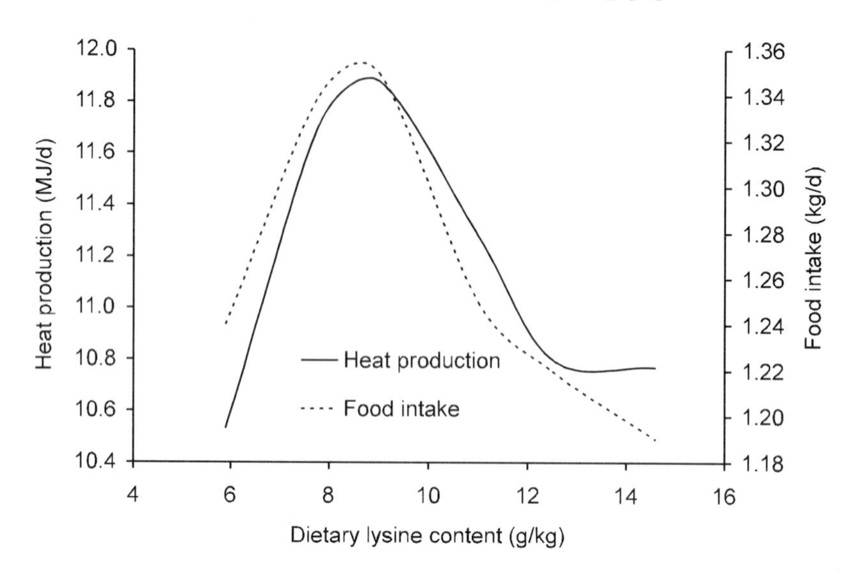

Figure 9.9 The relationship between heat production and food intake with different levels of dietary lysine concentration (after Ferguson and Gous, 1997)

and 11.0 g/kg at 30°C (Ferguson, 1997). From these data and Figure 9.9 it would appear that young pigs fed lysine deficient diets will increase feed intake in an attempt to compensate for the deficiency in the diet but the compensation will depend on the maximum amount of heat the animal can lose. The lower the temperature the more heat that can be lost and hence the greater the potential to maintain maximum protein retention. Conversely, as the environmental temperature increases above thermoneutrality, the more critical the correct dietary lysine concentration becomes, with lysine concentrations lower than 12.5 g/kg (0.87 g lysine/MJ ME) adversely affecting lean growth (Figure 9.10). Similar results were observed by Schenck *et al.* (1992b).

The similar responses in both THL and food intake to dietary lysine concentration and temperature suggest that the maximum amount of heat that the animal can dissipate is an important factor in regulating voluntary food intake: the lower the ambient temperature the greater the amount of heat the animal can lose hence enabling diets with lower lysine:ME ratios to be fed successfully at lower temperatures. Although some of the additional heat produced in animals grown at 18°C is a result of cold thermogenesis, the potential exists for such animals to lose more heat when fed a diet with a high heat increment (high protein, low fat diets) than those kept at higher temperatures.

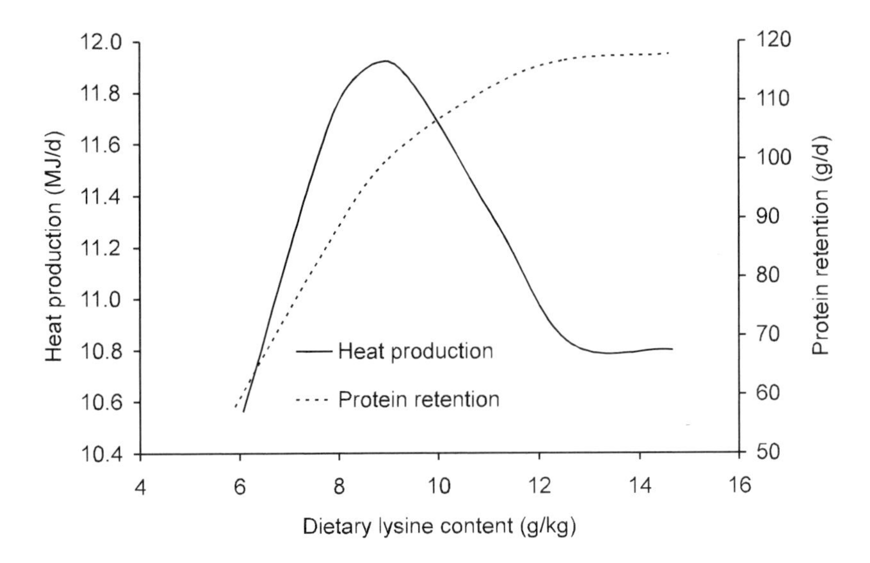

Figure 9.10 The relationship between heat production and protein retention with different levels of dietary lysine concentration (after Ferguson and Gous, 1997)

IMPLICATIONS FOR PREDICTING GROWTH AND FOOD INTAKE

As previously stated an important component of any model predicting voluntary food intake in growing pigs is defining how much heat an animal can dissipate to its environment and, in particular, the upper limit to the animal's capacity to lose heat. Therefore the problem is not in predicting how much heat an animal can produce but rather in predicting how much it can lose, and to what extent the maximum heat loss will constrain the desired food intake and hence reduce the growth rate below the potential of the animal. From the previous section it is apparent that animals fed an amino acid (or protein) -deficient diet will only achieve their maximum rate of protein retention if they can consume sufficient amounts of the limiting amino acid (or protein) to meet their daily requirements. This can only be achieved if they can lose the heat produced from depositing body tissue and from the heat increment of feeding that is generated by such feed. Therefore, assuming zero heat storage, the environmental constraint on the rate of heat loss becomes a constraint on the rate of heat produced. A corollary to the above is that the amount of heat an animal can lose in a hot environment will determine the rate of protein deposition. This underlines the importance of knowing the heat loss capability of the animal.

For modelling purposes, to achieve a reasonable level of accuracy of prediction, it is imperative that THL is adequately defined and that there are rules to cope with predicting the failure of the animal to grow to its potential because of either too much or too little heat. The latter scenario has been discussed by Verstegen *et al.* (1995) but the effect of high temperatures and/or low nutrient:energy ratios on depressing the rate of growth and food intake is still not well known. A number of attempts, both simple and complex, to define some of the relationships between temperature, feed composition and body composition have been made with varying degrees of success (Bruce and Clarke, 1979; Whittemore, 1983; Black, Campbell, Williams, James and Davies,1986 and Usry, Turner, Bridges and Nienaber, 1992). If it is assumed that for an animal in any given state the relationship between THL and temperature is linear then it is possible to determine the upper limit of the capacity of an animal to lose heat across a range of ambient temperatures (Ferguson *et al.*, 1994). The main determinants of the rate of heat loss from the animal to its environment are likely to include the state of the animal (such as surface area, body protein and fat content), external environmental factors (such as insulation, air movement, floor material, stocking density etc.) and the ambient temperature. Based on independent data from Campbell and Taverner (1988), Rinaldo and Le Dividich (1991), Arnold (1997) and Ferguson and Gous (1997) the following function was determined to describe such a relationship between THL and temperature across a range of temperatures from 12 to 35 °C:

$$\text{THL} = a - 23.4 \times \text{ET} \quad (\text{kJ/ kg}^{0.67}) \qquad \text{(Eq. 9.2)}$$

Where **a** ranges between 1650 to 2100 depending on body protein and fat content, and stocking density

ET = air temperature adjusted for insulation, floor material and air movement

With an estimate of the maximum rate of heat loss that is independent of dietary factors, such as protein and energy intake, it is then possible to ensure that the predicted energy intake does not exceed an amount that would cause heat production to exceed the upper limit of the animal's capability to lose heat. The lower the amino acid:energy ratio the greater the chance that heat production will exceed that which is possible to lose to the environment and the more likely food intake will drop. This is likely to be exacerbated as air temperature increases and by potentially high rates of lean tissue deposition associated with fast, lean growing strains of pigs. Once energy intake has been determined it is then a case of applying rules to partition available nutrients into protein and fat tissue.

Conclusions and practical implications

It can be concluded from the analysis of the five papers summarised in this review that nutrition × environment interactions play an important role in regulating food intake and hence daily growth rate. The prevailing environmental temperature will define the maximum amount of heat an animal can lose which, in turn, will define the upper limit to the amount of heat the animal can produce, and hence the food intake and growth responses of the animal. The results also show that pigs fed amino acid deficient diets will increase their food intake in an attempt to achieve daily amino acid intakes that will allow them to express their genetic potential. However, below a critical dietary amino acid concentration the animal can no longer compensate for the deficiency by consuming food at a higher rate and voluntary food intake will decline. The extent of the compensation will depend on the amount of heat the animal can lose which in turn is dependent on the environmental temperature. This would suggest that part of the variation in response to lysine observed in practice, and between experiments, is due to the different temperatures under which animals have been housed. The determination of amino acid requirements, and the formulation of diets for young fast, growing pigs, must take into account the effect of the environmental temperature on the growth response to amino acid. Ignoring the importance of the temperature × nutrition interaction can result in the improper feeding of the young pig and a consequential decline in growth, an increase in food intake, a decline in FCE and an increase in body fat content. An example of such a consequence at warm temperatures is shown in Table 9.2.

Thus, when considering the effects of feeding a diet designed for thermoneutral conditions (26 °C) containing 12.5 g lysine/ kg to young pigs (highlighted and bold text in Table 9.2) at two ambient temperatures, 26 °C and 30 °C. FCE would decline, at the higher temperature, from 658 to 585 g gain/kg feed and body fat weight would increase. However, if the lysine concentration increased to 14.6 g/kg (whilst maintaining an

Table 9.2 Comparison of performance (FCE, PCr, % fat gains and body fat) with different lysine concentrations at 26 and 30°C (after Ferguson, 1997)

		Diets		
		Lys = 11.0 g/kg *Lys:ME = 0.75 g/MJ*	*Lys = 12.5 g/kg* *Lys:ME = 0.87 g/MJ*	*Lys = 14.6 g/kg* *Lys:ME = 1.01 g/MJ*
			Proportion of requirement	
Temperature (°C)	*Variable*	*0.91*	*1.06*	*1.20*
26	FCE (g gain/kg feed)	562	658	671
	PR (g/d)	114	116	118
	% fat gain	17.6	13.6	11.2
	Body fat (g)	3.93	3.18	2.81
30	FCE (g gain/kg feed)	588	585	641
	PR (g/d)	103	115	117
	% fat gain	16.3	15.2	11.3
	Body fat (g)	3.58	3.62	2.86

ideal amino acid balance) there would be a marked improvement in the leanness of the animal at both temperatures with minimal effect on feed efficiency. A similar scenario would be observed if the lysine content were increased from 11.0 g lysine/kg to 12.5 g/kg when temperature increased. Therefore to circumvent the problem of feeding an animal according to its requirement for maximum carcass leanness under varying environmental conditions it is important to feed higher rather than lower levels of amino acids.

The objective of this paper was to demonstrate the effect nutrition × environment interactions have on food intake, growth and body composition in young pigs and to highlight the consequences of ignoring such interactions when trying to predict pig performance. Therefore, in designing feeds for different classes of pigs it is imperative that the interaction between the animal, the feed and the environment is considered. Failure to consider such an interaction would result in sub-optimal performance, something that is likely to prevail until more widespread use is made of simulation models. Theory-based simulation models have the potential to predict accurately the effects of genotype, feed composition and environment on food intake and growth in a systematic way and are therefore the most effective means available to solve the problem of estimating pig performance.

References

Adeola, O. 1995. Dietary lysine and threonine utilization by young pigs: Efficiency for carcass growth. *Canadian Journal of Animal Science* **75**:445-452.

Arnold, G.D., 1997. *Responses in growing pigs to lysine and threonine limiting feeds and environmental temperature.* M Sc Thesis, University of Natal, South Africa.

Batterham, E.S., Giles, L.R., and Dettman, E. 1985. Amino acid and energy interactions in growing pigs. 1. Effects of food intake, sex and live weight on the responses of growing pigs to lysine concentration. *Animal Production* **40**: 331-343.

Batterham, E.S., Andersen, L.M., Baigent, D.R. and White, E. 1990. Utilization of ileal digestible amino acids by growing pigs: effect of dietary lysine concentration on efficiency of lysine retention. *British Journal of Nutrition* **64**:81-94.

Black, J. L., Campbell, R. G., Williams, I. H., James, K. J. and Davies, G. T. 1986. Simulation of energy and amino acid utilisation in the pig. *Research and Development in Agriculture* **3**: 121-125.

Bruce, J. M. and Clark, J.J. 1979. Models of heat production and critical temperature for growing pigs. *Animal Production* **28**: 353-369.

Campbell, R. G. and Taverner, M.R. 1988. Relationships between energy intake and protein and energy metabolism, growth and body composition of pigs kept at 14 or 32 °C from 9 to 20 kg. *Livestock Production Science* **18**: 289-303.

Close, W.H. 1987. The influence of the thermal environment on the productivity of pigs. In *Pig Housing and the Environment,* pp 9 - 24. Edited by A.T. Smith and T.L.J. Lawrence. Animal Production Occasional Publication No 11, London.

Close, W. H. and Mount, L.E. 1978. The effects of nutrition and environmental temperature on the energy metabolism of the growing pig. 1. Heat loss and critical temperature. *British Journal of Nutrition* **40**: 413-421.

Dauncey,M.J., Ingram,D.L., Walters,D.E. and Legge,K.F., 1983. Evaluation of the effects of environmental temperature and nutrition on growth and development. *Journal of Agricultural Scence (Cambridge)* **101**:291-299.

Emmans, G.C., 1994. Effective energy: a concept of energy utilization applies across species. *British Journal of Nutrition* **71**: 801-821.

Emmans G.C. and Oldham, J.D., 1988. Modelling of growth and nutrition in different species. In: *Modelling of Livestock Production Systems.,* pp 13 - 21. Edited by S. Karver and J.A.M. Arendonk. Kluwer Academic Publishers, Dordrecht.

Ferguson, N.S., 1997. The effect of heat production on growth responses in young pigs fed a range of lysine diets. *Proceedings of the Eastern Nutrition Conference*, pp 109-124. University of Guelph, Guelph.

Ferguson, N.S. and Gous, R.M, 1997. The influence of heat production on voluntary food intake in growing pigs fed protein deficient diets. *Animal Science* **64**:365-378.

Ferguson N.S., Gous R.M. and Emmans, G.C., 1994. Preferred components for the construction of a new simulation model of growth, feed intake and nutrient requirements of growing pigs. *South African Journal of Animal Science* **24**:10-17.

Friesen, K.G., Nelssen, J.L., Goodband, R.D., Tokach, M.D., Unruh, J.A., Kropf, D.H. and Kerr, B.J. 1994. Influence of dietary lysine on growth and carcass composition of high-lean-growth gilts fed from 34 to 72 kilograms. *Journal of Animal Science* **72**: 1761-1770.

Fuller,M.F. and Boyne,A.W., 1971. The effects of environmental temperature on growth and metabolism of pigs given different amounts of food. 1. Nitrogen metabolism, growth and body composition. *British Journal of Nutrition* **25**:259-272.

Henry, Y., Colleaux, Y. and Seve, B. 1992. Effects of dietary level of lysine and of level and source of protein on feed intake, growth performance and plasma amino acid pattern in the finishing pig. *Journal of Animal Science* **70**:188-195.

Henry, Y., and Seve, B. 1993. Feed intake and dietary amino acid balance in growing pigs with special reference to lysine, tryptophan and threonine. *Pig News and Information* 14:35N.

Kyriazakis, I. and Emmans, G.C. 1991. Diet selection in pigs: Dietary choices made by growing pigs following a period of underfeeding with protein. *Animal Production* **52**:337-346.

Kyriazakis, I., Stamataris, C., Emmans, G.C. and Whittemore, C.T. 1991. The effects of food protein content on the performance of pigs previously given foods with low or moderate protein contents.. *Animal Production* **52**:165-173.

Le Dividich,J. and Noblet,J., 1986. Effect of dietary energy level on the performance of individually housed early-weaned piglets in relation to environmental temperature. Livestock Production Science **14**:255-263.

Meyr, R.O., Bucklin, R.A. and Fialho, F.B., 1998. Effects of increased dietary lysine (protein) level on performance and carcass characteristics of growing-finishing pigs reared in a hot, humid environment. *Transactions ASAE* **41**:447-452.

NRC. 1988. *Nutrient requirements of Swine (9th Ed.).* National Academy Press, Washington, DC.

Nienaber, J.A., Hahn, G.L., Korthals, R.L. and McDonald, T.P.,1993. Eating behaviour of swine as influenced by environmental temperature. In *Livestock Environment IV. Fourth International Symposium,* pp.909-916. Edited by E. Collins. University of Warwick, Coventry.

Rinaldo, D. and Le Dividich, J. 1991. Assessment of optimal temperature for performance and chemical body composition of growing pigs. *Livestock Production Science* **29**: 61-75.

Schenck, B.C., Stahly, T.S. and Cromwell, G.L. 1992a. Interactive effects of thermal environment and dietary amino acid and fat levels on rate, efficiency and composition of growth of weanling pigs. *Journal of Animal Science* **70**: 3791-3802.

Schenck, B.C., Stahly, T.S. and Cromwell, G.L. 1992b. Interactive effects of thermal environment and dietary amino acid and fat levels on rate and efficiency of growth of pigs housed in a conventional nursery. *Journal of Animal Science* **70**: 3803-3811.

Stahly,T.S., Cromwell,G.L. and Aviotti,M.P. 1979. The effects of environmental temperature and dietary lysine source and level on the performance and carcass characteristics of growing swine. *Journal of Animal Science* **49**:1242-1251.

Taylor, A.J., Cole, D.J.A. and Lewis, D. 1981. Amino acid requirements of growing pigs. 2. Identification of limiting amino acid(s) in a low-protein diet supplemented with lysine. *Animal Production* **33**: 87-97.

Usry, J.L., Turner, L.W., Bridges, T.C. and Nienaber, J.A., 1992. Modeling the physiological growth of swine Part III: Heat production and interaction with environment. *Transactions of the ASAE* **35**:1035-1042.

Verstegen,M.W.A., Brandsma,H.A. and Mateman,G., 1982. Feed requirements of growing pigs at low environmental temperatures. *Journal of Animal Science* **55**:88-94.

Verstegen, M. W. A., Close, W. H., Start, I. B. and Mount, L. E. 1973. The effects of environmental temperature and plane of nutrition on heat loss, energy retention and deposition of protein and fat in groups of growing pigs. *British Journal of Nutrition* **30**: 21-35.

Verstegen, M.W.A., de Greef, K.H. and Gerrits, W.J.J., 1995. Thermal requirements in pigs and modelling of the effects of coldness. In: *Modelling growth of the pig,* pp 123 - 135. Edited by P.J. Moughan, M.W.A. Verstegen. and M.I. Visser-Reyneveld. Wageningen Pers, Wageningen.

Whittemore, C. T. 1983. Development of recommended energy and protein allowances for growing pigs. *Agricultural Systems* **11**: 159-186.

Whittemore, C. T. 1993. *The science and practice of pig production.* Longman Scientific and Technical, Essex, United Kingdom.

Yen, H.T., Cole, D.J.A. and Lewis, D. 1986. Amino acid requirements of growing pigs. 7. The response of pigs from 25 to 55 kg live weight to dietary ideal protein. *Animal Production* **43**: 141-154.

10

NON-STARCH POLYSACCHARIDES IN PIG DIETS AND THEIR INFLUENCE ON INTESTINAL MICROFLORA, DIGESTIVE PHYSIOLOGY AND ENTERIC DISEASE

J.R. PLUSKE[1], D.W. PETHICK[1], Z. DURMIC[1], D.J. HAMPSON[1] AND B.P. MULLAN[2]

[1] Division of Veterinary and Biomedical Sciences, Murdoch University, Murdoch WA 6150, Australia; [2] Agriculture Western Australia, Locked Bag No. 4, Bentley Delivery Centre 6983, Western Australia

Introduction

In most feedstuffs, carbohydrates constitute 0.65-0.8 of the digestible dry matter. This fraction can be divided into a part that is digested by the endogenous enzymes of the mouth and gastrointestinal tract (e.g., α-amylase for starch, lactase for lactose) and a component that is digested predominately in the large intestine (dietary fibre; DF). The major components of DF pertinent to the pig are the non-starch polysaccharides, or NSP. The NSP fraction contained in diets fed to pigs is fermented largely by the commensal biota present in the caecum and colon with the production of metabolites, gases and microbial biomass, although some pre-caecal fermentation of NSP occurs. Microbial digestion requires the presence of a stable and diverse microflora in the large intestine that can be altered by many factors, with the source of DF being fed just one of these. It is the balance between the many and diverse microbial populations present in the large intestine and the nature of the material entering this organ, both exogenous and endogenous, that determines the stability of the microflora, and which will constitute the basis of this chapter.

This chapter commences with a brief description of DF, and then continues with the importance of NSP in maintaining a stable microflora within the large intestine. The effect of various dietary substrates on the balance of microbial populations will also be discussed. Physiological consequences of NSP will be covered, including a review of the anti-nutritive effects of NSP in pigs and the energetic contribution of hind-gut fermentation to metabolism in pigs. This is especially relevant given the increasing interest in the use of (cheaper) feed ingredients, such as food by-products, for inclusion in pig diets. Often, the nutritive value of such products is reduced due to the presence of high levels of NSP. This is also of interest given that cereals commonly included in pig diets, such as wheat and barley, contain NSP that may have anti-nutritive effects.

A section highlighting several non-nutritional aspects of NSP in pork production will also be presented. The final part of this chapter will examine research conducted largely by this group pertaining to the effects of NSP on the development of swine dysentery, a diarrhoeal disease that occurs in the large intestine of growing-finishing pigs. This section has been included as an example of how undigested carbohydrate can have effects in the pig that are not entirely related to its nutritional needs. Finally, this review will focus specifically on the influence of NSP in the growing and adult pig. The effects of NSP in younger pigs has been covered in the next chapter (Mosenthin, Hambrecht and Sauer, 1999).

Dietary fibre

A plethora of papers, reviews, books and monographs have been written defining "dietary fibre" (DF). In an article appearing in Lancet, Trowell, Southgate, Wolever, Leeds, Miguel and Jenkins (1976) redefined DF as "the plant polysaccharides and lignin which are resistant to hydrolysis by the digestive enzymes of man". Variations to this definition have occurred since and, no doubt, will continue to do so. For example, some authors (e.g., Englyst, 1989) advocate that because lignin is not a polysaccharide and behaves very differently *in vivo* to NSP, DF should be defined as the sum of monosaccharides released by acid hydrolysis of the NSP. Others (e.g., Baghurst, Baghurst and Record, 1996) state that "resistant starch" (RS, or starch that enters the large intestine of man) should be included in the definition of DF due to some of its properties being similar to those of NSP in the large intestine of monogastric animals. It is not the intention of this chapter to continue this discussion but to state that, regardless, the carbohydrates (which include the low molecular weight sugars, starch and various cell wall and storage NSP) are the most important energy-yielding ingredients for pigs and the indigenous microflora of the large intestine. In addition, they possess important and sometimes potent influences on aspects of production and health in modern-day pig production.

The NSP and lignin are the principal components of plant cell walls. In the past, the DF component of pig diets has been expressed as "crude fibre", with most pig diets containing between 30 and 100g crude fibre/kg. However, the term "crude fibre" is virtually redundant in non-ruminant nutrition because it provides no indication as to any possible physiological properties that NSP may have in the gastrointestinal tract of the pig. Even terms like neutral detergent fibre, while describing the cell wall component (as cellulose, hemicellulose and lignin) of a feed or diet, tend to ignore the soluble NSP that may have physico-chemical effects in the gastrointestinal tract.

The polysaccharides, because of their large size and structure, comprise an array of chemical and physical properties. Starch is present within plant cells as discrete granules, with amylopectin and amylose contained within each granule. The plant cell wall consists of a series of polysaccharides often associated and (or) substituted with proteins and phenolic compounds, in some cells together with the phenolic polymer lignin (Selvendran, 1984). The building blocks of cell wall polysaccharides are the *pentoses* (arabinose and xylose), *hexoses* (glucose, galactose and mannose), *6-deoxyhexoses* (rhamnose and fucose), and *uronic acids* (glucuronic and galacturonic acids). The major polysaccharides of plant cell walls are: cellulose, arabinoxylans, mixed linked ß(1→3; 1→4)-D-glucans (ß-glucans), xyloglucans, xylans, rhamnogalacturonans and arabinogalactans. The NSP are then generally defined as being either "soluble" or "insoluble". These terms provide an indication as to the solubility and behaviour of NSP *in vivo*. Lignins are branched networks of phenylpropane units and are partly linked to cell wall cellulose and non-cellulosic polysaccharides, and serve to cement and anchor the cellulose microfibrils and other matrix polysaccharides to prevent degradation and physical cell wall damage (Bach Knudsen, 1997).

Each of these categories can be subdivided into subgroups on the basis of their molecular structure, i.e., the component monosaccharides forming the backbone of the polymer, and the types of branching present. The configuration of the glycosidic linkages is extremely important because it determines the tertiary structure of the molecule that, in turn, determines the physico-chemical properties of the polysaccharides and their capacity to associate with other polysaccharide chains and with proteins (Asp, Schweizer, Southgate and Theander, 1992).

No endogenous enzymes capable of hydrolysing the non-α-glucans (i.e., NSP or lignin) have been demonstrated in the small intestine of the pig, even though considerable degradation of some NSP (e.g., ß-glucans from barley and oats) can occur prior to the caecum. Consequently, the large majority of undigested carbohydrate passes into the caecum and colon where it undergoes microbial digestion. Given the relative importance of the large intestine (i.e., caecum plus colon) over the small intestine to microbial fermentation in the pig, the following discussion will focus on general characteristics of the commensal biota in the large intestine, and how this can be altered by dietary means.

Microflora of the large intestine in the pig

INTRODUCTION

Pigs are non-ruminant, simple-stomached animals that do not rely on a symbiotic relationship with the microbial flora within their gastrointestinal tract to survive to the same degree as ruminants. As such, microbial populations are numerically fewer and also less diversified, meaning that they have less fermentative capacity (and ability) and do not cope as well with poorer quality foodstuffs as do ruminants, although this appears dependent upon the mature body size of the pig. Essentially, the nature of the carbohydrate polymers present and the degree of lignification influence the extent of microbial breakdown of NSP (Bach Knudsen, Jensen, Anderson and Hansen, 1991).

Bacteria that are present in the large intestine of pigs are fastidious and require specific environmental conditions, including correct pH, temperature, redox potential, osmolarity, anaerobiosis, endogenous secretions, enzyme activity and dry matter content to enable them to be actively involved in breakdown of NSP or other substrates. These factors can act to increase or decrease bacterial activity. Microbial interactions are also very important in this dense population, and these may act to promote or prevent growth of other bacteria. Facultative anaerobes utilize available O_2, enabling O_2-sensitive anaerobic organisms to survive. Certain species of bacteria produce various antibacterial substances such as colicines. Anaerobic organisms produce volatile fatty acids (VFA) that are inhibitory to bacterial growth by affecting pH and oxidation-reduction potential (E_h) levels (Simon and Gorbach, 1981).

The principal substrates for bacterial growth in the large intestine of the pig are dietary carbohydrates that have escaped digestion in the upper gastrointestinal tract. These carbohydrates may be starches, plant cell wall polysaccharides and host mucopolysaccharides together with various proteins, peptides, and other low-molecular weight carbohydrates. A wide range of bacterial polysaccharidases, glycosidases, proteases, and peptidases degrades these complex polymers to smaller oligomers and their component sugars and amino acids. In the large intestine, bacteria are able to metabolise these substrates to VFA (mainly acetic, propionic and butyric acids), hydroxy and dicarboxylic acids, combustible gases (H_2 and CH_4), H_2O, CO_2 and microbial biomass. In addition, smaller quantities of succinate, formate, H_2S and NH_3 are produced, as well as fermentation "heat" (Macfarlane and Gibson, 1997). The degree of fermentation depends primarily on the source of DF and the presence of nitrogen, vitamins and minerals that are essential for the overall nutrition of the microbial populations and for the synthesis of microbial proteins. With protein, this supply could be small if the digestion of dietary protein and of endogenous protein was close to completion before the ileum and large intestine.

The supply of fermentable carbohydrate is an important factor determining the growth of bacteria in the gut. Microbial metabolism in the hindgut is profoundly altered by the presence of fermentable carbohydrate. Bacteria account for 0.55 of faecal solids. This proportion is increased when the diet is supplemented with fermentable fibre, while non-fermentable fibre increases undegraded fibre and water contents (Stephen and Cummings, 1980). In lignified tissues, insoluble NSP is inaccessible to bacterial degradation. However, if soluble NSP are present, bacteria can degrade soluble NSP first and then gain access to degrade the insoluble fibre.

The conditions that exist in different regions of the gut result in the selection of a number of different microbial populations. The major factors influencing the survival and diversity of bacteria in the gut are:

- the presence of attachment sites
- the ability to tolerate the pH and E_h (i.e. oxidation-reduction potential) regimen
- resistance to bile acids and toxic compounds in the diet
- the ability to grow on the dietary and endogenous substrates present (Stewart, 1997)

The focus in this part of the chapter, however, relates specifically to how the type of diet fed, and in particular the NSP component of the diet, can alter microbial populations in the pig and hence influence metabolism and animal production. It is first necessary though to describe the microbiota present in the large intestine of the pig.

ESTABLISHMENT OF A "STABLE" MICROFLORA

At birth, the young pig is bacterium-free, but the multiplication of bacteria received either orally or via the rectum apparently starts within hours of parturition. During birth, the piglet successively encounters microorganisms in the maternal vagina, in the faeces, and in the environment (Sansom and Gleed, 1981). From the large number and variety of microorganisms encountered, conditions in the gut result in the selection of particular strains that will establish to form the commensal biota, while other organisms fail to colonise the gut. As such, the different sections of the gastrointestinal tract develop mixed populations of microorganisms, and these change constantly in response to external stimuli such as the type of diet fed. Indigenous colonic flora develop as a result of the influence of intestinal physiology on the interaction between the bacteria that contaminate the body. Sows are obviously the initial source of major bacterial groups for piglets but, as they are weaned, the microflora tend to change becoming more similar between littermates. During the growing-finishing phases, the intestinal flora become unstable and considerable variation occurs between individuals. Excellent reviews on the microbial development and populations in younger pigs are provided by Ewing and Cole (1994), Conway (1997) and Stewart (1997), and will not be discussed further in this chapter.

Indigenous intestinal microorganisms play significant roles in the gastrointestinal tract. They aid in digestion of food, produce essential vitamins, and help prevent pathogens from colonising the gastrointestinal tract (Meynell 1963; Freter, 1972; Savage, 1986; Wilson, Moore, Patel and Peramond, 1988). Protein degrading and amino acid fermenting bacteria occur in the colon, but saccharolytic species are numerically predominant (Finegold, Attebery and Sutter, 1974; Moore and Holdeman 1974). The principal saccharolytic bacteria are Gram-negative anaerobes from the genus *Bacteroides*. They hydrolyse dietary and endogenous polymers (e.g., mucins) that provide energy, carbon and macromolecular building blocks for the host. They also synthesise vitamins and conserve nitrogen. In exchange, flora compete directly with the host for nutrients and represses the absorptive capacities of the tissues. Whether bacteria compete with the host for this food or aid the animal in its utilisation is determined by the type of food consumed and the environmental conditions established by the physiology of the host.

GENERAL FEATURES OF THE LARGE INTESTINAL MICROFLORA

The large intestine of the pig is an extremely complex ecosystem that provides an excellent habitat for the establishment of a large number and diverse range of bacterial species. This is assisted by the large intestine having a large volume of digesta coupled with a longer residence time (up to 60 hours) that permits a very active bacterial fermentation of DF (Varel and Yen, 1997). Moreover, a slower transit time of digesta in the large intestine allows for considerable water resorption together with an increase in DM content as digesta is propelled distally. In general, the concentrations of bacteria of normal pigs are much higher in the caecum and colon than in more proximal portions of the tract (Jensen and Jorgensen, 1994), chiefly because the extensive retention of digesta in the hindgut provides a more stable environment for microbial proliferation and fermentation.

About 10^{10} cells per gram of gut content are present, with strict anaerobes the most common organisms isolated (Radecki and Yokoyama, 1991; Varel, 1987). As many as 500 bacterial species may be present, and many more have yet to be identified (Simon and Gorbach, 1981; Moore, Moore, Cato, Wilkins and Kornegay, 1987). Obligate anaerobes comprise 0.99 or more of the total bacterial numbers in the large bowel. The numerically predominant species are non-sporing anaerobes from genera *Bacteroides*, *Eubacterium*, *Fusobacterium* and *Bifidobacterium*. Gram-positive anaerobic cocci and ureolytic streptococci are also present (Robinson, Stromley and Varel, 1988; Levett 1990). Unlike the rumen, the large intestine does not contain protozoans or anaerobic fungi, which are active contributors to DF degradation in the rumen, especially cellulose and cellulose residues. Less methanogenic activity is also found, although considerable production (12.5 L/day *vs.* 1.4 L/day) was observed when 7-month-old castrate pigs were fed a high-fibre diet based on pea fibre and pectin compared to pigs fed a barley and wheat-starch diet (Jensen and Jørgensen, 1994).

Using anaerobic techniques for the isolation of bacteria from faeces, Salanitro, Blake and Muirhead (1977) found that ≈ 0.90 of the bacteria were Gram-positive and consisted of facultatively anaerobic streptococci, *Eubacterium* sp., *Clostridium* sp., and *Propionibacterium acnes*. The Gram-negative bacteria included *Treponema* (*Serpulina*) sp., *Selenomanas* sp., *Veillonella* sp., *Bacteroides* sp., and *E. coli*. Russell (1979) measured the types and distribution of anaerobic bacteria every 25-cm along the length of the large intestine in 20- to 25-week-old pigs. Bacteria were recovered from the colon wall, surface layer and contents and, as reported earlier by Salanitro *et al.* (1977), around 0.90 of the bacteria isolated were Gram-positive. Species included

Lactobacillus sp., *Streptococcus* sp., *Peptococcus* sp., *Peptostreptococcus* sp., *Eubacterium* sp., and *Megasphaera elsdenii*. Robinson, Whipp, Bucklin and Allison (1984) found that most (0.71) of the isolated bacteria from colonic epithelia in healthy, 7- to 9-week-old-pigs were Gram-positive, whereas in pigs infected with the intestinal spirochaete *Serpulina hyodysenteriae*, 0.88 of the isolates recovered from the epithelia were Gram-negative. In contrast to the results using faecal or colonic bacteria, studies using anaerobic techniques found that 0.78 of isolates recovered in the caecum of newly-weaned pigs were Gram-negative, with 0.35 and 0.21 of all isolated being represented by *Bacteroides ruminocola* and *Selenomonas ruminantium*, respectively (Robinson, Allison and Bucklin, 1981). This difference is most likely attributable to the undeveloped hindgut of the weaned pig and (or) the differential flow of undigested carbohydrate into the colon according to the type of substrates available to the microflora. Other bacteria found by Robinson *et al.* (1981) were characterized as *Lactobacillus acidophilus, Butyrivibrio fibrisolvens, Peptostreptococcus productus, Bacteroides uniformis* and *Eubacterium aerofaciens.* The difference that exists between bacterial populations attached to the epithelium and those present in the lumen and the predominant culturable bacteria recovered from colonic epithelia is mostly due to Gram-positive organisms, and includes *Streptococcus* sp., *L acidophilus, L fermentus, Bifidobacterium adolescentis* and *Coprococcus* spp. The minority of Gram-negative organisms comprise of *Bacteroides* sp., *Selenomonas ruminantium* and *Fusobacterium prausnitzii* (Robinson *et al.* 1984). Moore *et al.* (1987) reported that the most common isolates were streptococci, which accounted for > 0.25 of all isolates. The predominant groups detected by numerous workers are shown in Table 10.1.

Enumeration of anaerobic bacteria using selective media has allowed characterisation of the microflora in functional terms. Allison, Robinson, Bucklin and Booth (1979) used bacteria counts to develop population profiles, and demonstrated that differences existed between (1) caecal populations of littermates, (2) caecal and colonic populations from the same pig, and (3) luminal versus mucosal populations.

ADAPTIVE RESPONSE OF THE LARGE INTESTINE TO NSP

Whilst changes in microbial populations in response to diet have been well documented in ruminants such as sheep and cattle, less research has been conducted with the pig. This is somewhat surprising given that the omnivorous diet of the pig, together with the high number and immense diversity of bacterial populations in the large intestine of the pig, would be expected to result in a changing microflora as a consequence of dietary alteration. One of the most significant contributors to this area of research has been the group from the USDA-ARS, U.S. Meat Animal Research Centre, Clay Centre, Nebraska. In one of the first controlled investigations in this area, Varel, Pond, Pekas and Yen (1982) determined the effect of feeding low- or high-fibre (achieved with

Table 10.1 A comparison of bacterial isolates in the faeces or large intestine of pigs from different studies

Item	Study of:			
	Moore et al. (1987)	*Salanitro et al (1977)*	*Robinson et al. (1984)*	*Russell (1979)*
Number of pigs	12	6	2	4
Age/weight of pigs	68-114 kg	Sows/gilts	7-9 weeks	20-25 weeks
Area sampled	Faeces	Faeces	Proximal colon	Large intestine
Incubation time, days	5	7	7	3
No. of colonies picked per sample	50	60-70	NR[1]	NR
No. of isolates	1,871	381	83	188
Bacteria isolated, proportion				
Streptococcus sp.	0.29	0.44	0.47	0.15
Lactobacillus sp.	0.14	+[2]	0.08	0.21
Eubacterium sp.	0.12	0.36[3]	0.06	0.26
Fusobacterium sp.	0.12	0	0.14	0
Bacteroides sp.	0.08	+	0.22	0.05
Peptostreptococcus sp.	0.05	0	0	0.17
Bifidobacterium sp.	0.04	0	0	0.02
Selenomonas sp.	0.04	+	0	0
Clostridium sp.	0.03	0.07	0	0.04
Butyrivibrio sp.	0.03	0	0	<0.01
Escherichia sp.	0.02	+	0.01	0.02
Ruminococcus sp.	<0.01	0	0	0.02
Succinivibrio sp.	0.01	0	0	0
Leptotrichia sp.	<0.01	0	0	0
Propionibacterium sp.	<0.01	0.04	0	0.01
Coprococcus sp.	<0.01	0	0	0
Veillonellae sp.	<0.01	+	0	0
Megasphaerum sp.	0	0	0	0.04

1)NR: not reported; 2) +: less than 0.078 of the flora; 3) Some Bifidobacteria may have been grouped with Eubacteria (after Moore *et al.* 1987).

500g dehydrated alfalfa meal/kg) diets to genetically obese- or lean-genotype pigs on bacterial populations at 0, 3 and 8 weeks after commencement of feeding. These authors concluded that the microflora are initially suppressed when exposed to a "high-fibre" diet but later adapt, especially in lean pigs. The adaptation in lean pigs was thought to be a consequence of greater hypertrophy of the large intestine in response to high-fibre diets.

In studies conducted by Varel, Fryda and Robinson (1984a) and Varel, Pond and Yen (1984b), an anaerobic, cellulose-degrading, Gram-negative rod and a Gram-positive coccus, identified as *Bacteroides succinogenes* and *Ruminococcus flavefaciens* respectively, were isolated from pig (26-32 kg) large intestine. This is of interest given that these bacteria are the predominant cellulolytic bacteria found in the rumen. These authors estimated the most probable proportions of these bacteria to be 0.04 and 0.06 for pigs fed the low-fibre and high-fibre diets, respectively. Both Varel *et al.* (1982) and Varel *et al.* (1984b) reported that the number and activity of cellulose-degrading bacteria in the large intestine of growing pigs increased when fed a high-fibre diet such as alfalfa meal. Although Varel *et al.* (1984b) found higher levels of cellulase activity in faeces of pigs fed the high-fibre diet, the levels of cellulase did not necessarily concur with the numbers of cellulolytic bacteria present. Stewart (1997) commented that this may have been caused by changes in composition of the cellulolytic populations or that the substrate for cellulase, i.e., carboxymethyl-cellulose, may not have been entirely appropriate.

Another question of interest, which will be discussed later in this chapter, is whether bacterial activity in response to feeding NSP changes with age. To test this hypothesis, Varel and Pond (1985) fed 20 primiparous sows one of four diets containing increasing levels of alfalfa meal for a total of 98 days. Their results indicated that the cellulolytic flora were increased by prolonged feeding of high-fibre diets and may represent up to 0.10 of the culturable flora. In comparison to growing-finishing pigs, Varel and Pond (1985) calculated that there were 6.7 times more cellulolytic bacteria in the large intestine of sows fed a diet with 400g alfalfa meal/kg than in the large intestine of growing pigs fed 350g alfalfa meal/kg (from Varel *et al.* 1984b). This would at least partly explain why more mature pigs (i.e., sows *vs.* growing pigs) have a greater potential to degrade more cellulosic materials and can extract a greater amount of digestible energy (DE) from these sort of ingredients.

Furthermore, Varel, Robinson and Jung (1987) enumerated xylanolytic and celluloytic bacteria in 8-month-old gilts fed for 86 days on either a control diet or one containing 400g alfalfa meal/kg. Numbers of xylanolytic and celluloytic bacteria increased within three days of feeding the high-fibre diet to levels approaching 10^8/g intestinal contents, which suggests they are significant in the breakdown of NSP in the large bowel. Degradation of xylan and cellulose, measured by *in vitro* DM disappearance after inoculation with faecal samples, was greater for pigs adapted to the high-fibre diet. Varel *et al.* (1987) showed that *Bacteroides succinogenes* and *Ruminococcus flavefaciens*, as well as a species identified as *Bacteroides ruminocola*, were the major bacteria responsible for breakdown of xylan and cellulose. Butine and Leedle (1989) reported that carboxy-methylcellulose-hydrolysing bacteria contributed between 0.20-0.30 of the caecal and colonic biota, depending upon sampling site and chronological development (4 to 11 weeks). Furthermore, Butine and Leedle (1989) found an increase in the proportion of carboxymethyl-cellulose-degrading bacteria in pigs fed a "standard" as opposed to the high-fibre diets used by Varel *et al.* (1982; 1984; 1985; 1987). More recently, a new cellulolytic clostridial species (*Clostridium herbivorans*) capable of degrading plant cell walls to the same extent as the cellulolytic bacteria found earlier by

Varel and colleagues was discovered in pig faecal enrichment cultures (Varel and Pond, 1992; Varel, Tanner and Woese, 1995; Varel, Yen and Kreikemeier, 1995).

The length of time required for pigs to adapt to a diet higher in NSP content is an important issue. In one of the few studies conducted to address this, Longland, Low, Quelch and Bray (1993) fed growing pigs (25-45 kg) on either cereal or semi-purified basal diets supplemented with either high or low levels of sugar-beet pulp or cellulose. The apparent digestibility and retention of nitrogen and apparent digestibility of energy and NSP were measured during the second, fourth and sixth weeks of the trial. The authors concluded that adaptation to the diets in terms of nitrogen and energy balance occurs after seven days; however a longer period of time (three weeks) was recommended for stability of NSP monomers from cellulose, but not from sugar-beet pulp. This is supported by data from the earlier work of Varel and Pond (1985) who enumerated cellulolytic bacteria in the colon of growing pigs (Table 10.2), although the source of DF is obviously important as pigs fed corncobs showed no increase in numbers of cellulolytic bacteria.

Table 10.2 The number of cellulolytic bacteria obtained from faecal samples of sows fed various sources of dietary fibre (after Varel and Pond, 1985)

	Cellulolytic bacteria (x 10^8/g of dry weight) in these diets:				
Days on diet:	*Control*	*20% corncobs*	*40% alfalfa meal*	*96% alfalfa meal*	*SE*[1]
0	14.7	6.0	10.8	14.1	8.7
5	10.1	10.2	34.4	56.5	-
14	22.4	17.5	18.8	24.2	-
21	28.4	16.9	41.3	71.0	-
35	27.8	16.3	105.3	54.9	-
49	24.6	32.8	43.5	76.3	-
70	25.0	9.3	56.5	59.3	-
98	33.3	12.5	50.2	63.7	-
Overall	23.3[a]	15.2[a]	45.1[b]	52.5[b]	5.2

1) SE calculated from error mean square.
[a,b] Values in the same row not having the same superscript differ ($P < 0.05$).

Physiological effects of NSP in the large intestine of the pig

In general terms, soluble NSP increase intestinal transit time, delay gastric emptying, delay glucose absorption, increase pancreatic secretion and slow absorption (e.g., Low,

1985; Dierick, Vervaeke, Demeyer and Decuypere, 1989; Ellis, Roberts, Low and Morgan, 1995; Johansen, Bach Knudsen, Sandstrom and Skjoth, 1996; Nyachoti, de Lange, McBride and Schulze, 1997; Haberer, Schulz and Flachowsky, 1998b), whereas insoluble NSP decrease transit time, enhance water-holding capacity, and assist in faecal bulking (Low, 1985). In addition to these effects, there are documented influences of NSP on dry matter flow and endogenous losses in the gastrointestinal tract. For example, Just, Fernández and Jørgensen (1983) found that a 1% increase in the crude fibre content of the diet from cellulose depressed the digestibility of GE by 1.3% and depressed the utilisation of ME by 0.9%. Schulze, van Leeuwin, Verstegen and van den Berg (1995) reported that ileal dry matter flow and ileal nitrogen (both endogenous and exogenous sources) excretion in pigs at the ileum increased with increasing amounts of NDF in the diet. To balance this (inevitable) loss, Schulze *et al.* (1995) remarked that diets rich in NDF (≈ 200g/kg) should be supplemented with approximately 10g ileal-digestible protein/kg diet. Finally, and in agreement with many authors, Jørgensen, Zhao and Eggum (1996) reported a depressive effect of "dietary fibre" on ileal and, in some cases, faecal digestibility of nutrients.

Non-starch polysaccharides also have a profound effect on gut size and gut development, especially in the large intestine. For instance, Jørgensen *et al.* (1996) fed a low- (59 g dietary fibre/kg DM) and high-(268 g dietary fibre/kg DM) fibre diet to pigs between 45 and 120 kg, with the high-fibre diet being based on pea fibre and pectin. On a per kg empty body-weight basis, pigs fed the high-fibre diet had a heavier stomach, caecum and colon consistent with a greater water-holding capacity, than pigs fed the low-fibre diet (Table 10.3). A similar hypertrophy of gut tissues has been reported in growing pigs by, amongst others, Kass, Van Soest, Pond, Lewis and McDowell (1980), Stanogias and Pearce (1985), Pond, Jung and Varel (1988), Siba, Pethick and Hampson (1996) and Pluske, Pethick and Mullan (1998b). This, in turn, can impact on production characteristics such as dressing percentage. For example, Pluske *et al.* (1998b) fed sorghum-based diets substituted with a source of resistant starch (RS) or soluble NSP (guar gum) to pigs between 25 and 55 kg. These authors found a positive linear relationship between the weight (full or empty) of the large intestine and the daily intake of NSP+RS. In turn, carcass yield was reduced by 0.0025 units absolute for each additional gram of NSP+RS intake. In general, changes in the size and weight of gastrointestinal organs are believed to reflect a hypertrophy of particular tissues of the organ in response to the increased amount of work performed by these organs in drying, mixing, shaping, moving and expelling large quantities of undigested dietary residues. There are also marked increases in the number of crypts in the colon consistent with increased crypt cell production rate (McCullough, Ratcliffe, Mandir, Carr and Goodlad, 1998), an effect that requires the presence of a large intestinal microflora.

Table 10.3 Effects of level of inclusion of dietary fibre on gut characteristics and apparent ileal and faecal digestibilities (after Jørgensen *et al.*, 1996)

Item	*Level of Dietary Fibre*	
	Low	*High*
Body weight, kg	122	129
Contents of GI tract, g/kg	29	82
Empty body weight (EBW), kg	118	119
GI tract, g/kg EBW		
Stomach	5.7	7.3
Small intestine	15.6	16.2
Caecum	1.6	2.8
Colon	8.7	17.2
GI tract length, m/kg EBW		
Small intestine	0.15	0.16
Large intestine	0.05	0.06
Ileal digestibility of:		
Crude protein	0.76	0.64
Energy	0.87	0.58
NSP	0.09	0.01
Faecal digestibility of:		
Crude protein	0.92	0.74
Energy	0.94	0.82
NSP	0.59	0.77
Digesta dry weight at ileum, kg/d	0.270	1.073

Changes in organ size in response to feeding high levels of NSP are likely to have an impact on energy metabolism as visceral organs have a high rate of energy expenditure relative to their size. For example, Yen, Nienaber, Hill and Pekas (1989) found that organs drained by the hepatic-portal vein, including the large intestine, consumed a disproportionately high amount (≈ 0.25) of whole-body maintenance expressed as O_2 consumption, although it represents < 0.04 of body weight. Given this, then it is possible that some of the reductions in efficiency of feed use seen when pigs are fed "high fibre" diets may be related to increased basal heat production in addition to reduced energy (and amino acid) digestibility (Varel and Yen, 1997).

DIGESTION OF NSP IN THE LARGE INTESTINE

Digestibility of NSP varies according to the botanical origin of the plant material and the physiological stage (i.e., age/size) of the pig. The net disappearance of carbohydrates in the hindgut is positively related to the intake of DF, while the disappearance of protein tends to be independent of DF intake, indicating the importance of carbohydrates as a fuel for the microflora (Jørgensen *et al.* 1996). Bach Knudsen, Jensen and Hansen (1993) showed that 0.92 of the non-starch carbohydrates were fermented in the caecum and proximal and ascending colon of pigs. Any starch remaining at the end of the ileum is fermented fully by the end of the colon. Buraczewska, Schulz, Gdala and Grala (1988) reported that the degree of fermentation of NDF in the small intestine varied between 0.10 and 0.32 dependent upon the source of NDF, supporting the concept that pre-caecal digestion of NSP occurs and, depending on the NSP source, to a considerable extent (see later). Amongst others, Fernández, Jørgensen and Just (1986), Noblet and Shi (1993), Shi and Noblet (1993) and Noblet and Bach Knudsen (1997) reported digestibility to be higher in adult sows than growing pigs, with this difference being dependent upon the nature of the NSP source. For example, Noblet and Bach Knudsen (1997) reported that digestibility coefficients for sugar-beet pulp (SBP) between growing pigs (66 kg) and adult sows (239 kg) were negligible, for wheat bran moderate, but for maize fibre high (Table 10.4). Shi and Noblet (1993) reported that hindgut fermentation contributed 0.16 and 0.25 of total digestible energy in growing pigs (45 kg) and adult sows (208 kg), respectively; the authors suggested that this might be attributed to differences in chemical composition of the NSP fractions influencing the physico-chemical properties of the NSP in the gastrointestinal tract. This, in turn, must be related to microbial activity. For example, SBP has a high concentration of pectic substances (159g/kg DM) that enhance attachment of the microflora and, in addition, has a high water-holding capacity that increases the surface area available for attachment (Noblet and Bach Knudsen, 1997).

Table 10.4 Digestibility of NSP from three fibre sources (wheat bran, maize fibre, sugar-beet pulp) in growing pigs (66 kg) and adult sows (239 kg) (after Noblet and Bach Knudsen, 1997)

Item	*Wheat bran*		*Maize fibre*		*Sugar-beet pulp*	
	G[1]	*S*[2]	*G*	*S*	*G*	*S*
Weende crude fibre	0.35	0.38	0.45	0.79	0.78	0.82
NDF	0.48	0.51	0.39	0.81	0.84	0.89
ADF	0.26	0.33	0.38	0.82	0.88	0.93
Hemicellulose	0.58	0.62	0.38	0.82	0.88	0.93
NCP[3]	0.54	0.61	0.38	0.82	0.89	0.92
arabinose	0.37	0.45	0.48	0.86	0.96	0.97
xylose	0.62	0.67	0.31	0.81	0.48	0.55
uronic acids	0.32	0.35	0.43	0.80	0.95	0.97
Cellulose	0.25	0.32	0.38	0.82	0.87	0.91
Insoluble NSP	0.43	0.50	0.34	0.81	0.87	0.91
Total NSP	0.46	0.54	0.36	0.82	0.89	0.92
Dietary fibre	0.38	0.46	0.32	0.74	0.82	0.86

1) G = growing pigs; 2) S = adult sows; 3) NCP = non-cellulosic polysaccharides.

In another study comparing the digestibility of SBP with that of wheat straw added to the diets of gestating sows, Yan, Longland, Close, Sharpe and Keal (1995) reported higher digestibilities of NSP monomers from SBP than from wheat straw. Level of feeding, i.e., 1.0 or 1.5 times maintenance, had no effect on digestibility of NSP. This may be due to a more rapid transit rate through the gastrointestinal tract, in addition to greater lignification in wheat straw and a higher glucose and xylose ratio (Table 10.5). These data are consistent with the observations of Varel and colleagues showing higher numbers of cellulolytic bacteria in the colon of growing pigs fed various sources of NSP.

Table 10.5 The apparent digestibility of NSP constituents from diets containing sugar-beet pulp or wheat straw fed to pregnant sows at 1.0 or 1.5 times their estimated maintenance energy requirement (after Yan *et al.*, 1995)

	Sugar-beet pulp		*Wheat straw*		
NSP constituent	*1.0*	*1.5*	*1.0*	*1.5*	*SED*[1]
Arabinose	0.85[a]	0.85[a]	0.66[b]	0.63[b]	0.04
Xylose	0.62[a]	0.57[a]	0.43[b]	0.44[b]	0.05
Mannose	0.89[a]	0.88[a]	0.80[b]	0.81[b]	0.02
Galactose	0.84[a]	0.84[a]	0.67[b]	0.65[b]	0.02
Glucose	0.83[a]	0.79[a]	0.57[b]	0.52[b]	0.03
Uronic acids	0.93[a]	0.92[a]	0.66[b]	0.62[c]	0.02
Total NSP	0.82[a]	0.79[a]	0.55[b]	0.52[b]	0.03

1)Standard error of difference between means.
[a,b,c] Values in rows not having the same superscript differ ($P < 0.05$).

Although NSP are digested in the large intestine, it remains to be established whether they make a contribution towards the overall energy balance of the pig. Jørgensen *et al.* (1996) estimated that the relative value of energy derived from hindgut fermentation in growing pigs was 0.73 compared to digestion in the small intestine. McBurney and Sauer (1993) estimated that 0.155-0.295 of the ME was absorbed from the hindgut of pigs fed increasing levels of pea fibre. Similarly, Yen, Nienaber, Hill and Pond (1991) determined that absorption of VFA accounted for 0.238 of whole-animal heat production, whilst Rérat, Fislewicz, Guisi and Vaugelade (1987) estimated that absorbed VFA could provide 0.30 or 0.20 of maintenance energy requirement in 60-kg pigs after 12 or 24 hours, respectively, of food deprivation. Zhu, Fowler and Fuller (1993) calculated that in a cereal-based diet fed to growing pigs (45-48 kg), microbial fermentation contributed up to 0.18 of total DE. Inclusion of 300 g SBP /kg diet resulted in fermentation supplying up to 0.34 of total dietary DE. In general, fermentation in the large intestine yields less energy than that for carbohydrates digested and absorbed in the small intestine as monosaccharides due to additional losses as H_2, CH_4 and fermentation heat, together with a lower efficiency of utilisation of VFA in intermediary metabolism.

Viscosity-inducing properties of NSP in the pig

In order to investigate the effects of viscous-forming properties (if any) of NSP in the pig, it is first necessary to reconsider briefly the influence of NSP in the chicken. There appears little doubt that the soluble NSP component of barley (ß–glucans) and wheat (arabinoxylans) can influence the digestion of nutrients such as starch, protein and fat (e.g., Choct and Annison, 1992), the apparent metabolisable energy (AME) content (see Annison, 1993; Bedford and Schulze, 1998), as well as affecting litter quality (Classen, 1996). However, similar effects of NSP in pigs are less clear. In young chicks, the increase in viscosity in the small intestine associated with the solubilisation of ß–glucans and arabinoxylans is thought to be a major factor contributing to their anti-nutritive effects, although intact cell walls may also act as a physical barrier to the endogenous enzymes and hence reduce utilisation of nutrients that are encapsulated within endospermal cells (Burnett, 1966; Campbell and Bedford, 1992; Classen, 1996; Bedford and Schulze, 1998). Furthermore, it is reasonably well established that the addition of exogenous enzymes (e.g., xylanases, ß–glucanases) to cereal-based diets fed to young chicks increases the digestibility of these diets and enhances the AME value of the feedstuff(s).

Choct and Annison (1990) provided one such summary of the relationship between NSP and performance in poultry; a relationship between the efficiency of GE utilisation (expressed as the coefficient of AME/GE) in chicks (35 days) and the sum of the arabinoxylans plus ß–glucans in cereals was established (Figure 10.1). This relationship shows that an increasing concentration of NSP in cereals has a detrimental effect on the metabolisability of energy in young chicks. Although this relationship may hold true across cereals, within a cereal there is less evidence to support this negative relationship. Using Australian wheats for example, Wootton, Acone and Wills (1995) reported a *positive* relationship between AME content and total pentosan levels. Moreover, and using the extensive data compiled with Australian wheat fed to birds by Rogel, Annison, Bryden and Balnave (1987) and Annison (1991), the coefficient of determination for the relationship between pentosan level and AME is 0.26, i.e. only 0.26 of the variation in AME content could be explained by pentosan level.

Dusel, Kluge, Gläser, Simon, Hartmann, Lengerken and Jeroch (1997) and Dusel, Kluge, Jeroch and Simon (1998) attempted to establish relationships using German wheat between *in vitro* extract viscosity, wheat NSP, nutrient digestibility, and AME content in broiler chickens. Their extensive studies showed a close relationship between levels of soluble arabinoxylans in wheat and *in vitro* extract viscosity which, in turn,

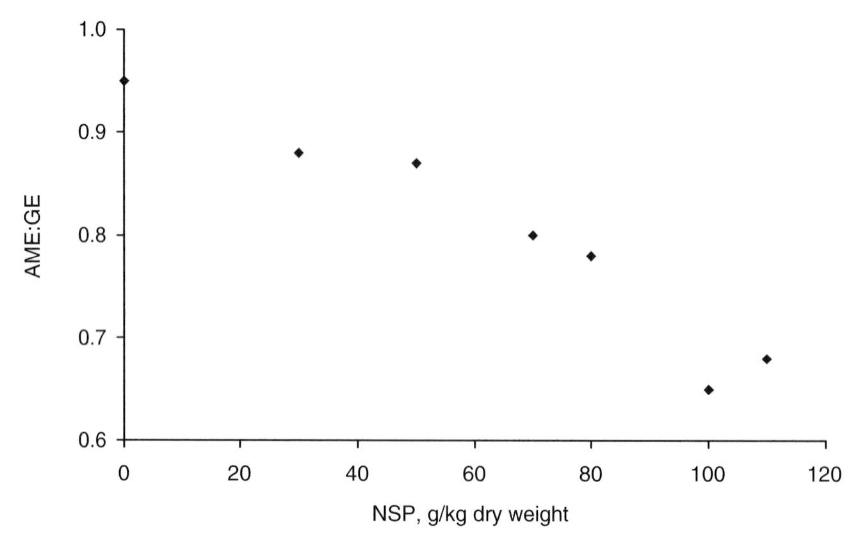

Figure 10.1 The relationship between energy metabolisability (AME/GE) of cereals and their NSP composition (arabinoxylans + ß-glucans; dry weight) ($r = -0.98$, $P < 0.001$) (After Choct and Annison, 1990).

was related to AME content. These results are in agreement to those of Annison (1991). Interestingly, Dusel *et al.* (1997) also reported a positive relationship between the rainfall that fell in the growing period and *in vitro* extract viscosity.

Wheat is also a major component in many pig diets, however very little work appears to have been conducted along the same lines as that undertaken in young birds. In one of the few studies relating *in vitro* viscosity measurements to nutrient digestibility in the pig, Lewis, McEvoy and McCracken (1998) fed 12 different varieties of wheat grown in the one location in Northern Ireland to growing pigs to investigate the relationship between wheat "quality" (as measured *in vitro* by extract viscosity) on ileal and faecal nutrient digestibility. These authors reported no significant relationships between viscosity and either ileal or faecal digestibility, suggesting that, at least in this study, viscosity measurements alone cannot provide a suitable basis for predicting the nutritive value of wheat for pigs. These data provide evidence that pigs and poultry differ in their responses to NSP from cereals.

DIGESTIBILITY OF NSP IN BARLEY AND WHEAT, AND THE USE OF ENZYMES

In the pig, concern is predominately with anti-nutritive effects of NSP contained in barley and wheat, although it is recognised that other cereals (e.g., rye, maize, sorghum, triticale) also contain varying levels of indigestible carbohydrate. Similar to what has happened in poultry, considerable research effort has been expended to enhance the utilisation of NSP in pig diets by the use of added enzymes, on the basis that NSP form viscous solutions in the small intestine of the pig. However, results in general have been equivocal and, where positive, have largely been restricted to the younger pig. Furthermore, accurate viscosity measurements are more difficult owing to limitations in animal numbers, and because the viscosity of most samples is so low that they often approach the detection limits of the apparatus (Bedford and Schulze, 1998).

Nevertheless, Bedford, Patience, Classen and Inborr (1992) reported that the addition of ß–glucanase to a barley-based diet fed to weanling pigs improved performance, although a xylanase added to a rye-based diet had no effect. Bedford *et al.* (1982) remarked that the digesta from weanling pigs was more "watery" compared to that found in chickens, suggesting that diffusional constraints, if any, would be less severe in pigs. In fact, these authors reported *increased* viscosities in the presence of an added pentosanase in rye-fed weanling pigs, but levels overall were far less than those reported for birds. Pigs differ physiologically from young chicks in that the digesta generally tends to have a lower DM (higher water) content. The DM content of the digesta in poultry is about 800 g/kg (Bedford, Classen and Campbell, 1991) compared to around 900 g/kg for pigs (Baas and Thacker, 1996). Since viscosity induced by ß–glucan has been reported to be logarithmically related to concentration (Bedford and Classen, 1992), simple dilution can essentially eliminate the viscosity problem and the associated constraints on luminal diffusion (Campbell and Bedford, 1992).

Graham, Fadel, Newman and Newman (1989) reported only small (but statistically significant) improvements in starch (0.926 to 0.943) and ß–glucan (0.957 to 0.971) digestibility in the presence of a ß–glucanase when added to a barley-based diet fed to 80-kg pigs. Other studies (Graham, Hesselman and Åman, 1986a; Graham, Hesselman, Jonsson and Åman, 1986b; Fadel, Newman, Newman and Graham, 1988) have also reported a high pre-caecal digestibility of mixed-linked ß–glucans from barley, a result thought to be a consequence of enhanced microbial activity and (or) residual ß–glucanase activity from the grain. This degradation is likely to play a role in rendering nutrients available for digestion in the small intestine, and may in part explain why the response to added ß–glucanase in barley-based diets for pigs is much less than that seen in chickens. In addition, it is thought that ß–glucan degradation increases with age (Graham *et al.* 1986b), an effect consistent with increased production benefits for younger (weaner) pigs (Bedford *et al.* 1992).

In contrast, Haberer, Schulz, Aulrich and Flachowsky (1998a) and Haberer *et al.* (1998b) fed growing pigs (26.5 kg) a diet based on barley, rye, wheat bran and soybean meal to study the effects of enzyme addition (xylanase + ß–glucanase) on digesta composition and nutrient and NSP disappearance prior to the caecum. The diet contained 189g NSP/kg DM (0.23 soluble) with 94 g arabinoxylan/kg DM, 45 g ß-glucan/kg DM and 29 g cellulose/kg DM. From their extensive work, there were no marked differences in nutrient content, NSP content or percentage of NSP fractions due to enzyme addition

(Haberer *et al.* 1998a), although insoluble ß–glucans showed a significant increase in disappearance rate in the small intestine in the presence of enzyme (Haberer *et al.* 1998b).

Baas and Thacker (1996) tested the efficacy of five commercial ß-glucanase preparations included in a hulless barley-based diet fed to growing-finishing pigs between 30 and 90 kg liveweight. The barley was of moderate viscosity (3.38). Supplementation of diets failed to improve average daily gain, feed intake or feed efficiency significantly. Nevertheless, anecdotal evidence and evidence presented by enzyme manufacturers suggests that in some instances the use of exogenous enzymes does provide a positive result.

The pig and chicken differ markedly in their digestive physiology, and this would also explain differences between these two species in their response to NSP. In chicks, the crop provides a comparatively beneficial environment for enzyme activity, at least for those enzymes operating in the pH optima 4-5 range. The stomach of the pig is more hostile with a lower pH, such that activity of most enzymes would be sub-optimal. For example, Baas and Thacker (1996) reported that across five commercial enzymes tested, 0.52 and 0.26 of the initial enzyme activity was detectable in duodenal digesta at 60 and 240 minutes after feeding, respectively. Although passage through the acidic stomach decreases enzyme activity, residual ß-glucanolytic activity does remain that provides the opportunity for these products to degrade ß-glucan in the small intestine and perhaps improve pig performance.

There is significant variation in the pre-caecal digestion of ß–glucan in cereals, with values for ß–glucan in oats ranging from 0.170-0.820 (Bach Knudsen and Hansen, 1991; Bach Knudsen *et al.* 1993; Johansen, Bach Knudsen, Wood and Fulcher, 1997) and in barley from 0.70-0.97 (Weltzien and Aherne, 1986; Fadel *et al.* 1988; 1989). Bach Knudsen and Canibe (1997) reported that it was primarily the soluble ß–glucan fraction that is degraded (0.67) compared with 0.21 for insoluble ß–glucan. This high level of degradation is caused presumably by the relatively high solubility of the linear ß–glucan and the high density of microorganisms present in the distal region of the small intestine. In contrast, Bach Knudsen and Canibe (1997) reported almost quantitative recovery of arabinoxylan from wheat and oats, an effect contributed to the cross-linking and highly branched structure of arabinoxylan that requires a more complex enzyme system for degradation. At the faecal level, there was nearly complete disappearance of soluble ß–glucan and arabinoxylan but considerable recovery of insoluble NSP.

The ability of the isolated soluble NSP sources to increase luminal viscosity has generally been considered a major determinant of delayed absorption. However, and using oat bran fed to pigs, Johansen *et al.* (1997) found low viscosities in jejunal contents, the region where absorption is most active; as suggested, dilution with pancreatic and bile juice may partly explain this, but the low and variable viscosity found in digesta of pigs was attributed to enclosure in intact cell walls leading to only partial solubilisation, and by depolymerisation of ß–glucan.

NSP IN VEGETABLE PROTEIN FEEDSTUFFS FOR PIGS

Focus to date on the anti-nutritive effects of NSP in pig production has focused largely on the cereal component. However, vegetable protein sources used in pig production (e.g., lupins, peas, beans, soybean, canola) also contain considerable levels of NSP that have the potential to exert physiological effects in the pig. In legumes for example, the NSP content varies from 270 to 420 g/kg DM in *Lupinus* species, from 160-200 g/kg DM in peas, and 173-181 g/kg DM in faba beans (Bach Knudsen, 1997; Gdala, 1998). The NSP in legumes is characterised by a high level of structural NSP, constituting 0.73-0.84, 0.27, and 0.25 of the total carbohydrate content in lupins, faba beans and peas, respectively (Gdala, 1998). Given the high level of NSP in these vegetable proteins, there is potential for them to influence gastrointestinal function in pigs, ranging from anti-nutritive activity to delayed gastric emptying.

Working with Australian lupins, Van Barneveld, Baker, Szarvas and Choct (1995a; 1995b) added 0, 120, 240 or 360 g lupin kernels/kg to sorghum-based diets and measured ileal and faecal energy digestibility, and the apparent digestibility of essential amino acids. The inclusion of lupin kernels at graded levels linearly decreased the ileal digestibility of dry matter, gross energy and digestible energy, but had no influence on faecal digestibility and faecal digestible energy content (Table 10.6). With respect to amino acids, decreases in apparent digestibility were observed for some amino acids (e.g., lysine) but not for others (e.g., valine). Using an NSP isolate extracted from *L. angustifolius*, Van Barneveld *et al.* (1995c) stated that decreases in ileal dry matter, energy and lysine digestibility were most likely attributable to increased viscosity, especially when the equivalent of > 120 g lupin kernels was included per kg diet. Lupins are commonly included in the diets of growing-finishing pigs in Australia at levels > 250g/kg, indicating that these observations are "real" effects that have a commercial bearing.

In addition, and although by definition not strictly a "non-starch polysaccharide", there is evidence that higher order oligosaccharides, i.e., α-galactosides (raffinose, stachyose, verbascose), contained in legumes exert anti-nutritive effects also. Gdala, Jansman, Buraczewska, Huisman and van Leeuwin (1997) used an exogenous α-galactosidase added to diets containing ≈ 300g lupins/kg and observed improvements in ileal oligosaccharide and amino acid digestibility. Van Barneveld, Olsen and Choct (1996) extracted 0.73 and 0.67 of oligosaccharides with ethanol from *Lupinus angustifolius* and *Lupinus albus* and reported an improvement in DE of 0.5 MJ and 0.7 MJ, respectively. In a follow-up study, Van Barneveld, Olsen and Choct (1997) investigated the apparent ileal digestibility of amino acids and found that ethanol-extraction improved the digestion of all amino acids by 0.05-0.10 absolute in *L. angustifolius*, and 0.05-0.08 absolute in *L. albus*. Given that oligosaccharides from lupins were not viscous forming, the negative effects of oligosaccharides were attributable to osmotic events occurring in the gastrointestinal tract.

Table 10.6 The influence of graded levels of lupin kernels on ileal and faecal digestibility in growing pigs (after Van Barneveld *et al.*, 1995b)

	Level of kernel inclusion, g/kg						
Item	*0*	*120*	*240*	*360*	*Diet*[1]	*Linear*[2]	*SEM*[3]
Ileal digestibility of:							
Dry matter	0.85	0.82	0.77	0.67	**	***	0.027
Gross energy	0.90	0.87	0.83	0.76	**	***	0.021
DE, MJ/kg	16.85	16.47	15.84	14.47	*	**	0.399
Faecal digestibility of:							
Dry matter	0.90	0.91	0.90	0.90	NS	NS	0.008
Gross energy	0.94	0.94	0.93	0.93	NS	NS	0.006
DE, MJ/kg	17.72	17.67	17.64	17.67	NS	NS	0.119

1) Comparison between the four diets; 2) Test for linear effects; 3) Standard error of the mean.
$P<0.05$; ** $P<0.01$; *** $P<0.001$; NS: not significant.

As with cereals, work has been conducted with pigs using enzymes to improve the digestibility of NSP in vegetable protein sources. To date, there is little information to support the notion that NSP in vegetable proteins is enhanced by enzyme supplementation (see review by Gdala, 1998). Needless to say, there is considerable scope for the design of specific enzyme "cocktails" to degrade the NSP component of vegetable protein sources used in pig production.

Non-nutritional effects of NSP

The inclusion of NSP in pig diets may have effects other than nutritional. In the case of sows for example, the use of high-fibre feeds has gained popularity because they may not only reduce diet cost, but also reduce the incidence of stereotypic behaviours (e.g., bar-biting and chain-chewing) and enhance reproductive performance. For example, in a study where pregnant gilts were fed different levels and sources of DF over two parities, Robert, Matte, Farmer, Girard and Martineau (1993) reported that provision of "high fibre" diets (either added wheat bran+corncobs, or oat hulls+oat) increased resting time by 10% (associated with decreased feeding motivation; Mroz, Partridge, Mitchell and Keal, 1986) and reduced the frequency of stereotypic activities. Matte, Robert, Girard, Farmer and Martineau (1994) found that sows spent 12.8% more time resting (> 3h/day) and showed a reduction of 50% in stereotypic behaviours when fed wheat bran and corncobs. Brouns, Edwards and English (1994) reported similar results. Matte *et al.* (1994) also reported a 20% increase in total litter weight at 3 days to 8 weeks of age from second-parity sows fed this diet. Earlier, Ewan, Crenshaw, Crenshaw, Cromwell, Easter, Nelssen, Miller, Pettigrew and Veum (1996) reported that wheat straw added to gestation diets increased the number of pigs born and weaned in a co-operative study using 141 sows over three parities. Münchow, Bergner, Seifert, Schönmuth and Baraband (1982) reported an extra 1.5 piglets born per sow when hydrolysed or pure straw meal replaced cereal in a gestation diet, while Mroz *et al.* (1986) reported higher ($P > 0.05$) numbers of piglets born alive, greater litter weights, and increased milk fat content, when oat hulls were added to the diet of pregnant sows. In contrast, Yan *et al.* (1995) found no improvement in these indices after sows had been fed sugar-beet pulp during pregnancy, although sow numbers used in this study were low.

Another benefit of feeding (fermentable) NSP relates to effects on the pH and NH_3 emission of slurry from pigs. Urea from urine is converted into NH_3 and CO_2 by the urease present in faeces, and this can present an environmental odour problem. Recent work by Canh, Aarnink, Verstegen and Schrama (1998) and Canh, Sutton, Aarnink, Verstegen, Schrama and Bakker (1998) has demonstrated that by shifting nitrogen excretion from the urine to bacterial protein in the faeces by feeding fermentable NSP sources such as soybean hulls and sugar-beet pulp, substantial decreases in slurry pH and NH_3 emissions can be achieved.

Interactions between dietary fibre and swine dysentery

The final section of this chapter examines the relationship between carbohydrates entering the large intestine and the expression of an economically important disease in pig production, swine dysentery (SD). Swine dysentery is an infectious transmissible typhlcolitis caused by the Gram-negative, anaerobic spirochaetal bacterium *Serpulina hyodysenteriae* that mainly affects growing-finishing pigs (Alexander and Taylor, 1969; Harris and Lysons, 1992). Pigs become affected with SD following ingestion of dysenteric faeces containing the spirochaete, with an inoculum of 10^5 colony forming units (CFU) being sufficient to produce SD. The organism displays a chemotactic response to mucin, and its active motility allows it to penetrate the mucus layer and to invade colonic crypts where it stimulates an outpouring of mucus into the lumen (Hampson and Trott, 1995). In severe cases, blood is seen in the faeces. As a consequence, colonic function, particularly H_2O, HCO_3^- and electrolyte absorption, is impaired, and this contributes to the diarrhoea observed.

Following failure of the spirochaete to produce disease in germ-free pigs (Meyer, Simon and Byerly, 1974), it became apparent that the aetiology of SD was complex. It was shown subsequently that the activity and composition of other components of the intestinal microbiota influenced both colonisation by *S. hyodysenteriae* and the severity of the associated disease (Hampson and Trott, 1995). In pigs with SD, the proportion of Gram-negative isolates associated with the colonic epithelium tends to increase (Robinson *et al.* 1984). Microorganisms identified as acting in synergy with the spirochaete and inducing SD in gnotobiotic pigs include *Bacteroides melaninogenicus, B. fragilis, Fusobacterium nucleatum, F. necrophorum, Clostridium* spp., *Listeria denitrificans*, and certain strains of *B. vulgatus* (Hampson and Trott, 1995). Although the mode of action of the associated bacteria is not known, they may act to facilitate colonisation of *S. hyodysenteriae* by providing various growth factors or essential nutrients for the spirochaete (Meyer, 1978), and (or) might increase lesion production in the large intestine (Whipp, Robinson, Harris, Glock, Mathews and Alexander, 1979).

To date, the control of SD in infected herds has relied heavily on the use of antimicrobials and (or) vaccines. This approach is expensive, resistant strains of *S. hyodysenteriae* have appeared, and the public is becoming increasingly concerned about the high level of usage of antimicrobials in animal production. Alternatively, the disease can be eradicated from piggeries by the combined use of medication, destocking, and disinfection. However, this strategy is not always successful, it is expensive and, even where it has worked, there remains the risk of reintroduction of infection.

Despite evidence that around 0.33 of pig herds in Western Australia are infected with the causative agent of SD, it is known that disease occurs much less commonly than this (Hampson, Cutler and Lee, 1992; Mhoma, Hampson and Robertson, 1992). The reasons for this are not clear, but Prohaszka and Lukacs (1984) reported that SD ceased to be a problem on a commercial pig unit in Eastern Europe when maize silage was used. This effect was thought to be mediated by the antibacterial effect of increased VFA production on *S. hyodysenteriae*. A similar finding but with a different bacterium (*Clostridium difficile*) was reported by May, Mackie, Fahey, Cremin and Garleb (1994). These authors added highly-fermentable fibre sources to an inoculum of pig faeces and observed increased VFA production, decreased gut pH, and suppressed growth and toxin elaboration by this bacterium that is responsible for pseudomembranous colitis in man.

Research at Murdoch University in Western Australia earlier this decade attempted to replicate the work of Prohaszka and Lukacs (1984); however the opposite effect was found. Feeding a diet based on cooked white rice and animal protein sources, which resulted in low fermentative activity in the hindgut and hence decreased production of VFA, completely protected growing pigs from SD following experimental infection with a virulent strain of *S. hyodysenteriae* (Siba *et al*., 1996). Subsequent experiments (Pluske, Siba, Pethick, Durmic, Mullan, and Hampson 1996; Pluske, Durmic, Pethick, Mullan and Hampson, 1998a) have confirmed these data and demonstrated that diets higher in fermentable carbohydrate sources, such as soluble NSP and resistant starch (RS), predispose experimentally-infected pigs to SD (Figure 10.2).

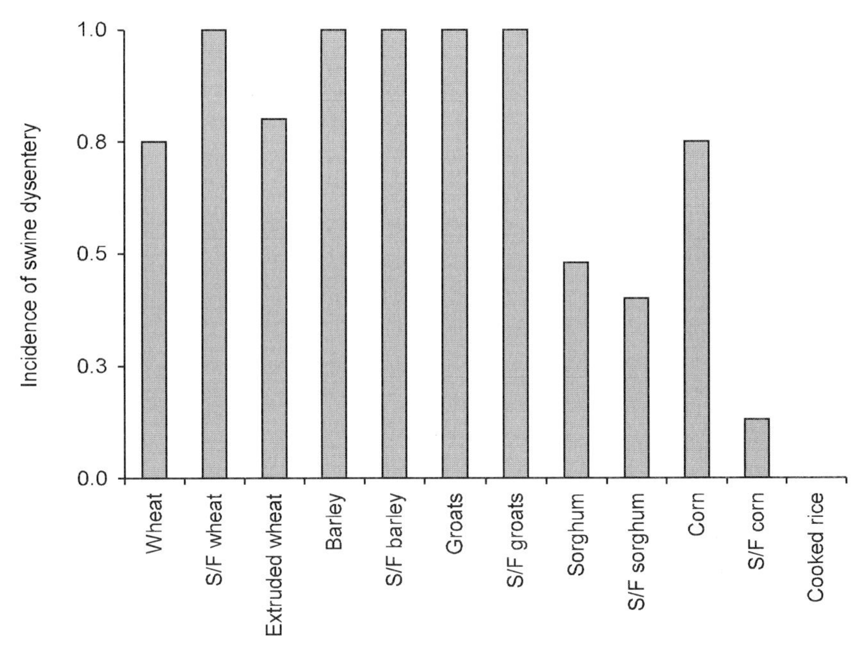

Figure 10.2 The incidence of swine dysentery in pigs fed a range of diets following experimental infection with *S. hyodysenteriae* (After Pluske *et al.* 1996).

Given the relationship between diet type and incidence of SD described in Figure 10.2, Pluske *et al.* (1996) demonstrated further that diets containing less than about 10 g soluble NSP/kg (≈ 50 g total NSP/kg) showed a reduced incidence of the disease (Figure 10.3). There was no correlation between the amount of insoluble NSP or the measured *in vitro* level of RS in diets and the incidence of SD. However, feeding some diets containing less than 10 g soluble NSP/kg (e.g., a diet based on steam-flaked sorghum) was still associated with a high incidence of disease. This was found to relate to higher levels of RS in these diets which appeared to exacerbate the clinical expression of SD (Pluske *et al.*, 1996). The role of these carbohydrates in the aetiology of SD has been studied further by adding them to a diet which completely prevents the development of the disease, i.e. cooked white rice with a supplement based on animal protein sources. In this way, the precise contribution of NSP to SD could be ascertained.

In a study by Pluske *et al.* (1998a), a source of soluble-NSP (guar gum, 100 g/kg), insoluble-NSP (oaten chaff, 70 g/kg), RS (200 g/kg, fed as Novelose™, a high-amylose maize starch, and equivalent to 76 g RS /kg), and a combination of soluble-NSP and RS (50 and 100 g/kg, respectively), was added to a diet containing cooked white rice and animal protein. A "control" diet containing rice and the animal protein supplement was also fed. Spirochaetes only colonised the colonic epithelium of pigs fed RS, guar

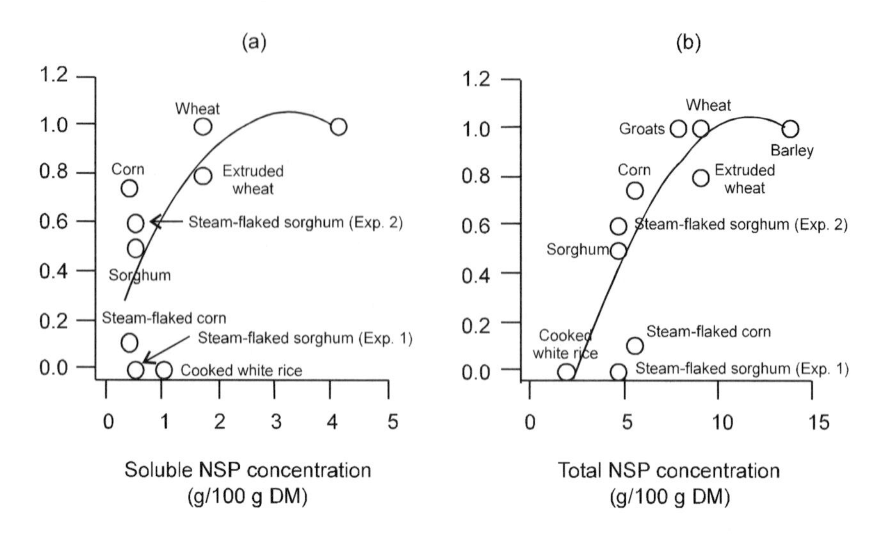

Figure 10.3 The relationship between the incidence of swine dysentery (proportion) (y-axis) and (*a*) soluble NSP concentration (x-axis) ($y = 0.0952 + 56.98x - 8.47x^2$, $R^2 = 0.561$, $P = 0.016$), and (*b*) total NSP concentration (x-axis) ($y = -0.5797 + 26.85x - 1.10x^2$, $R^2 = 0.712$, $P = 0.002$), in pigs fed different diets (After Pluske *et al.* 1996).

gum, and RS plus guar gum; however the clinical expression of SD only occurred when pigs were fed guar gum or RS plus guar gum. Expression of SD was associated with a lower pH in the colon and a heavier large intestine when expressed as a percentage of empty body weight. These data confirm our previous studies, and implicate the presence of soluble NSP in the clinical expression of SD. However, the contribution of RS to SD could not be confirmed despite spirochaetes colonising the epithelium of the large intestine.

From these series of experiments, we suggest a link between the more fermentable fraction of carbohydrates and the clinical expression of SD. Also, and given the data of Siba *et al.* (1996), the oligosaccharides contained in legumes such as lupins most likely play a role in the aetiology of SD. However, "classical" indices of the rate and extent of hindgut fermentation, such as pH, VFA pool size and ATP concentrations, have not necessarily supported our findings. There are several possible reasons for this phenomenon. No one index of fermentation can be considered in isolation and, instead, a suite of measurements is more appropriate to allow full interpretation of fermentation status. Using pH of the digesta as an indicator of fermentation can have limitations since the final pH will depend on the pKa of the VFA accumulating and, also, any factors which change the buffering capacity of the chyme will alter the final pH achieved at any given VFA concentration. The production of VFA allows the microorganisms in the hindgut to produce ATP by substrate level phosphorylation for cellular function. Given this logic, a high concentration (or total pool size) of VFA in combination with a lower concentration of ATP indicates that VFA is accumulating in the caecum and colon due to an imbalance between absorption and production. Guar gum, a soluble NSP, may have altered the functionality of the hindgut by reducing the efficiency of VFA absorption.

To date, our research has not unraveled the precise mechanism(s) whereby some diets are more protective than others, and in particular why a diet based on cooked white rice completely abrogates the disease. The protective effect of these diets may be attributable to a number of factors including: -

- an indirect effect on mucin production and (or) composition,
- altered viscosity,
- mobility of *S. hyodysenteriae* in the mucosal lining of the colonic epithelium,
- modifications to metabolic pathways,
- the ability of *S. hyodysenteriae* to express haemolysins and (or) lipopolysaccharides and effect inflammation of the epithelium, and (or)
- the dry matter content of colonic contents that may inhibit spirochaetal survival in the large intestine.

As an extension to this work with SD, research at Murdoch University in Australia is also investigating dietary means of controlling porcine intestinal spirochaetosis (PIS), a diarrhoeal disease of younger pigs associated with a mild colitis, and which is caused by *Serpulina pilosicoli* (Hampson, Robertson, Oxberry and Pethick, 1998).

MICROBIOLOGICAL INTERACTIONS AND SWINE DYSENTERY

Little or no research has been conducted examining possible interactions between *S. hyodysenteriae* and other microflora of the large intestine with the type of diet fed, and how this could then predispose pigs to SD. By definition, SD is a mixed synergistic infection (Meyer *et al.* 1974; 1975) and infective combination of resident microorganisms of low pathogenic potential, interacting with the spirochaete and with each other to represent an inherent pathogenic potential. Furthermore, it is well recognised that different components of the microflora of the large intestine can either enhance (Whipp *et al.* 1979) or reduce (Suenaga and Yamazaki, 1986) colonisation by *S.hyodysenteriae*, and subsequently can affect the incidence of SD. It is possible, therefore, that the protective effects of diets operated through some specific or unspecific alteration(s) in the resident microflora. In this regard, a component of our research investigating nutritional influences on SD has focused on changes in bacterial populations in response to different diets as a possible mechanism for unraveling the aetiology of SD under different nutritional conditions (Durmic, Pethick, Pluske and Hampson, 1998).

Durmic *et al.* (1998) showed that pigs fed a diet based on cooked white rice and animal protein sources, with a source of (largely) insoluble NSP added in the form of oaten chaff, had a lower total number of colonic bacteria, a predominance of Gram-positive populations, there was a failure of *S.hyodysenteriae* to colonise, and SD did not develop. Synergistic bacteria (*Fusobacterium necrophorum* and *Fus. nucleatum*), which have been reported to facilitate colonisation by *S.hyodysenteriae*, were found only among isolates from pigs fed a diet with added soluble NSP and RS. These pigs developed SD. However, these bacteria were not observed in pigs fed a diet with soluble NSP, and in which 0.80 of infected pigs came down with SD. These pigs also had a lower anaerobe count and a predominance of Gram-positive organisms. Addition of RS, however, increased total bacterial counts and stimulated growth of Gram-negative isolates in the colon. In one experiment this permitted colonisation by *S.hyodysenteriae* but clinical disease was not observed, whereas in another experiment both colonisation and clinical expression of SD were seen. In conclusion, these data are not sufficient to support a consistent, direct relationship between changes in the large intestinal microflora and protection against SD. However, these data suggest that different sources of carbohydrate fermented to varying degrees in the large intestine can generate changes in the large intestinal microbiota that, in turn, might have some influence on proliferation of *S.hyodysenteriae* and development of SD. This may occur as a result of a certain threshold of synergistic bacteria being reached, either as increased total numbers, or by an increase in their proportion at the expense of other bacteria.

THE USE OF ENZYMES TO REDUCE THE CLINICAL EXPRESSION OF SWINE DYSENTERY

Based on previous data, there is good reason to suspect that addition of in-feed enzymes to diets may ameliorate the incidence of SD by hydrolysing glycosidic linkages of soluble NSP prior to their passage into the large intestine. This hypothesis was tested in a study comprising a 2 x 2 factorial arrangement of treatments where wheat was fed to pigs either in extruded form (to reduce the contribution of RS to the expression of SD) or hammer-milled form, and where an exogenous xylanase was added or not added to the diet (Durmic, Pethick, Mullan, Schulze and Hampson, 1997). Pigs were infected with a virulent strain of *S. hyodysenteriae* at ≈ 25 kg and monitored for expression of the disease.

Both extrusion and enzymatic hydrolysis of wheat NSP increased pre-caecal starch digestion, as judged by reduced starch levels in the large intestine (Table 10.7). The addition of a xylanase preparation to the diet failed to reduce the incidence of SD. As stated by Durmic *et al.* (1997), the failure of combined extrusion and enzyme treatment effects to protect against SD might be related to the apparent increased fermentation in proximal areas of the hindgut, as judged by an increase in bacterial ATP concentrations.

A significant main effect of enzyme addition on digesta pH was seen but only in the distal part of the colon, such that pigs fed added xylanase had a higher pH than pigs not fed enzyme (6.68 *vs.* 6.35, $P = 0.017$). These data suggest that the enzyme was having some effect on fermentation, but that it was occurring towards the end of the large bowel. This may have allowed colonisation by *S. hyodysenteriae* in the anterior parts of the hindgut, with subsequent expression of SD. An appropriate combination of grain processing and dietary enzyme inclusion to obtain more complete protection against SD is currently under investigation.

Table 10.7 Production data, large intestinal fermentation indices, and incidence of swine dysentery in pigs fed wheat-based diets subject to different processing and addition of an exogenous xylanase preparation (after Durmic *et al.*, 1997)

	Diet type[1]					*Significance*[3]		
Item	*RW*	*ExtW*	*RW-Enz*	*ExtW-Enz*	*SED*[2]	*W*	*Enz*	*W x Enz*
Growth rate, g/day	427	430	489	423	33.8	NS	NS	*
Starch, mg/g								
Proximal colon	10.2	0.6	6.2	2.0	2.52	***	NS	*
Distal colon	7.2	0.2	2.1	0.0	2.97	**	NS	NS
pH								
Proximal colon	5.7	6.1	5.7	6.0	0.34	NS	NS	NS
Distal colon	6.1	6.6	6.6	6.8	0.30	**	*	NS
ATP, nmol/g								
Proximal colon	0.30	0.10	0.42	0.44	0.26	NS	*	NS
Distal colon	0.18	0.14	0.17	0.23	0.18	NS	NS	NS
Incidence of Swine Dysentery	0.667	0.333	1.00	1.00		-	-	-

1) RW: raw wheat; ExtW: extruded wheat; RW-Enz: raw wheat + enzyme; ExtW-Enz: extruded wheat plus enzyme; 2) Standard error of difference between means; 3) *** $P < 0.001$, ** $P < 0.01$, * $P < 0.05$, NS: not significant.

Conclusions

The maintenance of a stable microflora in the large intestine of the growing and adult pig is vital for the efficient conversion of undigested carbohydrates to sources of energy that the pig can use. It is also important for situations such as enteric disease, where an imbalance in microbial populations may predispose the pig to morbidity and, in some cases, mortality. There are a multitude of interacting influences, such as pH, redox potential and resistance to bile acids, which determine the balance of microbial populations in the hindgut. The type of diet fed to pigs undoubtedly has a dramatic influence on the balance of the microflora also, and this has been demonstrated in this chapter. In particular, it is evident that the pig, especially the adult pig, has considerable propensity to degrade NSP that enter the large intestine that, in turn, can contribute a significant proportion of maintenance energy requirements. Soluble NSP fed to pigs may also have anti-nutritive effects during their passage through the small intestine, although the literature is equivocal in this regard and suggests that this might only be a problem with the younger pig. Evidence to support the use of in-feed enzymes to counteract any anti-nutritive effects of NSP again is equivocal, although they may have a place in diets for young pigs and in cases where high levels of (largely) insoluble NSP are fed. Data were also presented showing that NSP in diets can have effects other than nutritional, such as modification of sow behaviour and odour emission. Finally, evidence has been presented to implicate the presence of fermentable sources of carbohydrate, i.e., soluble NSP and RS, in the clinical expression of SD. Although the exact mechanism(s) whereby these carbohydrates have such an effect have yet to be elucidated, it is suspected that changes in microbial populations and (or) numbers in the large intestine are implicated. This provides a good example whereby disruption of the "stable" microflora by nutrition has an important influence on gut function and health in the pig.

References

Alexander, T.J.L. and Taylor, D.J. (1969) The clinical signs, diagnosis and control of swine dysentery. *Veterinary Record,* **85,** 59-63.

Allison, M.J., Robinson, I.M., Bucklin, J.A. and Booth, G.D. (1979) Comparison of bacterial populations in the pig cecum and colon based upon enumeration with specific energy sources. *Applied and Environmental Microbiology,* **37,** 1142-1151.

Annison, G. (1991) Relationship between the levels of soluble non-starch polysaccharides and the apparent metabolizable energy of wheat assayed in broiler chickens. *Journal of Agricultural Food and Chemistry,* **39,** 1252-1256.

Annison, G. (1993) The role of wheat non-starch polysaccharides in broiler nutrition. *Australian Journal of Agricultural Research,* **44,** 405-422.

Asp, N.-G., Schweizer, T.F., Southgate, D.A.T. and Theander, O. (1992) Dietary fibre analysis. In *Dietary Fibre – A Component of Food - Nutritional Function in Health and Disease*, pp. 57-95. Edited by M. Eastwood, C. Edwards, J. Mauron and T. Schweizer. Springer, London.

Baas, T.C. and Thacker, P.A. (1996) Impact of gastric pH on dietary enzyme activity in swine fed ß-glucanase supplemented diets. *Canadian Journal of Animal Science,* **76,** 245-252.

Bach Knudsen, K.E. (1997) Carbohydrate and lignin contents of plant materials used in animal feeding. *Animal Feed Science and Technology,* **67,** 319-338.

Bach Knudsen, K.E. and Canibe, N. (1997) Digestion of carbohydrates in the small and large intestine of pigs fed on wheat or oat based rolls. In *Proceedings of the VIIth International Symposium on Digestive Physiology in Pigs*, pp. 562-566. Edited by J.-P. Laplace, C. Février and A. Barbeau. EAAP Publication No. 88, Saint Malo, France.

Bach Knudsen, K.E. and Hansen, I. (1991) Gastrointestinal implications in pigs of wheat and oat fractions. 1. Digestibility and bulking properties of polysaccharides and other major constituents. *British Journal of Nutrition,* **65,** 217-232.

Bach Knudsen, K.E., Jensen, B.B., Anderson, O. and Hansen, I. (1991) Gastrointestinal implications in pigs of wheat and oat fractions. 2. Microbial activity in the gastrointestinal tract. *British Journal of Nutrition,* **65,** 233-248.

Bach Knudsen, K.E., Jensen, B.B. and Hansen, I. (1993) Oat bran but not oat gum enhanced butyrate production in the large intestine of pigs. *Journal of Nutrition,* **123,** 1235-1247.

Baghurst, P.A., Baghurst, K.I. and Record, S.J. (1996) Dietary fibre, non-starch polysaccharides and resistant starch: A review. *Food Australia,* **48,** S1-S35.

Bedford, M.R. and Classen, H.L. (1992) Reduction of intestinal viscosity through manipulation of dietary rye and pentosanase concentration is effected through changes in the carbohydrate composition of the intestinal aqueous phase and results in improved growth rate and food conversion efficiency of broiler chicks. *Journal of Nutrition,* **122,** 560-569.

Bedford, M.R., Classen, H.L. and Campbell, G.L. (1991) The effect of pelleting, salt and pentosanase on the viscosity of intestinal contents and the performance of broilers fed rye. *Poultry Science,* **70,** 1571-1577.

Bedford, M.R., Patience, J.F., Classen, H.L. and Inborr, J. (1992) The effect of dietary enzyme supplementation of rye- and barley-based diets on digestion and subsequent performance in weanling pigs. *Canadian Journal of Animal Science,* **72,** 97-105.

Bedford, M.R. and Schulze, H. (1998) Exogenous enzymes for pigs and poultry. *Nutrition Research Reviews,* **11,** 91-114.

Brouns, F., Edwards, S.A. and English, P.R. (1994) Effect of dietary fibre and feeding system on activity and oral behaviour of group housed gilts. *Applied Animal Behavioural Science,* **39,** 215-222.

Buraczewska, L.E., Schulz, E., Gdala, J. and Grala, W. (1988) Ileal and total digestibility of NDF and ADF in the pig. In *Proceedings of the IVth International Symposium on Digestive Physiology in Pigs*, pp. 224-228. Edited by L. Buraczewska, S. Buraczewska, B. Pastuczewska and T. Zebrowska. Institute of Animal Physiology and Nutrition, Jablonna, Poland.

Burnett, G.S. (1966) Studies of viscosity as the probable factor involved in the improvement of certain barleys for chickens by enzyme supplementation. *British Poultry Science,* **7,** 55-75.

Butine, T.J. and Leedle, J.A.Z. (1989) Enumeration of selected anerobic bacterial groups in cecal and colonic contents of growing-finishing pigs. *Applied and Environmental Microbiology,* **55,** 1112-1116.

Campbell, G.L. and Bedford, M.R. (1992) Enzyme applications for monogastric feeds: A review. *Canadian Journal of Animal Science,* **72,** 449-466.

Canh, T.T., Aarnink, A.J.A., Verstegen, M.W.A. and Schrama, J.W. (1998) Influence of dietary factors on the pH and ammonia emissions of slurry from growing-finishing pigs. *Journal of Animal Science,* **76,** 1123-1130.

Canh, T.T., Sutton, A.L., Aarnink, A.J.A., Verstegen, M.W.A., Schrama, J.W. and Bakker, G.C.M. (1998) Dietary carbohydrates alter the fecal composition and pH and the ammonia emission from slurry of growing pigs. *Journal of Animal Science,* **76,** 1887-1895.

Choct, M. and Annison, G. (1990) Anti-nutritive activity of wheat pentosans in broiler diets. *British Poultry Science,* **31,** 811-821.

Choct, M. and Annison, G. (1992) The inhibition of nutrient digestion by wheat pentosans. *British Journal of Nutrition,* **67,** 123-132.

Classen, H.L. (1996) Cereal grain starch and exogenous enzymes in poultry diets. *Animal Feed Science and Technology,* **62,** 21-27.

Conway, P.L. (1997) Development of intestinal microbiota. In *Gastrointestinal Microbiology*, pp. 3-38, Volume 2. Edited by R.I. Mackie, B.A. White and R.E. Isaacson. Chapman and Hall Microbiology Series, Chapman and Hall, New York, NY.

Dierick, N.A., Vervaeke, I.J., Demeyer, D.I. and Decuypere, J.A. (1989) Approach to the energetic importance of fibre digestion in pigs. I. Importance of fermentation in the overall energy supply. *Animal Feed Science and Technology,* **23,** 141-167.

Durmic, Z., Pethick, D.W., Mullan, B.P., Schulze, H. and Hampson, D.J. (1997) The effects of extrusion and enzyme addition in wheat based diets on fermentation in the large intestine and expression of swine dysentery. In *Manipulating Pig Production VI*, p. 180. Edited by P.D. Cranwell. Australasian Pig Science Association, Werribee.

Durmic, Z., Pethick, D.W., Pluske, J.R. and Hampson, D.J. (1998) Influence of dietary fibre sources and levels of inclusion on the colonic microflora of pigs, and on the development of swine dysentery in experimentally-infected pigs. *Journal of Applied Microbiology,* **85,** 574-582.

Dusel, G., Kluge, H., Gläser, K., Simon, O., Hartmann, G., Lengerken, J.v. and Jeroch, H. (1997) An investigation into the variability of extract viscosity of wheat – relationship with the content of non-starch polysaccharide fractions and metabolisable energy for broiler chickens. *Archives of Animal Nutrition,* **50,** 121-135.

Dusel, G., Kluge, H., Jeroch, H. and Simon, O. (1998) Xylanase supplementation of wheat-based rations for broilers: influence of wheat characteristics. *Journal of Applied Poultry Research,* **7,** 119-131.

Ellis, P.R., Roberts, F.G., Low, A.G. and Morgan, L.M. (1995) The effect of high-molecular-weight guar gum on net apparent glucose absorption and net apparent insulin and gastric inhibitory polypeptide production in the growing pig – relationship to rheological changes in jejunal digesta. *British Journal of Nutrition,* **74,** 539-556.

Englyst, H. (1989) Classification and measurement of plant polysaccharides. *Animal Feed Science and Technology,* **23,** 27-42.

Ewan, R.C., Crenshaw, J.D., Crenshaw, T.D., Cromwell, G.L., Easter, R.A., Nelssen, J.L., Miller, E.R., Pettigrew, J.E. and Veum, T.L. (1996) Effect of addition of fiber to gestation diets on reproductive performance of sows. *Journal of Animal Science,* **74 (Suppl. 1),** 190.

Ewing, W.N. and Cole, D.J.A. (1994) *The Living Gut.* Context Publications, Leicester, England.

Fadel, J.G., Newman, C.W., Newman, R.K. and Graham, H. (1988) Effects of extrusion cooking of barley on ileal and fecal digestibilities of dietary components in pigs. *Canadian Journal of Animal Science,* **68,** 891-897.

Fadel, J.G., Newman, C.W., Newman, R.K. and Graham, H. (1989) Effects of baking hulless barley on the digestibility of dietary components as measured at the ileum and in the feces of pigs. *Journal of Nutrition,* **119,** 722-726.

Fernández, J.A., Jørgensen, H. and Just, A. (1986) Comparative digestibility experiments with growing pigs and adult sows. *Animal Production,* **43,** 127-132.

Finegold, S.N., Attebery, H.R. and Sutter, V.L. (1974) Effect of diet on human fecal flora: comparison of Japanese and American diets. *American Journal of Clinical Nutrition,* **29,** 1456-1469.

Freter, R. (1972) Interactions between mechanisms controlling the intestinal microflora. *American Journal of Clinical Nutrition,* **27,** 1409-1417.

Gdala, J. (1998) Composition, properties, and nutritive value of dietary fibre of legume seeds. A review. *Journal of Animal and Feed Sciences,* **7,** 131-149.

Gdala, J., Jansman, A.J.M., Buraczewska, L., Huisman, J. and van Leeuwin, P. (1997) The influence of α–galactosidase supplementation on the ileal digestibility of lupin seed carbohydrates and dietary protein in young pigs. *Animal Feed Science and Technology,* **67,** 115-125.

Graham, H., Fadel, J.G., Newman, C.W. and Newman, R.K. (1989) Effect of pelleting and ß-glucanase supplemenation on the ileal and fecal digestibility of a barley-based diet in the pig. *Journal of Animal Science,* **67,** 1293-1298.

Graham, H., Hesselman, K. and Åman, P. (1986a) The influence of wheat bran and sugar-beet pulp on the digestibility of dietary components in a cereal-based diet. *Journal of Nutrition,* **116,** 242-251.

Graham, H., Hesselman, K., Jonsson, E. and Åman, P. (1986b) Influence of ß-glucanase supplementation on digestion of a barley-based diet in the pig gastrointestinal tract. *Nutrition Reports International*, **34,** 1089-1094.

Haberer, B., Schulz, E., Aulrich, K. and Flachowsky, G. (1998a) Effects of ß-glucanase and xylanase supplementation in pigs fed a diet rich in nonstarch polysaccharides: composition of digesta in different prececal segments and postprandial time. *Journal of Animal Physiology and Animal Nutrition,* **78,** 84-94.

Haberer, B., Schulz, E. and Flachowsky, G. (1998b) Effects of ß-glucanase and xylanase supplementation in pigs fed a diet rich in nonstarch polysaccharides: disappearance and disappearance rate of nutrients including the nonstarch polysaccharides in stomach and small intestine. *Journal of Animal Physiology and Animal Nutrition,* **78,** 95-103.

Hampson, D.J., Cutler, R. and Lee, B.J. (1992) Isolation of virulent *Serpulina hyodysenteriae* from a pig in a herd free from clinical dysentery. *Veterinary Record,* **131,** 318-319.

Hampson, D.J., Robertson, I.D., Oxberry, S.L. and Pethick, D.W. (1998) Evaluation of vaccination and diet for the control of *Serpulina pilosicoli* infection (Porcine Intestinal Spirochaetosis). In *Proceedings of the 15th IPVS Congress*, p. 56. Birmingham, England.

Hampson, D.J. and Trott, D.J. (1995) Intestinal spirochaetal infections of pigs: an overview with an Australian perspective. In *Manipulating Pig Production V*, pp. 139-169. Edited by D.P. Hennessy and P.D. Cranwell. Australasian Pig Science Association, Werribee.

Harris, D.L. and Lysons, R.J. (1992) Swine dysentery. In *Diseases of Swine*, 7th Edition, pp. 599-616. Edited by A.D. Leman, B.E. Straw, W.L Mengeling, S. D'Allaire and D.J. Taylor. Iowa State University Press, Ames, IA.

Jensen, B.B. and Jørgensen, H. (1994) Effect of dietary fiber on microbial activity and microbial gas production in various regions of the gastrointestinal tract of pigs. *Applied and Environmental Microbiology,* **60,** 1897-1904.

Johansen, H.N., Bach Knudsen, K.E., Sandstrom, B. and Skjoth, F. (1996) Effects of varying content of soluble dietary fibre from wheat flour and oat milling fractions on gastric emptying in pigs. *British Journal of Nutrition,* **75,** 339-351.

Johansen, H.N., Bach Knudsen, K.E., Wood, P.J. and Fulcher, R.G. (1997) Physico-chemical properties and the degradation of oat bran polysaccharides in the gut of pigs. *Journal of the Science of Food and Agriculture,* **73,** 81-92.

Jørgensen, H., Zhao, X.-Q. and Eggum, B.O. (1996) The influence of dietary fibre and environmental temperature on the development of the gastrointestinal tract, digestibility, degree of fermentation in the hind-gut and energy metabolism in pigs. *British Journal of Nutrition,* **75,** 365-378.

Just, A., Fernández, J.A. and Jørgensen, H. (1983) The net energy value of diets for growth in pigs in relation to the fermentative processes in the digestive tract and the site of absorption of the nutrients. *Livestock Production Science,* **10,** 171-186.

Kass, M.L., Van Soest, P.J., Pond, W.G., Lewis, B. and McDowell, R.E. (1980) Utilization of dietary fiber from alfalfa by growing swine. I. Apparent digestibility of diet components in specific segments of the gastrointestinal tract. *Journal of Animal Science,* **50,** 175-191.

Levett, P. (1990) Anerobic bacteria. Open University Press, Milton Keynes, Philadelphia.

Lewis, F.J., McEvoy, J. and McCracken, K.J. (1998) Lack of relationship between wheat *in vitro* viscosity and digestibility parameters for pigs. *Animal Science,* **(Abstr.)**, 32.

Longland, A.C., Low, A.G., Quelch, D.B. and Bray, S.P. (1993) Adaptation to the digestion of non-starch polysaccharide in growing pigs fed on cereal or semi-purified basal diets. *British Journal of Nutrition,* **70,** 557-566.

Low, A.G. (1985) Role of dietary fibre in pig diets. In *Recent Advances in Animal Nutrition*, pp. 87-112. Edited by W. Haresign and D.J.A. Cole. Butterworths, London.

Macfarlane, G.T. and Gibson, G.R. (1997) Carbohydrate fermentation, energy transduction and gas metabolism in the human large intestine. In *Gastrointestinal Microbiology*, pp. 269-318, Volume 1. Edited by R.I. Mackie and B.A. White. Chapman and Hall Microbiology Series, Chapman and Hall, New York, NY.

Matte, J.J., Robert, S., Girard C.L., Farmer, C. and Martineau, G.-P. (1994) Effect of bulky diets based on wheat bran or oat hulls on reproductive performance of sows during their first two parities. *Journal of Animal Science,* **72,** 1754-1760.

May, T., Mackie, R.I., Fahey, Jr., G.C., Cremin, J.C. and Garleb, K.A. (1994) Effect of fiber source on short-chain fatty acid production and on the growth and toxin production by *Clostridium difficile. Scandinavian Journal of Gastroenterology,* **19,** 916-922.

McBurney, M.I. and Sauer, W.C. (1993) Fiber and large bowel energy absorption: validation of the integrated ileostomy-fermentation model using pigs. *Journal of Nutrition,* **123,** 721-727.

McCullough, J.S., Ratcliffe, B., Mandir, N., Carr, K.E. and Goodlad, R.A. (1998) Dietary fibre and intestinal microflora: effects on intestinal morphometry and crypt branching. *Gut,* **42,** 799-806.

Meyer, R.C. (1978) Swine dysentery – a perspective. *Advances in Veterinary Science and Comparative Medicine,* **22,** 133-158.

Meyer, R.C., Simon, J. and Byerly, C.S. (1974) The etiology of swine dysentery. I. Oral inoculation of germ-free swine with *Treponema hyodysenteriae* and *Vibrio coli*. *Veterinary Pathology,* **11,** 515-526.

Meyer, R.C., Simon, J. and Byerly, C.S. (1975) The etiology of swine dysentery. III. The role of selected gram-negative obligate anaerobes. *Veterinary Pathology,* **12,** 46-54.

Meynell, T.G. (1963) Antibacterial mechanisms of the mouse gut. II. Role of E_h and volatile fatty acids in the normal gut. *British Journal of Experimental Pathology,* **44,** 209-219.

Mhoma, J.R.L., Hampson, D.J. and Robertson, I.D. (1992) A serological survey to determine the prevalence of infection with *Treponema hyodysenteriae* in Western Australia. *Australian Veterinary Journal,* **69,** 81-84.

Moore, W.E.C. and Holdeman, L.V. (1974) Human fecal microflora: the normal flora of 20 Japanese-Hawaiians. *Applied Microbiology,* **27,** 961-979.

Moore, W.E.C., Moore, L.V.H., Cato, E.P., Wilkins, T.D. and Kornegay, E.T. (1987) Effect of high-fiber and high-oil diets on the fecal flora of swine. *Applied and Environmental Microbiology,* **53,** 1638-1644.

Mosenthin, R., Hambrecht, E. and Sauer, W.C. (1999) Utilisation of different fibres in piglet feeds. In: *Recent Advances in Animal Nutrition - 1999* (Eds. P.C.Garnsworthy and J. Wiseman). pp 227-256. Nottingham University Press, Nottingham, England.

Mroz, Z., Partridge, I.G., Mitchell, G. and Keal, H.D. (1986) The effect of oat hulls, added to the basal ration for pregnant sows, on reproductive performance, apparent digestibility, rate of passage and plasma parameters. *Journal of the Science of Food and Agriculture,* **37,** 239-247.

Münchow, M., Bergner, H., Seifert, G., Schönmuth, G. and Baraband, E. (1982) Investigation on the use of partly hydrolysed and untreated straw meal in the feeding of breeding sows. I. Influence on the birth weight and the number of piglets. *Archives Tierernaehrung,* **32,** 483-491.

Noblet, J. and Bach Knudsen, K.E. (1997) Comparative digestibility of wheat, maize and sugar beet pulp non-starch polysaccharides in adult sows and growing pigs. In *Proceedings of the VIIth International Symposium on Digestive Physiology in Pigs*, pp. 571-574. Edited by J.-P. Laplace, C. Février and A. Barbeau. EAAP Publication No. 88, Saint Malo, France.

Noblet, J. and Shi, X.S. (1993) Comparative digestibility of energy and nutrients in growing pigs fed ad libitum and adult sows fed at maintenance. *Livestock Production Science,* **34,** 137-152.

Nyachoti, C.M., de Lange, C.F.M., McBride, B.W. and Schulze, H. (1997) Significance of endogenous gut nitrogen losses in the nutrition of growing pigs: A review. *Canadian Journal of Animal Science,* **77,** 149-163.

Pluske, J.R., Durmic, Z., Pethick, D.W., Mullan, B.P. and Hampson, D.J. (1998a) Confirmation of the role of rapidly fermentable carbohydrates in the expression of swine dysentery in pigs after experimental infection. *Journal of Nutrition,* **128,** 1737-1744.

Pluske, J.R., Pethick, D.W. and Mullan, B.P. (1998b) Differential effects of feeding fermentable carbohydrate to growing pigs on performance, gut size, and slaughter characteristics. *Animal Science,* **67,** 147-156.

Pluske, J.R., Siba, P.M., Pethick, D.W., Durmic, Z., Mullan, B.P. and Hampson, D.J. (1996) The incidence of swine dysentery in pigs can be reduced by feeding diets that limit the amount of fermentable substrate entering the large intestine. *Journal of Nutrition,* **126,** 2920-2933.

Pond, W.G., Jung, H.G. and Varel, V.H. (1988) Effect of dietary fiber on young adult genetically lean, obese and contemporary pigs: Body weight, carcass measurements, organ weights and digesta content. *Journal of Animal Science,* **66,** 699-706.

Prohaszka, L. and Lukacs, K. (1984) Influence of the diet on the antibacterial effect of volatile fatty acids on the development of swine dysentery. *Zentralblatt für Veterinar Medizin B,* **31,** 779-785.

Radecki, S.V. and Yokoyama, M.T. (1991) Intestinal bacteria and their influence on swine nutrition. In *Swine Nutrition*, pp. 439-447. Edited by E.R. Miller, D.E. Ullrey and A.J. Lewis. Butterworths-Heinemann, Boston, MA.

Rérat, A., Fislewicz, M., Guisi, A. and Vaugelade, P. (1987) Influence of meal frequency on postprandial variations in the production and absorption of volatile fatty acids in the digestive tract of conscious pigs. *Journal of Animal Science,* **64,** 448-456.

Robert, S., Matte, J.J., Farmer, C., Girard, C.L. and Martineau, G.-P. (1993) High-fibre diets for sows: effects on stereotypies and adjunctive drinking. *Applied Animal Behavioural Science,* **37,** 297-309.

Robinson, I.M., Allison, M.J. and Bucklin, J.A. (1981) Characterization of the cecal bacteria of normal pigs. *Applied and Environmental Microbiology,* **41,** 950-955.

Robinson, I.M., Stromley, J.M. and Varel, V.H. (1988) Streptococcus intestinalis, a new species from the colons and feces of pigs. *International Journal of Systematic Bacteriology,* **38,** 245-248.

Robinson, I.M., Whipp, S.C., Bucklin, J.A. and Allison, M.J. (1984) Characterization of predominant bacteria from the colons of normal and dysenteric pigs. *Applied and Environmental Microbiology,* **48,** 964-969.

Rogel, A.M., Annison, E.F., Bryden, W.L. and Balnave, D. (1987) The digestion of wheat starch in broiler chickens. *Australian Journal of Agricultural Research,* **38,** 639-649.

Russell, E.G. (1979) Types and distribution of anaerobic bacteria in the large intestine of pigs. *Applied and Environmental Microbiology,* **37,** 187-193.

Salanitro, J.P., Blake, I.G. and Muirhead, P.A. (1977) Isolation and identification of fecal bacteria from adult swine. *Applied and Environmental Microbiology,* **33,** 79-84.

Sansom, B.F. and Gleed, P.T. (1981) The ingestion of sow's faeces by suckling piglets. *British Journal of Nutrition,* **46,** 451-456.

Savage, D.C. (1986) Gastrointestinal microflora in mammalian nutrition. *Annual Reviews of Nutrition,* **6,** 155-178.

Schulze, H., van Leeuwin, P., Verstegen, M.W.A. and van den Berg, J.W.O. (1995) Dietary level and source of neutral detergent fiber and ileal endogenous nitrogen flow in pigs. *Journal of Animal Science,* **73,** 441-448.

Selvendran, R.R. (1984) The plant cell wall as a source of dietary fibre: chemistry and structure. *American Journal of Clinical Nutrition,* **39,** 320-337.

Simon, T.L. and Gorbach, S.L. (1981) Intestinal flora in health and disease. In *Physiology of the Gastrointestinal Tract*, pp. 1361-1380. Edited by L.R. Johnson. Raven Press, New York, NY.

Stanogias, G. and Pearce, G.R. (1985) The digestion of fibre by pigs. I. The effects of amount and type of fibre on apparent digestibility, nitrogen balance and rate of passage. *British Journal of Nutrition,* **53,** 513-530.

Suenaga, I. and Yamazaki, T. (1986) Eliminating organisms against *Treponema hyodysenteriae* in the gut of mice. *Zentralblatt für Bakteriologie and Hygiene A,* **261,** 322-329.

Shi, X.S. and Noblet, J. (1993) Contribution of the hindgut to digestion of diets in growing pigs and adult sows: effect of diet composition. *Livestock Production Science,* **34,** 237-152.

Siba, P.M., Pethick, D.W. and Hampson, D.J. (1996) Pigs experimentally infected with *Serpulina hyodysenteriae* can be protected from developing swine dysentery by feeding them a highly digestible diet. *Epidemiology and Infection,* **116,** 207-216.

Stephen, A.M. and Cummings, J.H. (1980) Mechanisms of action of dietary fibre in the human colon. *Nature,* **284,** 283-284.

Stewart, C.S. (1997) Microorganisms in hindgut fermentors. In *Gastrointestinal Microbiology*, pp. 142-186, Volume 2. Edited by R.I. Mackie, B.A. White and R.E. Isaacson. Chapman and Hall Microbiology Series, Chapman and Hall, New York, NY.

Trowell, H., Southgate, D.A.T., Wolever, T.M.S., Leeds, A.R., Gassull, M.A. and Jenkins, D.J.A. (1976) Dietary fibre redefined. *Lancet,* **1,** 967.

Van Barneveld, R.J., Baker, J., Szarvas, S.R. and Choct, M. (1995a) Effect of lupin kernels on the apparent ileal digestibility of amino acids by growing pigs. In

Manipulating Pig Production V, p. 29. Edited by D.P. Hennessy and P.D. Cranwell. Australasian Pig Science Association, Werribee.

Van Barneveld, R.J., Baker, J., Szarvas, S.R. and Choct, M. (1995b) Effect of lupin kernels on the ileal and faecal digestibility of energy by growing pigs. In *Manipulating Pig Production V*, p. 30. Edited by D.P. Hennessy and P.D. Cranwell. Australasian Pig Science Association, Werribee.

Van Barneveld, R.J., Baker, J., Szarvas, S.R. and Choct, M. (1995c) Effect of lupin non-starch polysaccharides (NSP) on nutreint digestion and microbial activity in growing pigs. *Proceedings of the Nutrition Society of Australia,* **19,** 43.

Van Barneveld, R.J., Olsen, L.E. and Choct, M. (1996) Effect of lupin oligosaccharides on energy digestion in growing pigs. *Proceedings of the Nutrition Society of Australia,* **20,** 114.

Van Barneveld, R.J., Olsen, L.E. and Choct, M. (1997) Lupin oligosaccharides depress the apparent ileal digestion of amino acids by growing pigs. In *Manipulating Pig Production VI*, p. 230. Edited by P.D. Cranwell. Australasian Pig Science Association, Werribee.

Varel, V.H. (1987) Activity of fiber-degrading microorganisms in the pig large intestine. *Journal of Animal Science,* **65,** 488-496.

Varel, V.H., Fryda, S.J. and Robinson, I.M. (1984a) Cellulolytic bacteria from pig large intestine. *Applied and Environmental Microbiology,* **47,** 219-221.

Varel, V.H. and Pond, W.G. (1985) Enumeration and activity of cellulolytic bacteria from gestating swine fed various levels of dietary fiber. *Applied and Environmental Microbiology,* **49,** 858-862.

Varel, V.H. and Pond, W.G. (1992) Characteristics of a new cellulolytic *Clostridium* sp. isolated from pig intestinal tract. *Applied and Environmental Microbiology,* **58,** 1645-1649.

Varel, V.H., Pond, W.G., Pekas, J.C. and Yen, J.T. (1982) Influence of high-fiber diet on bacterial populations in gastrointestinal tracts of obese- and lean-genotype pigs. *Applied and Environmental Microbiology,* **44,** 107-112.

Varel, V.H., Pond, W.G. and Yen, J.T. (1984b) Influence of dietary fiber on the performance and cellulase activity of growing-finishing swine. *Journal of Animal Science,* **59,** 388-393.

Varel, V.H., Robinson, I.M. and Jung, H.-J.G. (1987) Influence of dietary fiber on xylanolytic and cellulolytic bacteria of adult pigs. *Applied and Environmental Microbiology,* **53,** 22-26.

Varel, V.H., Tanner, R.S. and Woese, C.R. (1995) *Clostridium herbivorans* sp. nov., a cellulolytic bacteria anaerobe from the pig intestinal tract. *International Journal of Systematic Microbiology,* **45,** 490-494.

Varel, V.H. and Yen, J.T. (1997) Microbial perspective on fiber utilization by swine. *Journal of Animal Science,* **75,** 2715-2722.

Varel, V.H., Yen, J.T. and Kreikemeier, K.K. (1995) Addition of cellulolytic clostridia to the bovine rumen and pig intestinal tract. *Applied and Environmental Microbiology,* **61,** 1116-1119.

Weltzien, E.M. and Aherne, F.X. (1986) The effects of anaerobic storage and processing of high-moisture barley on its ileal digestibility by, and performance of, growing pigs. *Canadian Journal of Animal Science,* **67,** 829-840.

Whipp, S.C., Robinson, I.M., Harris, D.L., Glock, R.D., Mathews, P.J. and Alexander, T.J.L. (1979) Pathogenic synergism between *Treponema hyodysenteriae* and other selected anaerobes in gnotobiotic pigs. *Infection and Immunity,* **26,** 1042-1047.

Wilson, K., Moore, L., Patel, M. and Peramond, P. (1988) Suppression of potential pathogens by a defined colonic microflora. *Microbiology and Ecology in Health and Disease,* **1,** 237-243.

Wootton, M., Acone, L. and Wills, R.B.H. (1995) Pentosan levels in Australian and North American feed wheats. *Australian Journal of Agricultural Research,* **46,** 389-392.

Yan, T., Longland, A.C., Close, W.H., Sharpe, C.E. and Keal, H.D. (1995) The digestion of dry matter and non-starch polysaccharides from diets containing plain sugar-beet pulp or wheat straw by pregnant sows. *Animal Science,* **61,** 305-309.

Yen, J.T., Nienaber, J.A., Hill, D.A. and Pekas, J.C. (1989) Oxygen consumption by portal vein-drained organs and by the whole animal in conscious growing swine. *Proceedings of the Society for Experimental Biology and Medicine,* **190,** 393-398.

Yen, J.T., Nienaber, J.A., Hill, D.A. and Pond, W.G. (1991) Potential contribution of absorbed volatile fatty acids to whole-animal energy requirement in conscious swine. *Journal of Animal Science,* **69,** 2001-2012.

Zhu, Q., Fowler, V.R. and Fuller, M.F. (1993) Assessment of fermentation in growing pigs given unmolassed sugar-beet pulp: a stoichiometric approach. *British Journal of Nutrition,* **69,** 511-525.

11

UTILISATION OF DIFFERENT FIBRES IN PIGLET FEEDS

R. MOSENTHIN[1], E. HAMBRECHT[2] and W.C. SAUER[3]

[1] *Hohenheim University, Institute of Animal Nutrition (450), D-70593 Stuttgart, Germany;* [2] *Nutreco R&D – Swine Research Centre, P.O.Box 240, NL-5830 AE Boxmeer, The Netherlands;* [3] *University of Alberta, Department of Agricultural, Food and Nutritional Science, Edmonton, Alberta, T6G 2P5, Canada*

Introduction

The role of high-fibre ingredients in piglet diets has been very little explored, partly because the microflora are thought to be of minor importance in young pigs. Creep feeds are usually relatively low in dietary fibre and the transition to high-fibre diets after weaning has often resulted in reduced feed intake; diarrhoea is a frequent problem during this time. The recent interest in fibre has resulted from the possibility that this component is involved in the aetiology of many diseases and possible health-modulating effects in the digestive tract. In addition, public concern of possible risks related to the inclusion of sub-therapeutic levels of antibiotics has intensified the search for specific attributes of dietary fibre as an alternative to support animal performance and health. This review will focus on various physiological, metabolic, microbial and nutritional aspects of dietary fibre (DF) in diets for pigs after weaning. Special attention will be paid to potential trophic and systemic effects of fermentation products from DF and the possible role of non-digestible oligosaccharides (NDO) as health-modulating substrates in the nutrition of piglets.

Definition and classification of dietary fibre

The definition of DF is still controversial and several have been suggested. Many use the terms crude fibre (CF), neutral detergent fibre (NDF), acid detergent fibre (ADF) or non-starch polysaccharides (NSP) interchangeably. To quote Trowell, Southgate, Wolever, Leeds, Miguel and Jenkins (1976) "dietary fibre may be defined as the sum of the polysaccharides and lignin which are not digested by the endogenous secretions of the gastrointestinal tract". This broad conceptual approach combines both chemical and physiological aspects of DF which proved to be too imprecise for analytical chemists concerned with devising routine methods for fibre estimation. A widely accepted chemical definition of DF is "the sum of all non-starch polysaccharides and lignin". However, this reductionist approach ignores many other food and feed components including starch resistant to amylase (resistant starch, RS), several NDO and some protein and lipid fractions (Englyst and Cummings, 1987) which, in the large intestine, behave similarly to some sources of NSP and might be included within the definition of fibre by those taking an holistic view. A classification of carbohydrates present in feedstuffs and as part of feed additives is given in Table 11.1.

Table 11.1 Classification of carbohydrates

Category	*Monomeric residues*	*Source*
Non-starch polysaccharides (NSP)		
Cell Wall NSP		
Cellulose	Glucose	Most feedstuffs
Mixed linked ß-glucans	Glucose	Barley, oats, rye
Arabinoxylans	Xylose, arabinose	Rye, wheat, barley
Arabinogalactans	Galactose, arabinose	Cereal by-products
Xyloglucans	Glucose, xylose	Cereal flours
Rhamnogalacturans	Uronic acids, rhamnose	Hulls of peas
Galactans	Galactose	Soya bean meal, sugar-beet pulp
Non-cell wall NSP		
Fructans	Fructose	rye
Mannans	Mannose	Coconut cake, palm cake
Pectins	Uronic acids, rhamnose	Sugar-beet pulp
Galactomannans	Galactose, mannose	Guar gum
Non-digestible oligosaccharides (NDO)		
a-Galacto-oligosaccharides	Galactose, glucose, fructose	Soya bean meal, peas, rapeseed meal
Fructo-oligosaccharides	Fructose	Cereals, feed additives
Transgalacto-oligosaccharides	Galactose, glucose	Feed additives, whey/milk products
Resistant starch (RS)		
Physical inaccessible starch	Glucose	Peas, faba beans
Native starch	Glucose	potatoes
Retrograded starch	Glucose	Heat-treated starch products

(After Bach Knudsen, 1997)

Starch may be resistant to hydrolysis in the small intestine for three reasons: the type RS1 includes physical inaccessible starch which is present in grains, seeds and legumes, type RS2 is ungelatinised starch primarily found in tubers, some peas and beans; type

RS3 is retrograded starch, an insoluble complex, which is formed when heated starch is cooled or dried which can result in recrystallisation of amylose (Bach Knudsen, 1997).

The term "oligosaccharide" is used in particular to denote a group of carbohydrates consisting of 2-10 sugar units. Oligosaccharides are natural constituents of various feedstuffs or they are produced under commercial conditions to be used as functional ingredients in feed and food. Since they escape enzymatic digestion and are therefore a potential substrate for the gastrointestinal microflora these carbohydrates are classified as NDO. These NDO may contain similar or different sugars, different linkages and can be linear or branched; they mainly include α-galacto-oligosaccharides (α-GOS), fructo-oligosaccharides (FOS) and transgalacto-oligosaccharides (TOS).

Finally, DF are also classified into soluble and insoluble fractions. This distinction governs their physico-chemical properties and their nutritional effects. Soluble fibres may create viscous conditions within the stomach and small intestine and may affect digestion and absorption whereas the more insoluble fibre fractions exert their effects usually in the lower part of the gastrointestinal tract (bulking effect). Consequently, many analytical procedures have been adapted to differentiate between soluble and insoluble fibre fractions. However, as was demonstrated by Graham and Åman (1991), this distinction is often designed to fit into an analytical procedure rather than to correspond to actual physiological conditions since the fibre complex is continuously modified during gastrointestinal transport.

The importance of describing DF in as much chemical and physical detail as possible needs to be emphasised since the lack of information makes comparisons of most studies on the effect of DF very difficult. In this review the term "dietary fibre" will be used in relation to a concept rather than to a specific substance and will not include NDO and RS which will be discussed separately.

Physiological, nutritional and metabolic effects of dietary fibre

GASTROINTESTINAL SECRETIONS

Studies with growing pigs revealed that the effects of DF on gastrointestinal secretions appear to be considerable. It should be emphasised, however, that no studies have been carried out so far with pigs after weaning. This is probably due to the lack of appropriate techniques for establishing permanent catheters in the secretory organs in order to collect digestive juices. However, the results and conclusions reported in this chapter may also apply to pigs after weaning.

It has been established from different studies with growing pigs (> 30 kg live weight) that various sources of DF stimulate gastric, biliary, pancreatic and intestinal secretions. For example, Zebrowska, Low and Zebrowska (1983) and Sambrook (1981) found significantly higher outputs of gastric, biliary and pancreatic secretions in pigs fed a barley-based diet containing a wide variety of types of DF, than in the same pigs when they received a semi-purified diet containing cellulose as the only source of DF. Although CF intakes were similar in both cases, NDF intakes were 180 and 50 g per day, respectively, which emphasises the large contribution of non-cellulosic components of DF in the barley-based diet. In agreement with results of the aforementioned studies, the inclusion of wheat bran in the diet increased the volume of secretion of pancreatic juice (Zebrowska and Low, 1987) and, in some cases, stimulated the secretion of protein and enzymes in pancreatic juice (Langlois, Corring and Février, 1987); although the results are interpreted as being related directly to changes in the DF content of the diets, it was suggested that other factors such as enzyme inhibitors in wheat bran might have been partly involved. Furthermore, wheat bran consumption induced an increase in the volume of secretion of gastric juice and increased pepsin activity which was attributed to vagal innervation as the predominant factor controlling the volume of gastric secretion (Korczyñski, Budzyñska and Zebrowska, 1997). Finally, DF from various oat mill fractions showed a stimulatory effect on secretion of gastrointestinal juices, but this effect was not correlated directly to the DF content of the diet (Johansen and Bach Knudsen, 1994a).

According to a recent report from Lizardo, Peiniau and Aumaître (1997), moderate amounts of NSP from sugar-beet pulp in diets for pigs after weaning maintain the activity of both pancreatic and intestinal enzymes which were measured in pancreatic and mucosal tissue of slaughtered animals. It was assumed that undigested residues of sugar-beet pulp or non-digested peptides in digesta could stimulate the activity of these enzymes.

Currently, no firm conclusions can be drawn as to which physical and chemical characteristics of DF and which sources of DF exert these effects since they were poorly defined in most studies. However, studies in pigs and humans indicate hormonal control as an additional mediator of gastrointestinal secretions. The presence of viscous fibre sources in the upper tract modified the secretion of hormones such as insulin, glucagon, gastric inhibitory polypeptide and possibly secretin and cholecystokinin probably through indirect mechanisms (Vahouny and Cassidy, 1985; Low, 1989; Roberfroid, 1993).

GUT MOTILITY AND TRANSIT TIME

Interest in transit time stems from the view that transit time of digesta through different sections of the gastrointestinal tract will determine the time available for digestion by host enzymes in the stomach and small intestine and for fermentation of undigested dietary residues in the large intestine. However, the statement of Low (1993) that there is a scarcity of information on the effect of DF on transit time in pigs after weaning is still valid. The results obtained in studies with growing pigs (>20 kg live weight) and the conclusions drawn from these studies may, in principle, also apply to the feeding of piglets with various sources of DF and will be considered subsequently.

Gut motility

There is strong evidence that high viscosity of digesta in the gut lumen decreases intestinal contractions, which facilitate the propulsion of digesta in the gastrointestinal tract, thereby inhibiting the mixing of dietary components with intestinal secretions and decreasing the flow rate of solid particles. Cherbut, Albina, Champ, Doublier and Lecannu (1990) showed a close relationship between viscosity of digesta and the gastrointestinal myoelectric activity and transit time in pigs. After consumption of diets containing 60 g of different sources of guar gums differing in viscosity/kg , the increase in gastrointestinal myoelectric activity was associated with an increase in the orocaecal transit time of liquids and solids. However, several authors pointed out that a direct relationship between meal viscosity and intestinal motor effects is difficult to establish since meal viscosity may be altered during transit in the gastrointestinal tract by dilution with gastric secretions, degree of acidification in the stomach and starch digestion in the small intestine (Rainbird and Low, 1986a; Edwards, Johnson and Read, 1987; Cherbut *et al*., 1990).

It still remains uncertain which specific attributes of DF affect gastrointestinal motility. According to Bardon and Fioramonti (1983), motility is largely related to physical properties rather than to the chemical composition of DF or fermentation products including volatile fatty acids (VFA). On the other hand, Cherbut, Salvador, Barry, Doulay and Delort-Laval (1991) suggested that non-fermentable fibre sources influence transit time and gut motility via mechanical effects whereas partly fermentable fibre sources may alter the rate of passage of digesta via their fermentation products.

Gastric emptying

Studies by Rainbird and Low (1986a, b) showed that soluble DF delayed gastric emptying of digesta but did not affect the emptying of dry matter and glucose in pigs fed semi-purified diets with various soluble DF sources including guar gum; it was concluded that the effect of soluble DF on gastric emptying was confined to the liquid phase. Similar results were obtained by Johansen, Bach Knudsen, Sandström and Skjøth (1996) after feeding oat mill fractions with varying contents of soluble DF in the form of ß-glucans. Higher levels of soluble DF showed a trend towards higher recoveries of digesta, the liquid phase and DF during the initial stage of gastric emptying.

Whether the slower emptying rate of liquids can be attributed to the water-retaining capacities of fibre and/or to the stimulatory effect of DF on gastric secretions (Low,

1989; Potkins, Lawrence and Thomlinson, 1991) has not yet been clarified. Rainbird and Sissons (1985) assumed that, in the case of guar gum, the retarding effects on gastric emptying possibly reflect an increased viscosity which, in turn, may affect gastric motility. In contrast to the aforementioned studies, an accelerated gastric emptying rate was observed when a proportion of barley was replaced by 50 g guar gum or pectin/kg (Potkins *et al.*, 1991). Purified cellulose delayed gastric emptying of both the liquid and solid phases of digesta, probably due to the high water-binding capacity (Johansen and Bach Knudsen, 1994b). However, these results are contradictory to previous reports on the effect of cellulose on gastric emptying published by Rainbird and Low (1986b). Furthermore, reducing the particle size of a barley-based diet accelerated (doubled) the rate of gastric emptying in one but not in another experiment (Potkins *et al.*, 1991).

It appears that gastric emptying cannot simply be related to the dietary level of soluble fibre or fibre in general. Other factors such as particle size, physical structure, starch digestibility, solubility of cell-wall polysaccharides and interactions between dietary components may interfere with the effect of DF, particularly in complex diets. The emptying rate is also influenced by feedback control from the duodenum via receptors sensitive to the osmolarity and acidity of digesta (Laplace, 1982). Ileal concentrations of VFA have been suggested as regulators of gastric emptying in pigs (Malbert, Montfort, Mathis, Guerin and Laplace, 1994).

Small and large intestinal transit time

The time interval for digesta to pass to the distal ileum can be described as a function of the rate of release from the stomach and the rate of passage in the small intestine with the ileocaecal valve controlling the release of digesta into the large intestine. However, the results obtained for different sources of native or purified DF are equivocal and show a high degree of variability. This may partly be related to problems in experimental techniques, in particular with regard to the suitability of markers used to estimate transit time for the liquid and solid phases of digesta (Latymer, Low, Fadden, Sambrook, Woodley and Keal, 1990).

For example, Potkins *et al.* (1991) reported a significant decrease in transit time in the small intestine for pectin and guar gum as soluble fibre sources, whereas other authors found no significant effect (Latymer *et al.*, 1990) or even an increase in orocaecal transit time when sugar-beet pulp was fed (Cherbut *et al.*, 1991). It remains questionable whether soluble DF may prevent the separation of liquid and solid phases during passage through the small intestine and thus could conceivably increase transit time to the distal ileum as was assumed by Low (1985). For insoluble fibre sources, Vahouny and Cassidy (1985) suggested an acceleration which was confirmed for alfalfa and oat husk meal by Den Hartog, Boon, Huisman, Van Leeuwen and Van Weerden (1985); this effect could not be found in studies by Latymer *et al.* (1990) and Potkins *et al.* (1991) using different sources of fibre-rich cereal by-products and cellulose.

Rate of passage through the large intestine has been shown to be the major factor on the rate of passage through the total gastrointestinal tract. Therefore, total transit time will primarily be determined by transit time through the large intestine and is often a function of the faecal bulking capacity of fibre. This in turn is determined by fibre degradability, water-binding capacity of fibre escaping degradation and bulk of microbial mass resulting from fibre fermentation. An inverse relationship between microbial degradability, water-binding capacity, faecal bulking capacity and particle size of DF and rate of passage was reported by several authors (Wrick *et al.*, 1983; Low, 1985; Graham and Åman, 1991; Cherbut, Aube, Mekki, Dubois, Lairon and Barry, 1997). This results in an increase in transit time in the large intestine for finely ground or easily fermentable soluble fibre, and a decrease (acceleration) for insoluble fibre with a particle size large enough to ensure intact cell wall structures to retain water and sufficient lignification to resist fermentation. Since soluble fibre sources with a high initial water-binding capacity may be fermented almost completely in the proximal part of the colon they will lose their water-binding capacity and will not contribute to faecal bulking whereas largely undegradable insoluble fibre sources are still able to exhibit some of their water-binding capacity (Potkins *et al.*, 1991; Bach Knudsen and Hansen, 1991). The resulting increase in faecal mass may decrease transit time via stimulation of mechanoreceptors (Cherbut *et al.*, 1997).

Overall transit time

There is one study on the effect of DF on overall transit time in weaner pigs. Schnabel, Bolduan and Güldenpenning (1983) reported that bran decreased transit time through the digestive tract and that transit time decreased over the period of four weeks after weaning (Table 11.2).

Table 11.2. Effect of increasing levels of crude fibre from wheat bran on overall transit time in piglets (initial BW 9.5 kg).

g/kg crude fibre (from bran)	*Transit time (h) weeks post weaning*			
	1	*2*	*3*	*4*
0	192	361	215	141
21	155	113	111	103
31	133	118	118	92
55	117	98	63	78

(After Schnabel, Bolduan and Güldenpenning, 1983)

Despite the fact that there are large individual variations in overall transit time (Latymer *et al.*, 1990; Stanogias and Pearce, 1985 a), some conclusions can be drawn. Firstly, the faster transit time of digesta through the total digestive tract from more insoluble fibre sources such as bran and oatmeal by-products is more likely to be related to differences in the rate of passage through the large intestine rather than to differences in gastric emptying or differences in small intestinal transit times, or both. Secondly, the sometimes faster gastric emptying rate and small intestinal transit time from more soluble fibre sources such as pectin and guar gum is likely to be compensated by, in view of similarities in transit times through the total digestive tract, delayed passage in the large intestine. Thirdly, the lack of effect of supplementary DF in reducing overall transit time in cereal-based diets which have a relatively high content of DF (Latymer *et al.*, 1990) may imply that there is a minimum transit time irrespective of the DF content of the diet (Low, 1993).

Finally, it has been pointed out that the transit time of digesta, in particular in the large intestine, should be of secondary importance to the total digestive capacity of DF, as composition of the undegraded insoluble DF is apparently the main limiting factor for fermentation (Vervaeke, Graham, Dierick, Demeyer and Decuypere, 1991).

DIGESTIBILITY OF DIETARY FIBRE

Ileal digestibility

There is growing evidence that the NSP fraction of piglet diets is modified in the small intestine. In studies by Inborr, Schmitz and Ahrens (1993), the soluble NSP fraction was enriched in all segments of the small intestine of piglets (11-13 kg live weight) fed a diet based on wheat (348 g/kg), barley (348 g/kg) and soybean meal (220 g/kg). As a result, negative digestibility coefficients were obtained ranging between –0.39 and –0.05. This is in agreement with results reported by Vervaeke *et al.* (1991) and is a consequence of NSP being solubilised by intestinal bacteria as they pass through the small intestine. According to Graham, Hesselman, Jonsson and Åman (1986), *Lactobacilli* are capable of hydrolysing pectins and ß-glucans. In a recent study with growing pigs, the molecular weight of soluble ß-glucans was reduced approximately 20-fold during passage through the small intestine despite the low digestibility of ß-glucans at the distal ileum (Johansen, Bach Knudsen, Wood and Fulcher, 1997). Vervaeke *et al.* (1991) pointed out that solubilisation could possibly lead to an underestimation of fibre degradation in the small intestine if inappropriate methods are used; it was reported that apparent ileal digestibilities of the soluble, insoluble and the total NSP fraction were –0.19, 0.21 and 0.18 , respectively, when feeding diets with increasing levels (15-90 g/kg) of sugar-beet pulp, alfalfa meal and wheat bran to pigs of 20 kg live weight. Somewhat lower apparent ileal digestibilities of total NSP in piglets were obtained by Gdala, Buraczewska, Jansman, Wasilewko and Van Leeuwen

(1994) with values ranging from 0.11 to 0.14 and by Inborr *et al.* (1993) of 0.05. Bach Knudsen (1997) concluded from a literature survey that approximately 0.20 of total NSP originating from different feedstuffs were fermented in the small intestine of growing pigs.

It is evident that differences in apparent ileal NSP digestibilities reflect differences in the source of NSP but are also affected by the age of the pigs. For example, the ileal digestibility of ß-glucans seems to increase with age (Graham, Löwgren, Pettersson and Åman, 1988) but is also influenced by the source and solubility of ß-glucans (Bach Knudsen and Hansen, 1991). In pigs of about 20 kg live weight, ileal digestibility of ß-glucans was reported to be 0.40 (Graham *et al.* 1988) which is about 0.1 units higher than the results obtained in pigs weighing between 10 and 13 kg (Inborr *et al.*, 1993). On the other hand, Li, Sauer, Huang and Gabert (1996) obtained ileal digestibilities for ß-glucans of about 0.80 in piglets (10 kg live weight) fed cereal-based diets.

Enrichment of soluble NSP monomers in ileal digesta of piglets has been reported for xylose and glucose, both components being predominant in the NSP fraction of cereals and lupin seeds (Gdala *et al.*, 1994), for uronic acids as major constituents of NSP in sugar-beet pulp (Inborr *et al.*, 1993) and for fucose (Gabert, Sauer, Mosenthin, Schmitz and Ahrens, 1995). The resulting negative ileal digestibilities of these monosaccharides may not only be attributed to the bacterial solubilisation of the NSP fraction but may be also due to the relatively high content of these sugars in endogenous secretions, both of bacterial and animal origin. For example, gastrointestinal mucin contains relatively high amounts of fucose and galactose (Neutra and Forstner, 1987).

The apparent ileal digestibilities of NSP monomers in piglets fed maize starch-based diets containing two different varieties of lupin seeds (350 g/kg) are presented in Table 11.3. The digestibility coefficients of the individual monosaccharides show a large variation both within and between varieties, the differences were significant for arabinose.

Table 11.3 Apparent ileal digestibility coefficients of NSP monomers in piglets (initial live weight 10 kg) fed corn starch-based semi-purified diets containing 350 g/kg lupin seeds

	Diet with	
Item	*L. luteus* c. v. *Amulet*	*L. angustifolius* c. v. *Saturn*
Arabinose	0.34 [a]	0.26 [b]
Xylose	-0.05	-0.02
Mannose	0.91	0.97
Galactose	0.38	0.27
Glucose	-0.01	-0.02
Uronic acids	0.27	0.28
Total NSP	0.11	0.14

[a,b] values within the same row with different superscripts differ ($p<0.05$)

(After Gdala, Buraczewska, Jansman, Wasilewko and Van Leeuwen, 1994)

Faecal digestibility

Studies by Longland, Carruthers and Low (1994) demonstrate the considerable capacity of piglets at both 32 and 56 d of age to digest NSP components from both cereals and sugar-beet pulp in the large intestine. The inclusion of 150 g sugar-beet pulp/kg in a cereal-based control diet resulted in a significant increase in the apparent faecal digestibility of total NSP at both 32 and 56 d of age. In addition, there was a significant improvement in the apparent faecal digestibility of total NSP from the diet with sugar-beet pulp but not from the cereal-based diet with an increase in piglet age (Table 11.4).

Table 11.4 Apparent faecal digestibility coefficients of NSP monomers in piglets 32 and 56 d old fed cereal-based soybean diets containing zero or 150 g sugar-beet pulp/kg

Item	*Age (d)*			
	32		*56*	
	sugar-beet pulp (g/kg)			
	Zero	*150*	*Zero*	*150*
Arabinose	0.49[a]	0.86[b]	0.51[a]	0.90[c]
Xylose	0.59[a]	0.54[a]	0.66[b]	0.69[b]
Mannose	0.59[a]	0.74[b]	0.62[a]	0.79[b]
Galactose	0.77[a]	0.79[a]	0.80[a]	0.84[a]
Glucose	0.53[a]	0.55[a]	0.55[a]	0.68[b]
Uronic acids	0.58[a]	0.81[b]	0.56[a]	0.89[c]
Total NSP	0.52[a]	0.72[b]	0.56[a]	0.78[c]

[a, b, c] values within the same row with different superscripts differ ($p<0.05$)
(After Longland, Carruthers and Low, 1994)

Studies by Vervaeke *et al.* (1991) support the results by Longland *et al.* (1994); similar faecal digestibilities for total NSP and NSP monomers were obtained in pigs of 20 kg live weight fed diets with increasing levels (15-90 g/kg) of beet pulp, alfalfa meal and wheat bran. In addition, only 0.1 of each soluble NSP component was recovered in faeces, suggesting a non-specific and highly fermentative capacity of the microflora in the large intestine for soluble NSP. However, the digestibility coefficients of the residues from insoluble NSP were 0.53 for xylose, 0.60 for glucose, 0.75 for arabinose and 0.82 for uronic acids in studies by Vervaeke *et al.* (1991) which is in the range of values reported by Longland *et al.* (1994), (Table 11.4). Vervaeke *et al.* (1991) concluded that cereal hemicelluloses and cellulose are the most resistant polysaccharides in cereals which was confirmed by Longland *et al.* (1994) who estimated that the NSP derived from cereal grains were only half as digestible as those from sugar-beet pulp.

These studies provide evidence of the early establishment of an efficient intestinal microflora and indicate that the fermentative capacity in young pigs may be higher than was previously thought.

EFFECT OF DIETARY FIBRE ON NUTRIENT DIGESTION AND ABSORPTION

The present level of knowledge makes it difficult to predict possible effects of specific types of DF on nutrient digestion and absorption in the small intestine of piglets. Little is known about the effect of DF on digestion and absorption of vitamins, carbohydrates and lipids in pigs after weaning. Recently, two studies with piglets with an initial live weight of 9 kg were carried out to determine the effect of level and source of fibre on apparent ileal crude protein and amino acid digestibilities (Li, Sauer and Hardin, 1994) and on ileal endogenous nitrogen flow (Schulze, Van Leeuwen, Verstegen and Van Den Berg, 1995).

In the studies by Li *et al.* (1994), the dietary inclusion of Solkafloc up to 133 g/kg, providing 168 g NDF/kg in the diet, did not affect apparent ileal crude protein and amino acid digestibilities. However, it was pointed out that the results, in which purified cellulose was the source of fibre, should not be extrapolated to other sources of fibre. Further consideration for this conclusion was provided by Schulze *et al.* (1995) who reported an increased loss of endogenous and exogenous nitrogen in ileal digesta of piglets fed either a purified NDF source or wheat bran or sunflower hulls. When NDF was part of a dietary ingredient, the total increase in nitrogen collected from the distal ileum was greater than from purified NDF. Li *et al.* (1994) also pointed out that purified sources of fibre such as Solkafloc are less likely to hinder physically the access of proteolytic enzymes to dietary protein than fibre in protein-containing ingredients. On the other hand, viscous sources of fibre in particular may impair digestion and absorption of nutrients in the following manner: (a) by reducing the movement and mixing of intestinal contents, thereby blocking enzyme-substrate interactions, (b) by forming an unstirred water layer, thereby creating a physical barrier to nutrient absorption (Anderson, Deakins, Floore, Smith and Whitis, 1990) and (c) by increasing the water-binding capacity which reduces the rate of diffusion of the products of digestion towards the mucosal surface (Leterme, Van Leeuwen, Théwis and Huisman, 1994). The latter authors reported a significant increase in the level of endogenous amino acids in ileal digesta of pigs (25 kg live weight) when fibre sources with a high water-binding capacity were compared with fibre sources with a low water-binding capacity. It can be concluded from these studies that not only was the re-absorption of endogenous amino acids reduced due to the bulking properties of fibre but also the digestion and absorption of exogenous amino acids from dietary protein.

It can be derived from various studies with growing and adult pigs that the higher faecal nitrogen excretion following the consumption of fibre is a combination of a greater flow of nitrogen to the large intestine and an increased excretion of microbial-bound nitrogen. However, it appears that these conclusions may not fully apply to piglets. According to Li *et al.* (1994), the inclusion of Solkafloc up to 133 g/kg in a diet for pigs after weaning did not affect apparent faecal protein and amino acid digestibilities which is in contrast to studies by Sauer, Mosenthin, Ahrens and Den Hartog (1991) with growing pigs who reported a significant decrease in apparent faecal crude protein and amino acid digestibilities with the inclusion of fibre from Alphafloc in the diet. This decrease resulted, in part, from a significant increase in the quantity of bacterial protein in faeces. Furthermore, supplementation of starter diets with 80 or 160 g peanut hulls /kg had only minor effects on apparent faecal crude protein digestibilities but led to a linear decrease in dry matter digestibilities due to a significant reduction in the digestibilities of NDF and ADF (Kornegay, Rhein-Welker, Lindemann and Wood, 1995). Despite a high fermentation rate of soluble NSP from sugar-beet pulp, the substitution of the cereal component of a piglet diet with 150 g sugar-beet pulp/kg did not affect apparent crude protein and gross energy digestibilities (Longland *et al.*, 1994). In contrast, the inclusion of oat hulls (150 g/kg), soybean hulls (150 g/kg) or alfalfa meal (200 g/kg) in a basal maize-soyabean meal diet decreased significantly the apparent digestibilities of crude protein, dry matter and energy but had only minor effects on mineral absorption and balance in piglets with an initial live weight of 9.7 kg (Moore, Kornegay, Grayson and Lindemann, 1988).

Eggum (1995) concluded from a comprehensive literature review that insoluble DF, because of its low degradability by the microflora, will increase bulk and faecal nitrogen excretion primary due to higher excretion of cell wall-bound protein, whereas soluble DF increases faecal bulk and nitrogen excretion due to greater excretion of microbial nitrogen. However, it appears that these effects cannot be predicted in piglets, and further investigations are warranted.

PRODUCTION ASPECTS

Growth trials

The utilisation of energy from DF in pigs has been of interest for decades. More than 30 years ago, the Agricultural Research Council (1967) concluded from data obtained from experiments with growing pigs that every 10 g increase in crude fibre/kg depresses energy digestibilities by 1.3 %, utilisation of metabolisable energy by 0.9 %, feed efficiency by 3 % and growth by 2 %. Fernández and Jørgensen (1986) came to similar conclusions. However, it still remains questionable wether these relationships also apply to piglets after weaning.

Several studies have been conducted to examine the utilisation of energy from DF in piglets. It can be concluded from most studies that DF originating from different fibre-rich feed ingredients such as soyabean hulls or peanut hulls (Kornegay *et al.*, 1995), sugar-beet pulp (Longland *et al.*, 1994), oat hulls, soya bean hulls or alfalfa meal (Moore *et al.*, 1988) or added as pure lignin (Valencia and Chavez, 1997) did not have major effects on production traits (feed intake, daily gain, feed efficiency). Other authors have even found that certain high-fibre ingredients such as canola meal (McIntosh, Baidoo, Aherne and Bowland, 1986) and destillers' grain (Richards, 1979) in piglet diets improved daily gain and feed efficiency. However, these ingredients were high in crude protein as well as in DF and this may have influenced the results.

Longland *et al.* (1994) concluded from a study, in which sugar-beet pulp (150 g/kg) was added at the expense of wheat and barley that the capacity of piglets four to eight weeks old to digest starch from cereals was somewhat limited. As a result, starch would escape digestion by host enzymes in the small intestine to be fermented in the large intestine. Such a shift from digestion to fermentation of starch, coupled with the higher fermentability of NSP from sugar-beet pulp than from cereals and a slightly higher digestibility of crude protein from the diet with sugar-beet pulp (possibly due to release of cell-wall associated protein from sugar-beet pulp, but not from cereal cell walls), may have led to the same performance by the piglets given the two diets. However, Longland *et al.* (1994) did not completely rule out that the live weight of the piglets may have been confounded by a heavier gut-fill due to the high water-binding capacity of sugar-beet pulp in the diet. Another explanation for the similar performance of piglets fed diets containing either oat hulls (150 g/kg), soya bean hulls (150 g/kg) or alfalfa meal (200 g/kg) included at the expense of maize and soya bean meal was given by Moore *et al.* (1988) who interpreted the subtle effects that these fibre sources had on apparent crude protein and energy digestibilities as of minor physiological and nutritional significance to the pigs. On the other hand, no differences in performance between piglets fed high- and low-fibre diets indicated that energy utilisation may have been increased since the lower energy concentration of the high-fibre diet was not compensated by a higher feed intake (Kornegay *et al.*, 1995); however, no further information that explained the improvement in energy utilisation was provided.

Contribution of volatile fatty acids to the energy supply of piglets

The net yield of dietary energy arising from fermentation mainly in the large intestine is expected to be lower than that of carbohydrates which are absorbed as monosaccharides in the small intestine. This difference results from additional losses as CH_4, H_2 and fermentation heat and from a lower efficiency of the utilisation of VFA in the intermediary metabolism of the pig. VFA, which include mainly acetate, propionate and butyrate, are rapidly absorbed and supply energy to local intestinal tissues; they are transported by the systemic circulation to and metabolised in peripheral tissues (e.g. Imoto and Namioka, 1978; Latymer and Woodley, 1984). An increase in the dietary level of crude fibre usually increases the amount of VFA produced in the large intestine (Gargallo and Zimmerman, 1981; Stanogias and Pearce, 1985b) but this is controversial (Argenzio and Southworth, 1974). Dierick, Vervaeke, Demeyer and Decuypere (1989), in a review of the literature, concluded that the average contribution of net energy from VFA to total net energy for maintenance was 0.15 in growing pigs ranging between 20 – 95 kg in live weight. However, it was considered that fermentation and production of VFA should be evaluated not in isolation but as an integrated part of the overall energy supply to the pig. The total impact of feeding DF on the energy supply of the pig was demonstrated by Giusi-Perier, Fiszlewicz and Rérat (1989). The additional daily supply of energy in the form of VFA, after increasing the level of cellulose or alfalfa meal in the diet, did not compensate fully for the decreased absorption of amino acids and glucose in the small intestine which in turn resulted in a lower energy utilisation.

In conclusion, it is unlikely that VFA will contribute significantly to the total energy supply of pigs, except for sows fed fibre-rich diets at maintenance level. The contribution of VFA to the energy requirement at maintenance is lowest for piglets capable of rapid growth. Since piglets held at zero gain or at maintenance are not productive animals, it is debatable whether such a small contribution of net energy from VFA to the overall energy supply is of economic advantage in the production of fast growing piglets. Finally, from a methodological point of view, a correct determination of the significance of VFA production to the energy supply of the pig can only be made through respiration trials and carcass analyses because only these approaches take into account the possible depressive effects of DF on nutrient absorption in the small intestine (Dierick *et al.*, 1989). In addition, growth parameters (e.g. daily gain, feed efficiency) may not be appropriate to determine the utilisation of energy from DF since the results may be confounded by differences in gut fill due to the bulking effect of some fibre sources.

Dietary fibre, intestinal microflora and microbial fermentation

MICROFLORA

Microbial fermentation occurs in all regions of the digestive tract to some degree. The extent is very dependent on the region of the gastrointestinal tract, the age of the pig and the composition of the diet (Cranwell, 1968). While the gastrointestinal tract is largely sterile at birth, the stomach and small intestine are colonised from about 3 hours after birth and bacteria are first detected in the large intestine at about 12 hours after birth (Smith and Jones, 1963). The concentration of the major bacterial groups in the various regions of the large intestine of piglets one to eight weeks old are presented in

Table 11.5. As can be seen, there is a considerable variation in the values reported by various authors; in some cases there is a range from 0 – 9.0 log colony-forming units/ g digesta. These large differences could be attributed to animal variation but could also reflect differences in techniques used for measuring the counts of anaerobic microbes (Conway, 1994).

Table 11.5 Range of concentrations of the major bacterial groups in different regions of the large intestine and in faeces of piglets 1 to 8 weeks of age (log colony-forming units per gram digesta)

Region	*Lactobacilli*	*Enterobacteria*	*Streptococci*	*Clostridia*	*Bacteroides*
Caecum	6.3–11.4	6.4–9.5	0–9.0	0–6.8	0–8.1
Colon	7.1–11.4	5.9–9.5	0–9.0	0–9.2	0–8.6
Rectum	6.8–9.7	5.5–9.0	5.2–9.0	0–9.0	5.2–8.6
Faeces	8.0–11.0	6.0–10.0	4.0–9.0	2.0–9.5	0–10.5

After Conway, 1994

It has been shown that microbes dominating the colon are Gram positive (Salanitro, Blake and Muirhead, 1977) while those dominating the caecum are Gram negative (Robinson, Allison and Bucklin, 1981). While *Lactobacilli* and *Streptococci* tend to dominate in the anterior sections of the digestive tract, *Bacteroides, Fusobacteria, Peptostreptococci* and *Eubacteria* become more important towards the distal sections of the intestinal tract (Savage, 1984). The density of microbial population in the caecum and colon amounts to 10^{10} – 10^{11} viable counts/g digesta, comprising more than 500 species of which only a few have been described in detail (Moore, Moore, Cato, Wilkins and Kornegay, 1987).

The increasing pressure for improving production capacity in the pig industry directs attention to possible ways of regulating the gastrointestinal microflora in order to enhance health and performance. It is usually recognised that the structure and function of the microflora can be influenced by environmental conditions within the gut ecosystem such as nutrient availability, pH, redox potential and microbial interactions. In addition, the composition of the microflora can be altered by modifying the diet. Jensen (1993) refers to diet composition as "probably the single most important control factor for microbial activity". In particular, diets high in DF promote the presence of cellulolytic bacteria without changing the total number of microorganisms (Varel and Pond, 1985; Moore *et al.*, 1987; Varel, 1987). Others (Jonsson and Henningsson, 1991) have shown a correlation between the diet of the piglet and the occurrence of faecal *Lactobacilli* with the ability to degrade mixed linked ß-glucans. Another approach is to use a range of NDO which cannot be degraded by host enzymes and hence can pass into the large intestine to stimulate the proliferation of a chosen microorganism such as

Bifidobacterium in order to maintain a desirable flora and suppress the growth of potential pathogenic microorganisms such as *E. coli*. Such a specific effect has not been identified with NSP, possibly due to the heterogeneity of the NSP fraction, but there is growing evidence that the unspecific stimulation of indigenous microorganisms may prevent the establishment of an undesirable microflora.

TROPHIC AND SYSTEMIC EFFECTS OF VOLATILE FATTY ACIDS

As was reviewed by Mosenthin and Hambrecht (1998), the production of VFA has several possible actions on the mucosa of the digestive tract. The VFA are readily absorbed by the mucosa of the caecum and colon, but only acetic acid reaches the systemic circulation in appreciable amounts via the liver (Cummings and Englyst, 1987) where it can be metabolised in muscle and adipose tissue. Propionic acid is metabolised in the liver (Eastwood, 1992) where it affects gluconeogenesis. The absorption of VFA from the colon stimulates sodium absorption from the intestinal lumen and thus provides an efficient mechanism for the re-absorption of water (Roediger, 1980). In this context, VFA must be seen as antidiarrhoeal and failure in their production leads to disturbances of salt and water balance. Since VFA are rapidly absorbed, they do not contribute to the osmotic load in the lumen of the large intestine (Bergmann, 1990). An additional effect of VFA production which is associated with a decrease in pH is binding of potentially toxic NH_3 with hydrogen to produce NH_4^+ that is non-diffusable (Younes, Garleb, Behr, Remsey and Demigne, 1995). This results in an increased faecal nitrogen excretion, lower blood urea levels and a reduction in urinary nitrogen excretion (Mosenthin, Sauer, Henkel, Ahrens and De Lange, 1992). These effects were demonstrated in experiments with NDO, RS and DF. In the case of DF, the higher faecal nitrogen excretion is quite often a combination of a higher flow of nitrogen to the large intestine and an increased excretion of nitrogen in the form of bacterial protein (Mosenthin, Sauer and Ahrens, 1994; Younes *et al*., 1995).

Of special interest is butyric acid which accounts for 0.70 of the total energy consumption of the colonocytes, and which may affect the regulation of gene expression and cell growth (Smith and German, 1995; Bingham, 1996). Butyrate represents an obligatory energy substrate for the epithelial cells of the colon and cannot be replaced by glucose or glutamine. In addition, there is evidence that the metabolic capacity of these cells is dependent on the content of DF in the diet (Darcy-Vrillon, Morel, Cherbuy, Bernard, Posho, Blachier, Meslin and Duée, 1993). Thus, the health of the epithelium of the colon may depend on the production of VFA, and in particular on the production of butyrate. A selective antimicrobial effect of dietary n-butyrate was observed in the intestine of piglets by increasing the number of *Lactobacilli* and decreasing the number of *E. coli* (Gálfi and Bokori, 1990). As a consequence, great effort has been expanded in attempting to enhance butyrate production in the large intestine of pigs by manipulating microbial fermentation. Studies with pigs weighing 35 – 45 kg revealed that arabinoxylans, but not ß-glucans, located in the cell walls of oat bran were responsible for an increase in butyrate production in the large intestine (Bach Knudsen, Jensen and Hansen, 1993). Furthermore, starch escaping digestion in the small intestine (RS) may yield high butyrate concentrations in contrast to fermentation of the more oxidised substrate pectin where acetate is the major end product (Englyst, Hay and McFarlane, 1987). The VFA produced during *in vitro* fermentation of different polysaccharide substrates are presented in Table 11.6. In conclusion, substrate redox state and availability, fermentation rate, transit time through the large intestine and the composition of the microflora may affect the absolute amounts and relative proportions of VFA produced by the microflora (Allison and McFarlane, 1989).

Table 11.6. *In vitro* production of VFA from fermentation of different polysaccharides

Item	*VFA (mg/mg polysaccharide fermented)*		
	Acetate	*Propionate*	*Butyrate*
Starch	0.25	0.13	0.21
Arabinogalactan	0.19	0.20	0.04
Xylan	0.42	0.10	0.02
Pectin	0.27	0.06	0.01

After Englyst, Hay and McFarlane, 1987

Although it is usually agreed that DF can modify the intestinal morphology by affecting appearance, length and number of villi, cell proliferation, mucosal cell division and absorptive functions, the results obtained are equivocal. The feeding of a high-fibre diet to piglets (initial live weight 14.3 kg) altered intestinal morphology and increased the rate of mucosal cell turnover in one study (Jin, Reynolds, Redmer, Caton and Crenshaw, 1994). As can be concluded from Figure 11.1, DF reaching the large intestine may not only affect the structure of this part of the digestive tract but may also affect gut structure in the small intestine through the influence of VFA (Jin *et al.*, 1994). Studies with sheep suggest that absorbed n-butyrate may stimulate the production of various messenger molecules such as insulin which, in turn, may enhance growth and development of the intestinal epithelium (Gálfi and Neogrády, 1996). In another study, however, measures of cellular proliferations in the mucosal epithelium of the caecum and colon of miniature pigs were not significantly correlated with the quantity of soluble DF consumed or with intestinal concentrations of VFA, including acetate and butyrate, or with volume or pH of digesta (Fleming, Fitch and De Vries, 1992). It was suggested that DF probably influences intestinal cell proliferation through many mechanisms, some of which are interactive. On the other hand, effects of VFA on epithelial cell proliferation may only be observed when the baseline concentrations of VFA are suppressed. There appears to be a limit for the stimulatory effect of VFA that cannot be exceeded by increasing the concentration of VFA through increased fermentation of DF (Sakata, 1987; Fleming *et al*., 1992). This was confirmed in studies by Moore *et al*. (1987), in which the addition of different fibre sources to a basal diet with a relatively high fibre content (90 g NDF/kg) did not affect intestinal morphology of piglets.

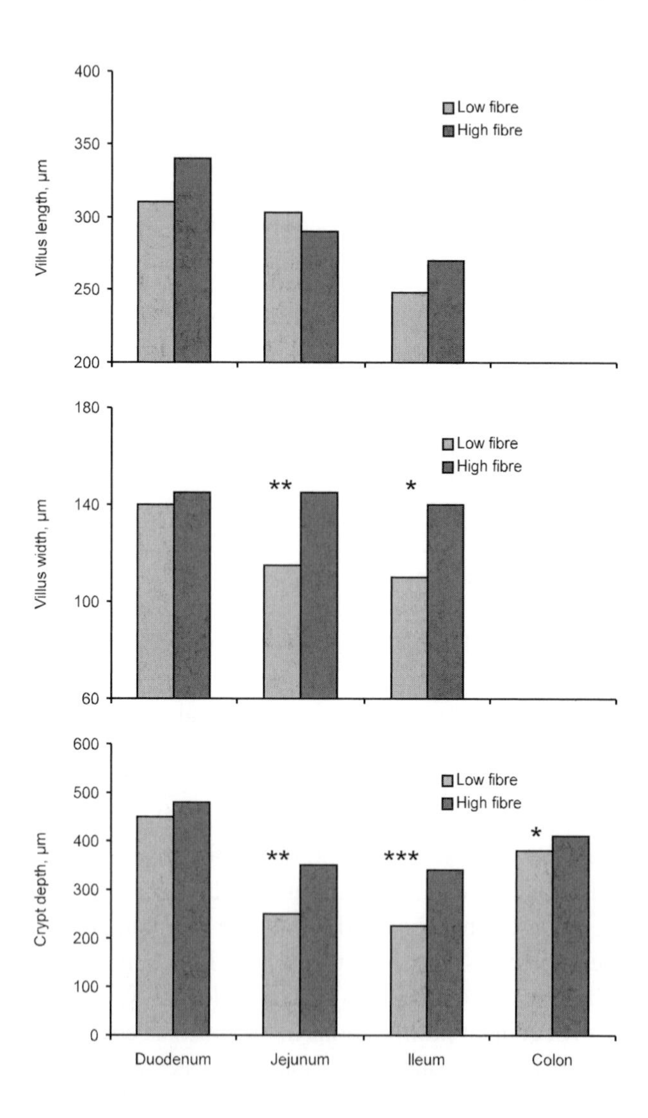

Figure 11.1 Effect of low- and high-fibre diets on gut structure in piglets (after Jin, Reynolds, Redmer, Caton and Crenshaw, 1994)

Most studies have shown an increase in visceral weights with the inclusion of DF in the diet of growing pigs (e.g. Pond, Varel, Dickson and Hascheck, 1989; Jørgensen, Zhao and Eggum, 1996) and piglets (Hambrecht, 1998), (Table 11.7), whereas Jin *et al.* (1994) failed to show this effect. Both Jin *et al.* (1994) and Hambrecht (1998) exposed the piglets over a period of 14 days to high-fibre diets before the measurements were taken. Thus, it can be questioned if the time period was too short to measure stimulatory effects of DF on visceral weights as was suggested by Jin *et al.* (1994). Hambrecht (1998) assumed that no response to DF is most likely due to a low production rate of VFA from a rather slowly and poorly fermentable fibre source or when the baseline concentrations of VFA from the basal diet are relatively high.

Table 11.7 Effect of non-starch polysaccharides (NSP) on visceral weights of piglets 2 weeks after weaning.

		Level of NSP in the diet	
		43 g/kg	*138 g/kg*
Body weight	kg	13.3	13.2
Empty body weight	kg	11.8	11.2
Stomach	g	95[a]	128[b]
Proximal small intestine	g	336[a]	387[b]
Distal small intestine	g	299[a]	361[b]
Caecum	g	26[a]	92[b]
Proximal colon	g	57[a]	89[b]
Distal colon	g	35[a]	54[b]
Rectum	g	20[a]	29[b]

[a,b] values within the same row with different superscripts differ ($p<0.05$)
After Hambrecht, 1998

Microbial, physiological and nutritional aspects of non-digestible oligosaccharides

In recent years, it has been recognised that many NDO have the ability to promote specifically the proliferation of *Bifidobacteria* in the large intestine, since NDO escape enzymatic digestion by host enzymes. These NDO may potentially be regarded as prebiotics, recently defined as "non-digestible food ingredients that beneficially affect the host by selectively stimulating the growth and/or activity of one or a limited number of bacteria in the colon, and thus improve health" (Gibson and Roberfroid, 1995). The different classes and main characteristics of NDO which are present in plant material or can be commercially produced were described in detail by Mul and Perry (1994). A limited number of NDO have been included in diets for pigs after weaning, with FOS, TOS and α-GOS as the most frequently used NDO.

MICROFLORA AND MICROBIAL METABOLITES

It can be concluded from *in vitro* and *in vivo* studies in different species including humans that NDO are usually readily fermented and act as selective growth substances for *Bifidobacterium,* so called "bifidus growth factors" (e.g. Roiberfroid, Gibson and Delzenne, 1993). *Bifidobacterium* and *Lactobacillus* species suppress the growth of pathogenic and putrefactive bacteria primarily by stimulating the production of acetate and lactate, which decreases the pH and may reduce the incidence of diarrhoea (Modler, McKellar and Yaguchi, 1990). In addition, supplementation of diets with FOS increased epithelial cell proliferation in the caecum and colon of neonatal pigs (Howard, Kerley, Gordon, Pace and Garleb, 1993).

However, studies with weanling pigs revealed that the effects of different sources of NDO on the microflora and their metabolites are equivocal. Galactan, a polymer of D- and L-galactose, decreased the pH and the concentration of *E. coli* in ileal digesta of newly-weaned pigs challenged orally with *E. coli*; it was suggested that the decrease in pH was responsible for the decline in *E. coli* because they are more sensitive to a low pH (Mathew, Sutton, Scheidt, Patterson, Kelly and Meyerholtz, 1993). Supplementation of diets for weanling pigs (initial live weight 10 kg) with FOS or TOS resulted in a lower number of aerobic bacteria which was probably due to a lower content of *E. coli*, the most dominant group of aerobic bacteria in ileal digesta (Houdijk, Hartemink, Van Laere, Williams, Bosch, Verstegen and Tamminga, 1997); a significant drop in pH was observed, but no changes in the total amount of VFA in ileal digesta. Propionate levels, probably formed from lactate, were increased rather than acetate levels and the pH in faeces was increased significantly due to the supplementation of the basal diet with FOS or TOS. No effect on the pH in faeces was obtained in a similar study with piglets which were heavier (initial live weight 15 kg) and fed diets with lower levels of FOS and TOS (Houdijk, Bosch, Verstegen and Berenpas, 1998). In agreement with the aforementioned study, several research groups have failed to detect an efffect of NDO on the composition and activity of the microflora, the level of microbial metabolites including VFA and NH_3, the pH, and/or the incidence of diarrhoea. In these studies, the pigs were weaned between 21 to 28 d of age and fed diets supplemented with TOS, α-GOS, gluco-oligosaccharides or lacitol (Gabert, Sauer, Mosenthin, Schmitz and Ahrens, 1995) or with galactosyl lactose, a trisaccharide produced by the enzymatic conversion of lactose (Mathew, Robbins, Chattin and Quigley, 1997) or with the sucrose thermal oligosaccharide caramel which is produced by thermal treatment of amorphous anhydrous acidified sucrose (Orban, Patterson, Adeola, Sutton and Richards, 1997), or with FOS (Farnworth, Modler, Jones, Cave, Yamazaki and

Rao, 1992; Howard *et al.*, 1993). Recently, oligosaccharides with mannose as the primary carbohydrate derived from yeast cell walls have been shown to improve health and performance in piglets (Spring and Privulescu, 1998). This was attributed to a reduction in the number of pathogens in the digestive tract, stimulation of the immune system and binding of mycotoxins. Since these mannan-oligosaccharides are associated with the yeast cell wall, other yeast metabolites may contribute to some of the effects described.

DIGESTIBILITY

Although NDO are resistant to digestion by host enzymes, Graham and Åman (1986) reported that the apparent ileal digestibility of fructans from Jerusalem artichoke tubers was 0.51. Furthermore, data from Gabert *et al.* (1995) and Petkevicius, Bach Knudsen, Nansen, Roepstorff, Skjøth and Jensen (1997) indicated that most of the TOS, gluco-oligosaccharides, lacitol and inulin were degraded at the distal ileum. However, in studies by Gabert *et al.* (1995), the fermentation of TOS and gluco-oligosaccharides was not correlated with changes in acetate concentrations in ileal digesta. Thus, it remains open if, due the relatively low levels of NDO in the diets (2 g/kg), the growth of *Bifidobacterium* and heterofermentative *Lactobacillus* species was promoted in this study.

There was no major effect of various sources and levels of NDO in diets for pigs after weaning on ileal or faecal nutrient and energy digestibilities (e.g. Gabert *et al.*, 1995; Houdijk *et al.*, 1997; Mathew *et al.*, 1997). Even changes in microbial ecology as observed in studies by Houdijk *et al.* (1997), did not affect nutrient and energy digestibilities.

PRODUCTION ASPECTS

Supplementation of diets for early-weaned pigs with 2.5 or 5g FOS/kg increased daily gain and reduced the incidence of diarrhoea (Modler *et al.*, 1990). According to Mul and Perry (1994), positive effects of NDO on performance of piglets are demonstrated best under farm conditions. Results provided by Japanese producers showed improvements in daily gain and feed efficiency between 4 to 6 % and in most cases a reduction in the incidence and severity of diarrhoea during the post weaning period. This may explain why under more standardised conditions at experimental research stations, there is often a failure to show an improvement in performance or incidence of diarrhoea when diets supplemented with different sources and levels of NDO were fed (Farnworth *et al.*, 1992; Mathew *et al.*, 1997; Orban *et al.*, 1997; Houdijk *et al.*, 1997, 1998).

It appears that there is not a clear and predictable relationship between supplementation of NDO to piglet feeds and the effects on digestive processes in the intestinal tract and on over-all performance of piglets. There may be several factors involved that partly explain the lack of response to dietary supplementation of NDO. Firstly, cereals and legumes which are commonly used in diets for pigs after weaning contribute certain levels of NDO to the diet (up to 30 g/kg), depending on factors such as variety and harvesting and processing conditions (Houdijk *et al.*, 1997). These relatively high levels may partly explain the lack of response to supplemented NDO because of masking or dilution effects (Gabert *et al.*, 1995). Secondly, well-kept animals that are not challenged with pathogenic bacteria or certain stress factors (e.g. housing conditions, feeding regimen) will hardly respond to supplementation of NDO. Thirdly, it cannot be ruled out completely that *E. coli* and several species of *Clostridium* do not metabolise NDO such as FOS (Modler *et al.*, 1990). Recent studies by Hartemink and Rombouts (1997) revealed that a significant proportion of different sources of NDO including FOS and TOS were fermented by other species than *Bifidobacterium* including species from *Clostridium* and *E. coli*. None of the NDO that were tested were selective for *Bifidobacteria*. This may explain increased flatulence and other abdominal discomfort after consumption of higher levels of NDO in humans and may stimulate research in pigs.

Interaction between dietary fibre and intestinal diseases

An interesting approach to prevent intestinal infections in pigs and weanling pigs has recently been introduced (McDonald, Pluske, Pethick and Hampson, 1996; Pluske, Siba, Pethick, Durmic, Mullan and Hampson, 1996; Siba, Pethick and Hampson, 1996 - see chapter 10). According to these studies, both the clinical expression of swine dysentery following experimental infection with *S. hyodysenteriae* and the occurrence of post-weaning diarrhoea caused by the proliferation of certain serotypes of strongly ß-haemolytic *E. coli* in the small intestine could be reduced by feeding diets low in rapidly fermentable fibre. Soluble NSP in guar gum, NDO in legumes and RS in feedstuff probably play a role in the aetiology of swine dysentery and colibacillosis (post-weaning diarrhoea). The expression of these diseases originating in both the small and large intestine could be controlled by feeding diets of low fermentability and high ileal digestibility. Diets based on cooked white rice and animal protein that contained low levels of soluble NSP (less than 15 g/kg) and low levels of RS inhibited or even prevented the development of swine dysentery and colibacillosis.

However, the mechanisms by which rapidly-fermentable fibre sources stimulate the proliferation of pathogenic bacteria is not known yet. They may have increased the viscosity of the intestinal contents, thereby prolonging intestinal transit times and increasing the availability of substrate for growth for pathogenic bacteria in the small intestine (Hampson, Pethick, Pluske, Mullan, Durmic, McDonald and Siba, 1997). As was suggested by Siba *et al.* (1996), the protective effect of diets with a low content of rapidly fermentable fibre sources may be an indirect effect of mucin production, factors affecting the mobility of *S. hyodysenteriae* in the mucosal lining of the epithelium of the colon and/or a higher dry matter content of colonic contents that may inhibit spirochaetal survival in the large intestine.

On the other hand, inclusion of fibre in the diet has been suggested as a means of ameliorating post-weaning diarrhoea by enhancing the development of intestinal function which, in turn, may decrease the transit time of digesta, and as such may reduce the availability of substrate for bacterial growth (Bolduan, Jung, Schnabel and Schneider, 1988). In addition, Göransson, Lange and Lönnroth (1995) reported a lower incidence of post-weaning diarrhoea in piglets after feeding high levels of NSP from guar gum. This area of research certainly warrants further investigations.

Conclusions

Due to the limited number of studies carried out with piglets after weaning, the effects of DF on gastrointestinal secretions, gut motility, transit time, absorption of nutrients and gut hypertrophy have been identified but not quantified and the results are inconclusive. Enrichment of soluble monomers of NSP in ileal digesta, as a consequence of NSP being solubilised by bacteria in the small intestine, has been reported and there are also studies that provide evidence of an early establishment of an active microflora in the large intestine which is able to ferment both soluble and non-soluble NSP to a considerable degree. The microbial products of carbohydrate fermentation in the large intestine are VFA, which mainly include acetate, propionate and butyrate. These VFA barely contribute to the maintenance energy requirements of piglets, are partly metabolised in the epithelial cells of the colon, can inhibit growth of pathogenic microorganisms and promote electrolyte absorption by moderately lowering the pH. In addition from being an important fuel for colonocytes, butyrate has been shown to regulate gene expression and epithelial cell proliferation. Thus, the development and health of the epithelial cells in the colon may depend on the production of VFA, in particular on the production of butyrate.

There is growing interest in the use of a range of NDO which cannot be degraded by host enzymes and hence pass into the large intestine where a limited range of microbes can ferment these compounds. The most notable effect of NDO such as FOS and TOS is their specific stimulation of one or a limited number of bacterial species (e.g. *Bifidobacteria*). They thereby assist in maintaining a desirable flora that suppresses the growth of pathogenic microorganisms. It should be emphasised, however, that there is not a predictable relationship between the inclusion of specific NDO in diets for piglets and their effects on digestive processes and over-all performance.

An interesting approach to prevent intestinal infections in piglets has been recently introduced. Piglets fed soluble sources of NSP (e.g. guar gum) and RS were challenged with different strains of pathogenic bacteria. As a result, the intestinal tract was colonised by these pathogenic bacteria and the piglets developed dysentery whereas piglets fed diets low in soluble fibre and RS content were protected against swine dysentery. However, factors responsible for the protective effect of diets with a low content of fermentable fibre have not yet been identified.

In conclusion, a number of questions remain to be answered. The complex relationships between the source and level of DF, the microflora and host metabolism are, particularly in piglets, largely unexplained. Furthermore, the energy value of VFA produced from fibre in the large intestine still has to be adequately described. Recent studies on the effect of DF on intestinal growth, cell proliferation and gut morphology in pigs revealed that DF probably influences these processes through several interactive mechanisms, some of which include luminal factors whose mode of action needs to be clarified. Finally, more information is needed on the relationship between fibre composition and the physico-chemical properties, including water-binding capacity, viscosity, cation-exchange capacity, adsorption characteristics and fermentability which, in turn, determine gastrointestinal activities.

References

Agricultural Research Council (1967) The Nutrient Requirements of Farm Livestock, No 3: Pigs. Agricultural Research Council, London.

Allison, C. and MacFarlane, G.T. (1989) *Applied and Environmental Microbiology*, **55**, 2894-2898.

Anderson, J.W., Deakins, D.A., Floore, T.L., Smith, B.M and Whitis, S.E. (1990) *Food Science Nutrition*, **29**, 95-147.

Argenzio, R.A. and Southworth, M. (1975) *American Journal of Physiology*, **228**, 454-460.

Bach Knudsen, K.E. (1997) In Proceedings of the International Symposium "Non-digestible oligosaccharides: Healthy Food for the Colon?" – 1997, Addendum. Edited by R. Hartemink. The Graduate School VLAG, Wageningen Institute of Animal Sciences, Wageningen.

Bach Knudsen, K.E. and Hansen, I. (1991) *British Journal of Nutrition*, **65**, 217-232.

Bach Knudsen, K.E., Jensen, B.B. and Hansen, I. (1993) *The Journal of Nutrition*, **123**, 1235-1247.

Bardon, T. and Fioramonti, J. (1983) *British Journal of Nutrition*, **50**, 685-690.

Bergman, E.N. (1990) *Physiological Reviews*, **70**, 567-590.

Bingham, S.A. (1996) *Nutrition Research Reviews*, **9**, 197-239.

Bolduan, G., Jung, H., Schnabel, E. and Schneider, R. (1988) *Pig News and Information*, **9**, 381-385.

Cherbut, C., Albina, E., Champ, M., Doublier, J.L. and Lecannu, G. (1990) *Digestion*, **46**, 205-213.

Cherbut, C., Salvador, V., Barry, J.-L., Doulay, F. and Delort-Laval, J. (1991) *Food Hydrocolloids*, **5**, 15-22.

Cherbut, C., Aube, A.-C., Mekki, N., Dubois, C., Lairon, D. and Barry, J.-L. (1997) *The British Journal of Nutrition*, **77**, 33-46.

Conway, P.L. (1994) In: Proceedings of the VIth International Symposium on Digestive Physiology in Pigs (Vol. 2) – 1994, pp 231-240. Edited by W.-B. Souffrant and H. Hagemeister. EAAP-Publication No. 80, Dummerstorf.

Cranwell, P.D. (1968) *Nutrition Abstracts and Reviews*, **38**, 721-730.

Cummings, J.H. and Englyst, H.N. (1987) *American Journal of Clinical Nutrition*, **45**, 1243-1255.

Darcy-Vrillon, B., Morel, M.T., Cherbuy, C., Bernard, F., Posho, L., Blachier, F., Meslin, J.-G. and Duée, P.H. (1993) *The Journal of Nutrition*, **123**, 234-243.

Den Hartog, L.A., Boon, P.J., Huisman, J., Van Leeuwen, P. and Van Weerden, E.J. (1985) In: Proceedings of the 3rd International Seminar on Digestive Physiology in the Pig - 1985, pp 199-202. Edited by A. Just, H. Jørgensen and J. Fernández. Copenhagen, National Institute of Animal Science, Copenhagen.

Dierick, N.A., Vervaeke, I.J., Demeyer, D.I. and Decuypere, J.A. (1989) *Animal Feed Science and Technology*, **23**, 141-167.

Eastwood, M.A. (1992) *Annual Reviews in Nutrition*, **12**, 19-35.

Edwards, C.A., Johnson, I.T. and Read, N.W. (1988), *European Journal of Clinical Nutrition*, **42**, 307-312.

Eggum, B.O. (1995) *Archives of Animal Nutrition*, **48**, 89-95.

Englyst, H.N. and Cummings, J.H. (1987) *The American Journal of Clinical Nutrition*, **45**, 423-431.

Englyst, H.N., Hay, S. and McFarlane, G.T. (1987) *FEMS Microbiology Ecology*, **45**, 163-171.

Farnworth, E.R., Modler, H.W., Jones, J.D., Cave, N., Yamazaki, H. and Rao, A.V. (1992) *Canadian Journal of Animal Science*, **72**, 977-980.

Férnandez, J.A. and Jørgensen, H. (1986) *Livestock Production Science*, **15**, 53-71.

Fleming, S.E., Fitch, M.D. and de Vries, S. (1992) *Journal of Nutrition*, **122**, 906-916.

Gabert, V.M., Sauer, W.C., Mosenthin, R., Schmitz, M. and Ahrens, F. (1995) *Canadian Journal of Animal Science*, **75**, 99-107.

Gálfi, P. and Bokori, J. (1990) *Acta Veterinaria Hungarica*, **38**, 3-17.

Gálfi, P. and Neogrády, S. (1996) In 4th International Feed Production Conference – 1996. Edited by G. Piva, Piacenza.

Gargallo, J. and Zimmerman, D.R. (1981) *Journal of Animal Science*, **53**, 1286-1291.

Gdala, J., Buraczewska, L., Jansman, A.M.J., Wasilewko, J. and Van Leeuwen, P. (1994) In Proceedings of the VIth International Symposium on Digestive Physiology in Pigs (Vol. 1) – 1994, pp 93-96. Edited by W.-B. Souffrant and H. Hagemeister. EAAP-Publication No. 80, Dummerstorf.

Gibson, G.R. and Roberfroid, B.M. (1995) *Journal of Nutrition*, **125**, 1401-1412.

Giusi-Perier, A., Fiszlewicz, M. and Rérat, A. (1989) *Journal of Animal Science*, **67**, 386-402.

Göransson, L. (1997) *Recent Advances in Animal Nutrition – 1997*, pp 45-56. Edited by P.C. Garnsworthy and J. Wiseman. Nottingham University Press, Nottingham.

Graham, H. and Åman, P. (1986) *Food Chemistry*, **22**, 67-76.

Graham, H. and Åman, P. (1991) *Animal Feed Science and Technology*, **32**, 143-158.

Graham, H., Hesselman, K., Jonsson, E and Åman, P. (1986) *Nutrition Reports International*, **34**, 1089-1096.

Graham, H., Löwgren, W., Pettersson, D. and Åman, P. (1988) *Nutrition Reports International,* **38**, 1073-1079.

Hambrecht, E. (1998) Effect of non-starch polysaccharides on performance, incidence of diarrhoea and gut growth in weaned pigs. MSc. Thesis, Institute of Animal Nutrition, Hohenheim University, Stuttgart.

Hampson, D.J., Pethick, D.W., Pluske, J.R., Mullan, B.P., Durmic, Z., McDonald, D.E. and Siba, P.M. (1997) *Feed Mix*, **5**, 8-12.

Hartemink, R. and Rombouts, F.M. (1997) In Proceedings of the International Symposium "Non-digestible oligosaccharides: Healthy Food for the Colon?" – 1997, pp 57-66. Edited by R. Hartemink. The Graduate School VLAG, Wageningen Institute of Animal Sciences, Wageningen.

Houdijk, J.G.M., Bosch, M.W., Verstegen, M.W.A. and Berenpas, H.J. (1998) *Animal Feed Science and Technology*, **71**, 35-48.

Houdijk, J.G.M., Hartemink, R., Van Laere, K.M.J., Williams, B.A., Bosch, M.W., Verstegen, M.W.A. and Tamminga, S. (1997) In: Proceedings of the International Symposium "Non-digestible oligosaccharides: Healthy Food for the Colon?" – 1997, pp 69-78. Edited by R. Hartemink. The Graduate School VLAG, Wageningen Institute of Animal Sciences, Wageningen.

Howard, M.D., Kerley, M.S., Gordon, D.T., Pace, L.W. and Garleb, K.A. (1993) *Journal of Animal Science*, **71**, (Suppl. 1), 177.

Imoto, S. and Namioka, S. (1978) *Journal of Animal Science*, **47**, 467-478.

Inborr, J., Schmitz, M. and Ahrens, F. (1993) *Animal Feed Science and Technology*, **44**, 113-127.

Jensen, B.B. (1993) In: Proceedings of the 44th Annual Meeting of the EAAP, **3**, Aarhus, Denmark.

Jin, L., Reynolds, L.P., Redmer, D.A., Caton, J.S. and Crenshaw, J.D. (1994) *Journal of Animal Science*, **72**, 2270-2278.

Johansen, H.N. and Bach Knudsen, K.E. (1994 a) *British Journal of Nutrition*, **72**, 299-313.

Johansen, H.N. and Bach Knudsen, K.E. (1994 b) *British Journal of Nutrition*, **72**, 717-729.

Johansen, H.N., Bach Knudsen, K.E., Sandström, B. and Skjøth, F. (1996) *British Journal of Nutrition*, **75,** 339-351.

Johansen, H.N., Bach Knudsen, K.E., Wood, P.J. and Fulcher, R.G. (1997) *Journal of the Science of Food and Agriculture*, **173**, 81-92.

Jonsson, E. and Hemmingsson, S. (1991) *The Journal of Applied Bacteriology*, **70**, 512-516.

Jørgensen, H., Zhao, X.-Q. and Eggum, B.O. (1996) *British Journal of Nutrition*, **75**, 365-378.

Korczyñski, W., Budzyñska, M. and Zebrowska, T. (1997) In Proceedings of the VIIth International Symposium on Digestive Physiology in Pigs – 1997, pp 613-616. Edited by J.-P. Laplace, C. Février and A. Barbeau. INRA, Saint Malo.

Kornegay, E.T., Rhein-Welker, D., Lindemann, M.D. and Wood, C.M. (1995) *Journal of Animal Science*, **73**, 1381-1389.

Langlois, A., Corring, T. and Février, C. (1987) *Reproduction, Nutrition, Development*, **27**, 929-939.

Laplace, J.P. (1982) In: Digestive Phyisology in the Pig. Les Colloques de l'INRA 12 – 1982, pp 29-44. INRA, Paris.

Latymer, E.A. and Woodley, S.C. (1984) The Proceedings of the Nutrition Society, **43**, 22A.

Latymer, E.A., Low, A.G., Fadden, K., Sambrook, I.E., Woodley, S.C. and Keal, H.D. (1990) *Archives of Animal Nutrition*, **40**, 667-680.

Leterme, P., Van Leeuwen, P., Théwis, A. and Huisman, J. (1994) In Proceedings of the VIth International Symposium on Digestive Physiology in Pigs (Vol. 1) – 1994, pp 67-70. Edited by W.-B. Souffrant and H. Hagemeister. EAAP-Publication No. 80, Dummerstorf.

Li, S., Sauer, W.C. and Hardin, R.T. (1994) *Canadian Journal of Animal Science*, **74**, 327-333.

Li, S., Sauer, W.C., Huang, S.X. and Gabert, V.M. (1996) *Journal of Animal Science*, **74**, 1649-1656.

Lizardo, R., Peiniau, J. and Aumaître, A. (1997) In Proceedings of the VIIth International Symposium on Digestive Physiology in Pigs – 1997, pp 630-633. Edited by J.-P. Laplace, C. Février and A. Barbeau. INRA, Saint Malo.

Longland, A.C., Carruthers, J. and Low, A.G. (1994) *Animal Production*, **58**, 405-410.

Low, A.G. (1985) In: *Recent Advances in Animal Nutrition – 1985*, 87-112. Edited by W. Haresign and D.J.A. Cole. Butterworths, London.

Low, A.G. (1989) *Animal Feed Science and Technology,* **23**, 55-65.

Low, A.G. (1993) In: *Recent Developments in Pig Nutrition - 2*, pp 137-162. Edited by D.J.A. Cole, W. Haresign and P.C. Garnsworthy, Nottingham University Press, Nottingham.

Malbert, C.H., Montfort, I., Mathis, C., Guérin, S. and Laplace, J.P. (1994) In: Proceedings of the VIth International Symposium on Digestive Physiology in Pigs (Vol. II) – 1994, pp 283-286. Edited by W.-B. Souffrant and H. Hagemeister. EAAP-Publication No. 80, Dummerstorf.

Mathew, A.G., Robbins, C.M., Chattin, S.E. and Quigley, J.D. (1997) *Journal of Animal Science*, **75**, 1009-1016.

Mathew, A.G., Sutton, A.L., Scheidt, A.B., Patterson, J.A., Kelly, D.T. and Meyerholtz, K.A. (1993) *Journal of Animal Science*, **71**, 1503-1509.

McDonald, D.E., Pluske, J.R., Pethick, D.W. and Hampson, D.J. (1996) Proceeding of the Nutrition Society of Australia, **20**, 136.

McIntosh, M.K., Baidoo, S.K., Aherne, F.X. and Bowland, J.P. (1986) *Canadian Journal of Animal Science*, **66**, 1051-1056.

Modler, H.W., McKellar, R.C. and Yaguchi, M. (1990) *Canadian Institute of Food Science and Technology*, **23**, 29-41.

Moore, R.J., Kornegay, E.T., Grayson, R.L. and Lindemann, M.D. (1988) *Journal of Animal Science*, **66**, 1570-1579.

Moore, W.E.C., Moore, L.V.H., Cato, E.P., Wilkins, T.D. and Kornegay, E.T. (1987) *Applied and Environmental Microbiology*, **53**, 1638-1644.

Mosenthin, R. and Hambrecht, E. (1998) In: Proceedings Symposium Series 1 of the 8th World Conference on Animal Production – 1998, 78-91. Seoul National University, Seoul.

Mosenthin, R., Sauer, W.C. and Ahrens, F. (1994) *Journal of Nutrition*, **124**, 1222-1229.

Mosenthin, R., Sauer, W.C., Henkel, H., Ahrens, F. and De Lange, C.F.M. (1992) *Journal of Animal Science*, **70**, 3467-3472.

Mul, A.J. and Perry, F.G. (1994) In: *Recent advances in Animal Nutrition – 1994*, pp 57-79. Edited by P.C. Garnsworthy and D.J.A. Cole. Nottingham University Press, Nottingham.

Neutra, M.R. and Forstner, J.F. (1987) In *Physiology of the Gastrointestinal Tract – 1987*, pp 975-1009. Edited by L.R. Johnson. 2nd edition Raven Press, New York.

Orban, J.I., Patterson, J.A., Adeola, O., Sutton, A.L. and Richards, G.N. (1997) *Journal of Animal Science*, **75**, 170-175.

Petkevicius, S., Bach Knudsen, K.E., Nansen, P., Roepstorff, A., Skjøth, F. and Jensen, K. (1997) *Parasitology*, **114**, 555-568.

Pluske, J.R., Siba. P.M., Pethick, D.W., Durmic, Z., Mullan, B.P. and Hampson, D.J. (1996) *Journal of Nutrition*, **126**, 2920-2933.

Pond, W.G., Varel, V.H., Dickson, J.S. and Haschek, W.M. (1989) *Journal of Animal Science*, **67**, 716-723.

Potkins, Z.V., Lawrence, T.L.J. and Thomlinson, J.R. (1991) *British Journal of Nutrition*, **65**, 391-413.

Rainbird, A.L. and Sissons, J.W. (1985) In Proceedings of the 3rd International Seminar on Digestive Physiology in the pig – 1985, pp 69-71. Edited by A. Just, H. Jørgensen and J. Fernández. Copenhagen, National Institute of Animal Science, Copenhagen.

Rainbird, A.L. and Low, A.G. (1986 a) *The British Journal of Animal Nutrition*, **55**, 87-98.

Rainbird, A.L. and Low, A.G. (1986 b) *The British Journal of Animal Nutrition*, **55**, 111-121.

Richards, G. (1979) In: Proceedings of the 27th swine short course – 1979, pp 30-34. Texas Agricultural Experimental Station, Labbouk.

Roberfroid, M., Gibson, G.R. and Delzenne, N. (1993) *Nutrition Reviews*, **51**, 137-146.

Roberfroid, M. (1993) *Critical Reviews in Food Science and Nutrition*, **33**, 103-148.

Robinson, I.M., Allison, M.J. and Bucklin, J.A. (1981) *Applied and Environmental Microbiology*, **41**, 950-955.

Roediger, W.E.W. (1980) *Gut*, **21**, 793-798.

Sakata, T. (1987) *British Journal of Nutrition*, **58**, 95-103.

Salanitro, J.P., Blake, I.G. and Muirhead, P.A. (1977) *Applied and Environmental Microbiology*, **33**, 79-84.

Sambrook, I.E. (1981) *Journal of the Science of Food and Agriculture*, **32**, 781-791.

Sauer, W.C., Mosenthin, R., Ahrens, F. and Den Hartog, L.A. (1991) *Journal of Animal Science*, **69**, 4070-4077.

Savage, D.S. (1984) In: *The Germ-free Animal in Biomedical Research – 1984*, pp 119-140. Edited by M.E. Coates and B.E. Gustafson. Laboratory Animals Ltd., London.

Schnabel, E., Bolduan, G. and Güldenpenning, A. (1983) *Archives of Animal Nutrition*, **33**, 371-377.

Schulze, H., van Leeuwen, P., Verstegen, M.W.A. and van den Berg, J.W.O. (1995) *Journal of Animal Science*, **73**, 441-448.

Siba, P.M., Pethick, D.W. and Hampson, D.J. (1996) *Epidemiology and Infection*, **116**, 207-216.

Smith, H.W. and Jones, J.E.T. (1963) *The Journal of Pathology and Bacteriology*, **86**, 387-412.

Smith, J.G. and German, J.B. (1995) *Food Technology*, **11**, 87-90.

Spring, P. and Privulescu, M. (1998) In: Proceedings Pre-Conference Symposium of the 8th World Conference on Animal Production – 1998, pp 21-28. Seoul National University, Seoul.

Stanogias, G. and Pearce, G.R. (1985 a) *British Journal of Nutrition*, **53**, 513-530.

Stanogias, G. and Pearce, G.R. (1985 b) *British Journal of Nutrition*, **53**, 531-536.

Trowell, H., Southgate, D.A.T., Wolever, T.M.S., Leeds, A.R., Gassull, M.A. and Jenkins, D.J.A. (1976) *The Lancet*, **1**, 967.

Vahouny, G.V. and Cassidy, M.M. (1985) Proceedings of the Society for Experimental Biology and Medicine, **180**, 432-446.

Valencia, Z. and Chavez, E.R. (1997) *Nutrition Research*, **17**, 1517-1527.

Varel, V.H. and Pond, W.G. (1985) *Applied and Environmental Microbiology*, **49**, 858-862.

Varel, V.H. (1987) *Journal of Animal Science,* **65**, 488-496.

Vervaeke, I.J., Graham, H., Dierick, N.A., Demeyer, D.I. and Decuypere, J.A. (1991) *Animal Feed Science and Technology*, **32**, 55-61.

Younes, H., Garleb, K., Behr, S., Rémésy, C. and Demigné, C. (1995) *Journal of Nutrition*, **125**, 1010-1016.

Zebrowska, T., Low, A.G. and Zebrowska, H. (1983) *British Journal of Nutrition*, **49**, 401-410.

Zebrowska, T. and Low, G. (1987) *The Journal of Nutrition*, **117**, 1212-1216.

12

FIBRE IN DIETS OF SOWS

M-C MEUNIER-SALAÜN
INRA Station de Recherches Porcines 35590 St-Gilles - France

Introduction

Sows are commonly fed to maintain a relatively constant body condition throughout the reproductive cycle for good health and optimal performance. This involves a restriction of feed intake during gestation to prevent excessive body weight gain and fat deposition, which could adversely affect the progress of farrowing and lead to problems during lactation; culling as a consequence of poor lactation performance or problems with locomotion is also possible. Excessive underfeeding in pregnant sows leads however to low body fat reserves at farrowing or at weaning leading to a delayed return to oestrus and low conception rate. On the other hand, lactating sows are usually fed *ad libitum*. During lactation, it seems that there is a critical feeding level below which reproductive performance is affected (weaning to oestrus interval, rate of return to oestrus, pregnancy rate) in relation to the loss of body fat during lactation or the absolute level of fat reserves at weaning. Accordingly, high feeding levels in lactating sows are necessary to ensure that the high energy and nutrient requirement for maintenance and milk production are met (Dourmad *et al.*, 1994).

Pregnant sows typically receive their whole daily feed allowance in one or two short concentrate meals. It supplies requirements for maintenance, and a small additional allowance for growth of maternal tissue and conceptus, assumed to be adequate for good health and performance. However, these conditions might not fulfil the other needs of the sow, especially feeding motivation. Indeed, the level of feed provided corresponds to about 0.5 to 0.6 of the voluntary intake (Brouns *et al.*, 1991), which results in a low level of satiety and a reduction of feeding behaviour expression.

The low feeding level has been linked to the occurrence of stereotyped activities (Appleby and Lawrence, 1987, Terlouw *et al.*, 1991; Lawrence and Terlouw, 1993; Spoolder *et al.*, 1995) which are behavioural patterns performed repetitively in fixed order and without any function (Odberg, 1978; Wiepkema *et al.*, 1983). These activities are more prevalent in the immediate post-feeding period (Rushen, 1984, 1985). The development of stereotypies due to food restriction has been attributed either to the limited nutrient and energy supply but also to reduced access to a foraging substrate in stall-housed sow or group housed sows. Spontaneous foraging behaviour is then directed towards less appropriate substrates such as pen components. The behavioural repertoire is channelled into a few simple behavioural sequences and persists through the sensitisation of the neural controls of behaviour (Figure 12.1 - Lawrence and Terlouw, 1993; Dantzer, 1986).

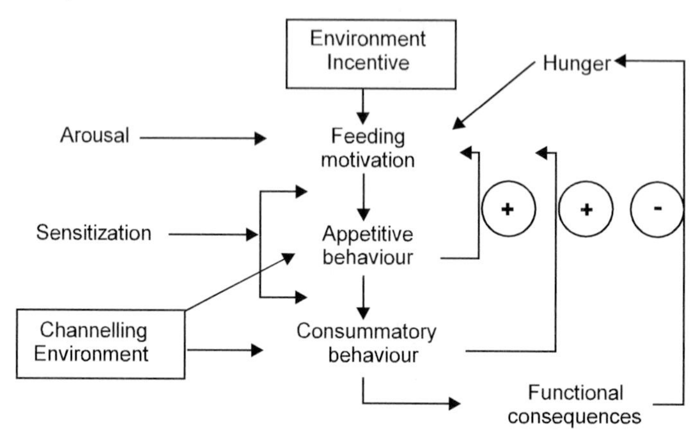

Figure 12.1 Motivational processes in feeding behaviour (after Lawrence and Terlouw, 1993)

The occurrence of stereotypic behaviour has been considered to reflect heightened feeding motivation after feeding and interpreted as an indication of impaired welfare (Wiepkema *et al.*, 1983). In addition, restricted feeding could lead to feeding competition in group-fed sows. The lack of individual access or a limited access to the feed can result in unequal intake between sows with detrimental effects on body reserves of low-ranking sows (Hansen *et al.*, 1982, Csermely and Wood-Gush, 1986, Signoret *et al.*, 1995).

Thus, feed restriction applied during gestation in sows has lead to doubts about whether their welfare is adequate. Intensive production is under increasing pressure of environmental and welfare legislation by government and local authorities, reflecting negative view of husbandry practices by consumers and a strong pressure exerted by animal welfare organisations. On the basis of the knowledge to date, a report was adopted in September 1997 by the welfare section of the Scientific Veterinary Committee and submitted to the European Council, with recommendations to improve the directive 91/630/EEC laying down minimum standards for the protection of pigs. One of these recommendations focused on the use of bulky or high fibre food to reduce hunger as well to provide for the need to chew.

Use of fibre in feeding strategy in sows

Many studies have shown that increasing the amount of feed or energy lowers behavioural activity level and especially stereotypic behaviour (Appleby and Lawrence, 1987; Terlouw *et al.*, 1991; Spoolder *et al.*, 1995; Bergeron and Gonyou, 1997). However, the high intake capacity of pregnant sows limits any attempt at offering a conventional diet *ad libitum* because of concomitant obesity, and detrimental effects at farrowing and during lactation. An alternative way of satisfying feeding motivation whilst maintaining sows on a restricted energy supply was suggested by providing diets with additional roughage. Different ways have been considered to supply fibrous components, either available in the environment, within a rack or on the floor, or incorporated into diets.

Considering the first option, wet chopped straw or oat hulls introduced in the trough of pregnant sows in addition to a basal diet increased the resting time but did not decrease the stereotyped behaviour whilst active (Fraser, 1975; Broom and Potter, 1984). On the other hand the provision of straw on the floor reduced the incidence of excessive or stereotypic chain and bar manipulation (Fraser, 1975; Spoolder *et al.*, 1995). Nevertheless, Spoolder *et al.* (1995) recorded a straw manipulation level in sows fed on a restricted basis as high as the level of manipulation of pen components exhibited by sows housed without access to straw. In addition, straw manipulation was performed in a persistent and compulsive way, suggesting that it is far from clear whether such activity represents a significant improvement in terms of welfare over the stereotypic manipulation of pen components.

The second approach consists of including high levels of fibrous ingredients in the diet, which allows increased feed supply without increasing energy and nutrient allowances. Furthermore, sows showed high fermentative capacity in digestion of fibrous diet (Noblet and Shi, 1993). The use of fibre in sow diets has mainly been investigated with regard to their nutritional and metabolic effects, and their consequences on performance (Etienne, 1987; Close, 1993). More recently, significant attention has been given to the effects of fibrous diets on behavioural activities. Feeding motivation has been measured on the basis of various criteria including the appetitive and behavioural sequences of the feeding activity during consumption, the feeding rate or the response level in operant conditioning procedures.

Use of fibrous diets in sows

EFFECTS ON BEHAVIOURAL ACTIVITY

Although studies concerning the effects of fibrous diets on the behaviour of pregnant sows have used different methodological procedures, as illustrated in Table 12.1, some conclusions may be drawn. They include a reduced general activity, especially the occurrence of oral stereotyped activity after the end of the meal. Sows fed fibrous diets spent more time in ingesting their daily ration, exhibited stronger mastication movements and consequently a lower rate of intake when compared to sows fed concentrate diets. Feeding motivation measured in operant conditioning procedures was also reduced with the supply of fibrous diets. Aggressive interactions associated with feed restriction could be also reduced in group-housed sows.

Nevertheless, the responses of sows when fed fibrous diets also depends on factors related to animals, feeding conditions or to the characteristics of the diet supplied.

Stereotypic behaviour

High fibre diets based either on wheat bran and maize cobs (CF 100 g/kg DM) or oat hulls and oats (CF 204 g/kg DM) compared to low fibre diet (CF 22 g/kg DM) reduced the occurrence of stereotyped chain manipulating over 24h in the first two parities (Robert *et al.*, 1993). However, duration of stereotypies around mealtime was reduced only in sows fed fibrous diets in the second parity. Moreover, fibrous diets slowed down the increase of repetitive activities between parities 1 and 2 (15% to less than 9%; Figure 12.2). These results suggest that high fibre diets are more effective in reducing stereotypic behaviour when the females are young and when they receive this feed for a long period (Robert *et al.*, 1993). This could be related to the fact that stereotypic behaviour becomes more rigid and frequent gradually over successive parities (Dantzer, 1986).

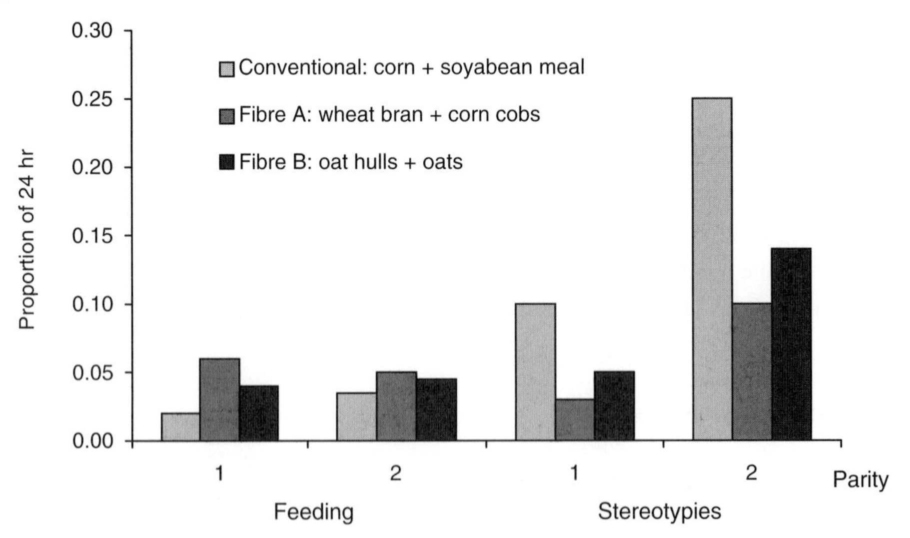

Figure 12.2 Effects of fibrous diets on the feeding behaviour and stereotypies performed by pregnant sows over the first two parities (proportion of 24 hr) (after Robert *et al.*, 1993)

Table 12.1 Review of the effects of fibrous diets supplied during pregnancy on behaviour of sows

Study,/worker	*Parity*	*Housing conditions*	*Fibrous diet supply Period (1)*	*Major effects of fibrous diet compared to conventional diet*				
				Lying time	*Non feeding oral activities (2)*	*Eating time*	*Agression*	*Operant feeding motivation*
Robert *et al.*., 1993	first -second	stalled concrete slatted floor	gestation	>	<	>		
Martin and Edwards, 1994	multiparous	outdoor-grouped feed line on floor	latin square design 2 weeks period		<	>	=	
Brouns and al, 1994	first	indoor-grouped straw bedded pen	gestation	>	<	>		
Brouns and Edwards, 1994	multiparous	indoor-housed straw bedded pen	gestation *ad libitum*				<	
Robert *et al.*, 1997	first	stalled concrete slatted floor	gestation		<	>		<
Braund *et al.*, 1998	multiparous	outdoor housed	latin square design 2 weeks period	>	<	>		
Ramonet *et al.*, 1999	multiparous	stalled concrete floor	latin square design 3 weeks period	>	<	>		

(1) In all studies, the experimental fibrous diets supply the same energy as the conventional diet, except when feed is provided *ad libitum*
(2) The non feeding oral activities include various motor acts implicated in the ingestive and foraging behaviour but performed after all food had been consumed. It covers chewing, rooting, biting, licking towards available substrate (bars, trough, drinker, floor, bedding) or self-directed (vacuum chewing, sucking). Such activities are identified as stereotypies when they are expressed in fixed sequences of motor acts and performed repetitively.

The reduction of stereotypic behaviour is stronger when the rate of inclusion of fibrous components leads to a concentration of crude fibre approaching 200 g/kg diet (Robert *et al.*, 1993, 1997b; Ramonet *et al.*, 1999), illustrated in Table 12.2. However these few studies need further investigations before concluding.

Table 12.2 Effects of the incorporation rate of fibre in diets on the behaviour of multiparous pregnant sows housed in stalls (after Ramonet *et al.*, 1998).

	Diet			
	Low	*Medium*	*High*	
g Crude Fibre/kg	*33*	*106*	*181*	*RSD(2)*
Behaviour (1)				
Feeding activity				
prehension	21.9[y]	25.5[y]	34.0[x]	6.0
mastication	7.4[z]	16.8[y]	42.6[x]	8.6
Drinking activity	4.7	4.7	3.5	3.3
Oral activities				
Total	53.7[x]	41.5[y]	14.6[z]	8.6
on trough	5.4[xy]	6.6x[x]	3.0[y]	4.1
on bars and floor substrates	6.8	6.4	4.1	5.0
self-directed	41.5[x]	28.4[y]	7.5[z]	7.8
Lying	11.1	10.0	4.2	5.3
Ingestion				
Feeding time, min	16.4[y]	24.3[y]	51.6[x]	5.1
Feeding rate, g/min	151.6[x]	119.6[y]	66.6[z]	13.8
Daily water intake, l/d; wk 3	17.0	13.9	9.7	12.5

(1) Behavioural recordings over the hour following the feed distribution, expressed in percent of total scan records; averaged adjusted-means values for the three 21 d periods of experimental diets supply.
(2) RSD : Residual Standard Deviation, values with different letters differed significantly between diets ($P<0.05$).

The manner in which the fibrous diet is supplied also modulates oral activity after the meal, as shown by Brouns *et al.* (1994) who recorded a reduced incidence of oral behaviours in feed-restricted sows fed a diet containing 500 g unmolassed sugar-beet pulp per kg of diet (11.3MJ DE, 107g CF/kg) compared to a conventional diet (13.0 MJ DE, 44g CF/kg); both diets were offered once daily in individual feeding stalls. In addition the oral behaviour that still occurred did not have the appearance of stereotypies. The same fibrous diet supplied *ad libitum* from a single space hopper reduced the occurrence of oral behaviour further and the level of behavioural manipulation of straw available on the floor. Feed-related behaviours were mostly redirected towards straw in feed-restricted sows raised in bedded pens (Spoolder *et al.*, 1995) indicating that the occurrence of stereotypies can be related to a combined lack of a sufficient amount of food to produce satiety and to a frustration of feeding/foraging behaviour.

It has been suggested that part of the effects of fibrous diets on the occurrence of rooting and redirected foraging behaviours performed in stereotypic situations, may result from the greater length of time taken to eat the increased volume of fibrous meal and then a shorter period of time available to perform such behaviour. Nevertheless, the reduction of post-feeding stereotypic activities with fibrous diet was still significant once the time spent eating was removed (Robert *et al.*, 1997b, Ramonet *et al.*, 1999).

The reduction of stereotypic behaviour has also been related to the increased bulk of food eaten. Robert *et al.* (1997b) compared isoenergetic diets (29.8 MJ DE/day) and reported a clear effect for the oat hulls and oat high–fibre diet (204 g CF) compared with the less bulky wheat bran and corn cobs high-fibre diet (101 g CF), in reducing post-feeding stereotypies which were higher in the conventional diet (22 g CF/kg). On the other hand, an oat-hull-based diet with a lower energy content (22 MJ DE/day) appeared less effective than one balanced for energy by adding fat (29.8 MJ DE/day) in reducing stereotypies after the meal, increasing resting time over 24h and decreasing the pre-feeding stereotypies. Results indicate that, in the short term, the greater volume of feed is responsible for the decrease in feeding motivation while, in the long term, both bulk and dietary energy have significant effects. Bergeron and Gonyou (1997) evaluated the effects of energy level of diets providing a similar volume of food in pregnant sows. A control diet based on barley and soybean meal (14.0 MJ DE/kg) was compared with a dry-fat and soybean meal diet formulated to provide 1.7 times as much digestible energy (23.7 MJ DE/kg) while having a similar density (0.59g/cm^3); an increased resting behaviour in the post-feeding period and reduced incidence of repetitive oral behaviours in sows fed the high–energy diet was observed. On the other hand, these behaviours were not evident in sows fed the control diet within the trough which provided more opportunities to forage at meal time, showing that a lack of foraging has less influence than a lack of energy in the development of stereotypies.

Finally, both studies suggest that nutrient and energy intake is an essential factor modulating the degree of satiety and point out that adding fibre in the diets of sows will be effective only if the requirements are met.

Feeding behaviour

The increased feeding time exhibited by sows fed fibrous diet is greater than expected solely by the increased quantity given to provide the same energy supply as the conventional diet (Brouns *et al.*, 1995). Martin and Edwards (1994) reported feeding time increased by 33% and 25% in outdoor-housed sows fed a high fibre diet (11,5 MJ DE/kg, 76 g CF), as compared to a conventional diet (13 MJ DE/kg, 41 g CF/kg) provided either at the same energy level or at the same daily quantity supply respectively.

Similar results have been obtained with individually–housed sows studied by these authors.

The analysis of feeding behaviour exhibited by individually-housed pregnant sows indicated a high proportion of mastication relatively to prehension when sows were fed a high fibre diet, based mainly on sugar-beet pulp (Ramonet *et al.*, 1999). Brouns *et al.* (1997) also showed that sows fed a sugar-beet pulp diet consumed their ration with pauses within the overall meal and with more chewing time.

Investigations on the effect of various fibrous ingredients used in bulky diets on feeding behaviour are currently limited, with most studies in restrict-fed sows using mixed sources of fibre. When diets are offered *ad libitum*, Brouns *et al.* (1995) showed lower feeding time as a proportion of active time and also lower voluntary feed intake of a diet with a high inclusion level of unmolassed sugar-beet pulp when compared to those formulated with inclusion of either barley straw, oat husks, malt culms, rice bran or wheat bran (Table 12.3). In addition, voluntary feed intake decreased from 50 to 30 kg/day when the inclusion level of sugar beet pulp increased from 400 to 650g/kg of feed. The effect on reducing voluntary feed intake of inclusion of sugar-beet pulp in the diet could be attributed to some properties of the sugar-beet pulp such as a lower palatability or specific physical and/or metabolic effects linked to its high water-holding capacity.

Table 12.3 Influence of fibrous feed ingredients on behaviour and performance in stall housed pregnant sows (after Brouns *et al.*, 1995).

Diet (1)	*Sugar beet pulp*	*Barley straw*	*Oat husks*	*Malt culms*	*Rice bran*	*Wheat bran*
g/kg (2)	*650*	*357.3*	*369.3*	*455.2*	*612.9*	*669.1*
NDF, g/kg DM	455.0	796.0	652.1	536.8	402.5	397.9
Voluntary intake, DE kg/day	2.3	6.4	7.7	6.8	7.6	7.1
MJ/day	24.6	57.0	79.0	70.1	70.8	72.9
Behaviour; min/24h						
feeding	12.8	15.6	18.5	13.4	13.9	15.1
lying (3)	1168	1194	1225	1203	1216	1243
standing (3)	96	59	2	59	53	36
Change 3-weeks period						
in weight, kg (3)	2.2	41.8	47.7	46.7	10.8	39.2
in backfat, mm (3)	-3.5	0.5	1.3	1.0	2.7	3.2

(1) Diets were formulated to contain the same calculated concentration of Digestible Energy (9.5 MJ/kg), minimum concentration of lysine (5g/kg) and crude protein (85g/kg). Barley and soya-bean meal were used in all diets
(2) content of fibre source in each diet, expressed as g/kg diet
(3) significant differences between the sugar-beet diet and the others diets, $P<0.05$

The lower eating rate recorded with fibrous diets has been considered as a measure of motivation level (Terlouw *et al.*, 1991). However, this criterion could also reflect potential lower palatability and/or longer feeding time owing to physical and metabolic processes during digestion associated with the nature of the fibrous components in diets (Brouns *et al.*, 1995, 1997). A diet containing fibrous components with high water-holding capacity may result in a marked increase in gastric distension and intestinal fill with local and central feedback limiting the voluntary intake (Cherbut *et al.*, 1988; Lepionka *et al.*, 1997). Sows consume a sugar-beet pulp diet much more slowly than the control diet, offered either as the daily food allowance or a subsequent test portion (Brouns *et al.*, 1997). Nevertheless, the feeding rate for the test diet in sows fed a sugar beet-pulp diet increases over the twelve experimental days, as illustrated in Figure 12.3. Increasing the interval between ration and test diet to 2 hours decreased further the feeding rate for the test diet by sows supplied with sugar beet pulp at meal. These results suggest that the sows adapt to some extent to the physical and /or metabolic effects of such diet and indicate that a sugar beet pulp diet causes a rapid feeling of satiety, which disappears gradually as digestion continues.

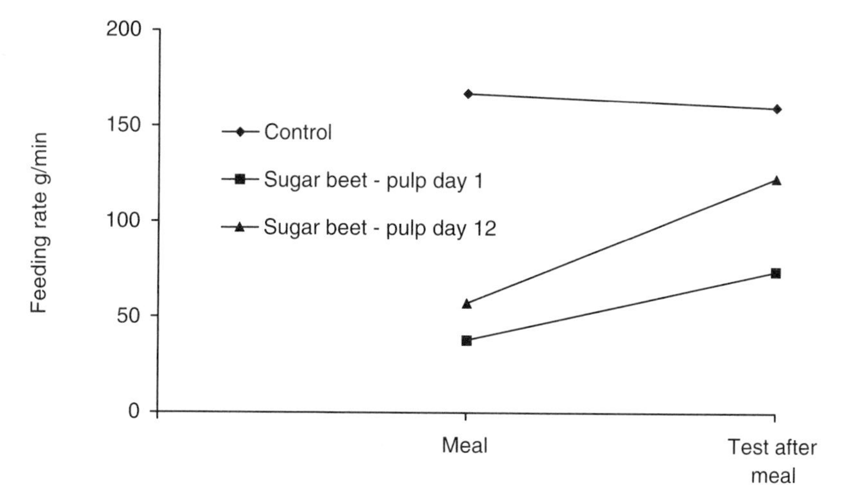

Figure 12.3 Feeding rate (g/min) of sows given a control or 500 g/kg sugar pulp diet (SBP) when offered the daily food allowance (meal) and subsequent standard test portion immediately after daily food allowance (test) the first day and 12 days later (after Brouns et al., 1997)

Operant conditioning feeding motivation

The operant conditioning procedure, requires pigs to press a paddle to obtain a feeding-related stimulus (food reward). The level of feeding motivation is evaluated according to the number of food reward obtained (Lawrence and Illius, 1989). Using a progressive ratio reward schedule, Robert *et al.* (1997b) reported a reduction in the number of food rewards obtained by sows fed high fibre diets (based on wheat-bran and oat hulls respectively) compared with sows receiving a concentrate (maize and soybean) diets. However, only high bulky diets (oat hulls and oats) compared with less bulky diet (wheat bran and maize cobs) were effective in reducing the feeding motivation, as described in Figure 12.4. These results suggest that the positive effects of fibrous diets on the motivation level increase as the volume of the ration increases and is maximised with increased degree of bulkiness of the diet.

Figure 12.4 Effects of high fibre diet supplied to pregnant sows on the feeding motivation measured in operant conditioning test (after Robert *et al.*, 1997b)

In the procedure of a second-order schedule of food reinforcement where the pig is trained to work not for food but for the presentation of food reward (Day *et al.*, 1996a), a decrease in the response level was recorded in growing male pigs with increased water holding capacity of diets (Day *et al.*, 1996b). The number of conditioned stimuli earned per unit time was inversely proportional to the water holding capacity of isoenergetic diets only after the meal, with no relationship being recorded before the meal. This suggests that feeding motivation will return to a level dictated by the nutrient intake as gastrointestinal distension declines after the meal. Thus the supply of bulky diets to sows could increase the satiety level in the short term but would not solve the sustained feeding motivation over a longer period of time. As pointed out by Savory *et al.* (1996) in broiler breeders, an animal exposed to qualitative food restriction, could still be "metabolically hungry " or feeding motivated, despite having a full stomach. This point needs to be evaluated in pregnant sows and its relevance from the welfare point of view established.

EFFECTS ON PHYSIOLOGICAL STATE

Fibrous diets induce higher chewing activity in sows during feed intake; this activity has been related to an increase in saliva and gastric secretions and has been associated with greater subjective satiety score in human subjects (Kritchevsky, 1988). Investigations on blood concentrations of glucose and insulin after feed distribution showed lower and more uniform levels in gilts or sows fed fibrous diets than in animals fed concentrate diets (Robert *et al.*, 1997a). In dry sows fed a basal diet supplemented with similar amounts of energy from three fibre sources (starch, sugar-beet pulp or wheat bran), Brouns *et al.* (1994a) reported higher acetate blood concentrations and more uniform glucose and insulin post feeding levels with the sugar beet pulp diet. Results of both studies indicate a more constant nutrient supply with fibrous diets, which should be related to a lower degree of hunger (Houpt *et al.*, 1983) and might maintain satiety for a longer period. Higher circulating levels of volatile fatty acids were also reported in pigs fed high sugar beet-pulp diets (Brouns *et al.*, 1994b) in association with a greater microbial fermentation in the hindgut which has been shown to depress appetite in ruminants (Farmingham and Whyte, 1993).

The increase of chewing with fibrous diet could be associated to physical and/or chemical characteristics of diets, such as the need to reduce particle size before swallowing, or high water-holding capacity of the fibrous components. In the case of ad libitum provision of bulky food, Day *et al.* (1996b) pointed out the need to ascertain whether the increased lying period in sows fed bulky fibrous diets represents a reduced level of feeding motivation or abdominal discomfort induced by high gastric distension with increased water holding capacity of diet.

The signs of chronic stress, associated with the occurrence of stereotypies, can be evaluated through the activation of the hypothalamic-pituitary-adrenal axis or immune function. Mac Glone and Fullwood (1996) did not show any influence of fibrous diets on the immune criteria (humoral and cellular systems) of restrict-fed pregnant sows in comparison with sows fed concentrate diets, in spite of differences in behaviour and feed intake. In addition, Spoolder *et al.* (1996) reported that the provision of straw as bedding to feed-restricted sows had no apparent effect on the response to ACTH challenge. Data are still limited and further work should be done in this area.

EFFECTS ON PERFORMANCE

Reese (1997) reviewed data on the overall effects of fibre diets supplied to sows during gestation on their reproductive performance and concluded there were potential beneficial effects (Table 12.4). Nevertheless, results described large variations between studies related to various parameters of the experimental diets such as the nature and the incorporation rate of the fibre sources, together with the amino acid, vitamin and trace mineral contents of diets.

Table 12.4 Average effect of fibrous diets supplied during gestation on reproductive performance of sows (after Reese, 1997)

	Fibre effect (1) (Fibre – Control)	*Number of studies*	*Number of litters (F/C)*
Sow traits			
Gestation weight gain, kg	-2.7	21	650/871
Lactation weight loss, kg	-1.3	18	593/790
Lactation feed/day, g	+ 272	17	571/742
Litter traits			
number pigs born alive, n	+0.3	24	738/1069
number pigs weaned, n	+0.3	24	732/1051
average pig birth weight, g	-91	24	762/1093
average pig weaning weight, g	+409	24	756/1075

(1) the fibre effect was estimated considering the difference between the means values in fibre and control diets respectively; means were weighted by the number of litters produced in each study

In restrict-fed sows, Matte *et al.* (1994) reported no marked effects of two high fibre diets (wheat bran and maize cobs, 107 g CF/kg - Oats hulls and oats, 204 g CF/kg) over two parities on the number of pigs born alive, on pre weaning mortality or on the weaning-to-oestrus interval, compared to a maize and soybean meal diet. The growth of the litter was increased by approximately 20% during the second parity when sows were supplied during the first and second parities with a diet based on wheat bran and maize cobs. In both parities, feed intake during lactation was 5% higher for sows fed on diets based on oats and oat hulls; the authors pointed out that results were partly due to an overestimation of the nutritive value and a subsequent compensation of pregnancy under-supply of energy by a higher energy intake during lactation; they concluded that a lack of knowledge of the availability of nutrients could result in imprecision in formulation and determination of the long term effects. Thus, additional research is needed to provide conclusive evidence that fibre source influences reproductive performance of sows.

Conclusion and research issues

There is some evidence that fibrous diets are effective in reducing the occurrence of sterotypic behaviour and in increasing the time spent in feeding activity in sows, suggesting an improvement of the welfare status of sows but only if diets meet the nutritional requirements of the sow. This chapter shows that further investigations are still needed to evaluate the long-term effects over successive parities. It is not clear whether stereotypies are reduced because of increased satiety or because sows have more opportunities to perform foraging behaviour on an appropriate substrate. The measurement of feeding motivation from the feeding rate is complex. As pointed out by Brouns *et al.* (1997), it is not yet known if there is a physiological variable governing satiety, e.g. gastrointestinal distension and metabolic changes, acting independently or as single variables on the feeding rate. A pig eating a bulky diet may remain metabolically hungry but be physically incapable of further ingestion. Further knowledge is necessary on the relationships between the behavioural activity and the metabolic and/or digestive effects of diets according to the fibrous components of diets. In terms of psychological responses, e.g. feeling of security and control of the environment, channelled by the activation of the pituitary-adrenal axis, previous studies are not conclusive and have yet to be actively pursued to evaluate the impact of fibre on the psychological processes implicated in feeding motivation. The reviewed effects of fibrous diets on reproductive performances of restrict-fed pregnant sows suggest a positive influence, although high variations are observed.

Moreover, there is currently considerable debate over whether or not the occurrence of stereotypies themselves constitutes a welfare problem (Mason, 1991). They are considered to be indicative of a welfare problem, because they are often prevalent in environments judged to be poor, develop from frustrated motivation and sometimes correlate with other signs of poor welfare. Some authors have even devised scales to gauge whether particular levels of stereotypies indicate unacceptably poor welfare (Broom and Johnson, 1993). Stereotypies are not a straightforward indicator of an animal's emotional state and should rather be treated as a sign of difficulties of copying with the environment.

The alternative feeding strategy by providing fibrous diets has some practical drawbacks, which could limit its use. It includes the low nutritional value of the fibrous ingredients increasing the volume of feed supply and consequently handling and storage constraints of bulky diets. From the environmental point of view, treatment of larger volume of solid manure produced with such diets has to be considered. These points show that more information must be obtained to formulate properly and feed adequate amount of fibrous diets in order to ensure the real improvement of welfare and to satisfy the environmental and economical requirements.

References

Appleby, M.C. and Lawrence, A.B. (1987) Food restriction as a cause of stereotypic behaviour in tethered gilts. *Animal Production* **45**, 103-110.

Bergeron, R. & Gonyou, H.W. (1997). Effects of increasing energy intake and foraging behaviours on the development of stereotypies in pregnant sows. *Applied Animal Behaviour Science* **53,** 259-270.

Broom, D.M., Potter, M.J., 1984. Factors affecting the occurrence of stereotypies in stall-housed dry sows. In Proceedings International Congress on Applied Ethology of farm Animals – 1984. pp 229-231. Edited by J. Unshelm, G. van Putten and K. Zeeb. Kiel, KTLB, Darmstadt.

Broom, D. and Johnson, K.G. (1993) *Stress and Animal Welfare*. Chapman and hall, London.

Brouns, F., Edwards, S.A. and English, P.R. (1991) Fibrous raw materials in sow diets : effects on voluntary food intake, digestibility and diurnal activity patterns. *Animal Production* **52**, 598.

Brouns, F. & Edwards, S.A. (1994). Social rank and feeding behaviour of group-housed sows fed competitively or ad libitum. *Applied Animal Behaviour Science* **39,** 225-235.

Brouns, F., Edwards, S.A. & English, P.R. (1992). Feeding motivation of sows fed a sugar-beet pulp diet. *Animal Production* **54,** 486-487.

Brouns, F., Edwards, S.A. & English, P.R. (1994). Metabolic effects of fibrous ingredients in pig diets. *Animal Production* **58,** 467

Brouns, F., Edwards, S.A. and English, P.R. (1994). Effect of dietary fibre and feeding system on activity and oral behaviour of group housed gilts. *Applied Animal Behaviour Science* **39**, 215-223.

Brouns, F., Edwards, S.A. and English, P.R. (1995). Influence of fibrous feed ingredients on voluntary intake of dry sows. *Animal Feed Science and Technology* **54,** 301-313.

Brouns, F., Edwards, S.A. and English, P.R. (1997). The effect of dietary inclusion of sugar-beet pulp on the feeding behaviour of dry sows. *Animal Science* **65,** 129-133.

Close, W.H. (1993) Fibrous diets for pigs. In Animal production in developing countries1983, pp 107-117. Edited by M. Gill and E. Owen. *British Society Animal Production Occasional Publication*, **16**.

Czermely, D and Wood-Gush, D.G.M. (1986) Agonistic behaviour in grouped sows . *Biology of behaviour* **11**, 244-252.

Cherbut, C., Barry, J.L., Wyers, M. and Delort-Laval, J. (1988). Effect of the nature of dietary fibre on transit time and faecal excretion in the growing pig. *Animal Feed Science and Technology* **20,** 327-333.

Dantzer, R. (1986). Behavioral, physiological and functional aspects of stereotyped behavior: a review and a re-interpretation. *Journal of Animal Science* **62,** 1776-1786.

Day, J.E., Kyriazakis, I. and Lawrence, A.B. (1996a). The use of second-order schedule to measure feeding motivation in the pig. *Applied Animal Behaviour Science* **50,** 15-31.

Day, J.E., Kyriazakis, I. and Lawrence, A.B. (1996b) The use of a second-order schedule to assess the effect of food bulk on the feeding motivation of growing pigs. *Animal Science* **63,** 447-455.

Dourmad, J.Y., Etienne, M., Prunier, A. and Noblet, J. (1994). The effect of energy and protein intake of sows on their longevity : a review. *Livestock Production Science* **40**, 87-97.

Etienne, M. (1987) Utilization of high fibre feeds and cereals by sows : a review. *Livestock Production Science* **16**, 229-242.

Farningham, D.A.H. and Whyte, C.C. (1993) The role of propionate and acetate in the control of feed intake in sheep. *British Journal of Nutrition* **70**, 37-46.

Fraser, D (1975) The effect of straw on the behaviour of sows in tether stalls. *Animal Production* **21**, 59-68.

Hansen, L.L., Hagelso, A.M. and Masden, A. (1982). Behavioural results and performance of bacon pigs fed ad libitum from one or several self feeders. *Applied Animal Ethology* **8**, 307-333.

Houpt, K.A. (1982) Gastrointestinal factors in hunger and satiety. *Neuroscience and Biobehavioural Reviews* **6**, 145-164.

Houpt, K.A., Baldwin, B., Houpt, R. and Hills, F. (1983) Humoral and cardiovascular responses to feeding in pigs. *American Journal of Physiology* **244,** R274-R284.

Kritchevsky, D. (1988) Dietary fibre. *Annual Reviews Nutrition* **8**, 301-328.

Lawrence, A.B., Appleby, M.C. and Mac Leod, H.A. (1988). Measuring hunger in the pig using operant conditioning: the effect of food restriction. *Animal Production* **47,** 131-137.

Lawrence, A.B. and Illius, A.W. (1989). Methodology for measuring hunger and food needs using operant conditioning in the pig. *Applied Animal Behaviour Science* **24,** 273-285.

Lawrence, A.B. and Terlouw, E.M. (1993). A review of behavioral factors involved in the development and continued performance of stereotypic behaviors in pigs. *Journal of Animal Science* **71,** 2815-2825.

Lepionka, L., Malbert, C.H. and Laplace, J.P. (1997). Proximal gastric distension modifies ingestion rate in pigs. *Reproduction Nutrition Development* **37,** 449-457.

McGlone, J.J. and Fullwood, D. (1996). Stereotyped oral/nasal/facial behaviors, reproduction and immunity of created sows: effects of high dietary fiber and rearing environment. *Journal of Animal Science* **74,** 131.

Martin, J.E. and Edwards, S.A. (1994). Feeding behaviour of outdoor sows: the effects of diet quantity and type. *Applied Animal Behaviour Science* **41**, 63-74.

Mason, G.J. (1991). Stereotypies: a critical review. *Animal Behaviour* **41,** 1015-1037.

Matte, J.J., Robert, S., Girard, C.L., Farmer, C. and Martineau, G.P. (1994). Effect of bulky diets based on wheat bran or oat hulls on reproductive performance of sows during their first two parities. *Journal of Animal Science* **72,** 1754-1760.

Noblet, J. & Shi, X.S. (1993). Comparative digestibility of energy and nutrients in growing pigs fed ad libitum and adult sows fed at maintenance. *Livestock Production Science* **34,** 137-152.

Odberg, F.O. (1978). Abnormal behaviors : sterotypes. In : Proceedings1st world congress Ethology and Applied Zootechnic. p 475-480. Madrid, Spain.

Ramonet, Y., Bolduc, J., Robert, S., Bergeron, R. and Meunier-Salaün, M.C. (1998). Feeding motivation and stereotypies in pregnant sows fed increasing levels of fibre and/or food. *In Proceedings of the 32nd congress of the ISAE* – 1998, pp 181. Edited by I. Veissier and A. Boissy. INRA Press, Clermont-Ferrand.

Ramonet, Y., Meunier-Salaün, M.C. and Dourmad, J.Y. (1999) High fiber diets in pregnant sows : digestive utilization and effects on the behavior of the animals. *Journal Animal Science* **77,** 591-599.

Reese, D.E. (1997) Dietary fiber in sows gestation diets reviewed. *Feedstuffs* **23**, Nebraska Swine report.

Robert, S., Matte, J.J., Farmer, C., Girard, C.L. & Martineau, G.P. (1993). High-fibre diets for sows: effects on stereotypies and adjunctive drinking. *Applied Animal Behaviour Science* **37,** 297-309.

Robert, S., Rushen, J. and Farmer, C. (1997a) Effets d'un ajout de fibres végétales au régime alimentaire des cochettes sur le comportement, le rythme cardiaque et les concentrations sanguines de glucose et d'insuline au moment du repas. *Journées de la Recherche Porcine en France* **27**, 161-167.

Robert, S., Rushen, J. & Farmer, C. (1997b). Both energy content and bulk of feed affect stereotypic behaviour, heart rate and feeding motivation of female pigs. *Applied Animal Behaviour Science* **54**, 161-171.

Rushen, J. (1984). Stereotyped behaviour, adjunctive drinking and the feeding periods of tethered sows. *Animal Behaviour* **32**, 1059-1067.

Rushen, J.P. (1985). Stereotypies, aggression and the feeding schedules of tethered sows. *Applied Animal Behaviour Science* **14**, 137-147.

Savory, C., Hocking, P.M., Mann, J.S. and Maxwell, M.H. (1996) Is broiler welfare improved by using qualitative rather than quantitative food restriction to limit growth rate? *Animal Welfare* **5**, 105-127.

Signoret, J.P., Ramonet, Y. and Vieuille-Thomas, C. (1995) L'élevage en plein-air des truies gestantes : problèmes posés par les relations sociales. J*ournées de la Recherche Porcine en France* **27**, 11-18.

Shi, X.S. and Noblet, J. (1993). Contribution of the hingut to digestion of diets in growing pigs and adult sows: effect of diet composition. *Livestock Production Science* **34,** 237-252.

Spoolder, H.A.M., Burbidge, J.A., Edwards, S.A., Simmins, P.H. and Lawrence, A.B. (1995). Provision of straw as a foraging substrate reduces the development of excessive chain and bar manipulation in food restricted sows. *Applied Animal Behaviour Science* **43**, 249-262.

Spoolder, H.A.M., Burbidge, J.A., Edwards, S.A., Simmins, P.H. and Lawrence, A.B. (1996) Effects of food level and straw bedding during pregnancy on sow performance and responses to an ACTH challenge. *Livestock Production Science* **47**, 51-57.

Terlouw, E.M.C., Lawrence, A.B. and Illius, A.W. (1991). Influences of feeding level and physical restriction on development of stereotypies in sows. *Animal Behaviour* **42**, 981-991.

Wiepkema, P.R., Broom, D.M., Duncan, I.J.H. and van Putten, G. (1983). Abnormal behaviours in farm animals. Commission of the European Communities Report, Brussels, 16pp..

LIST OF PARTICIPANTS

The thirty-third University of Nottingham Feed Manufacturers Conference was organised by the following committee:

Mr B. Ashington
Dr M.R. Bedford *(Finnfeeds International)*
Miss J. Bell *(Pye Milk Products)*
Dr D.J.A. Cole *(Nottingham Nutrition International)*
Mr P.W. Garland *(BOCM Pauls Ltd.)*
Dr M. Marsden *(ABN)*
Dr J. O'Grady *(IAWS Group Plc)*
Dr S.A. Papasolomontos *(Dalgety Agriculture)*
Mr J.R. Pickford
Dr S.J. Taylor *(Nutec Ltd)*
Mr J. Twigge *(Trouw Nutrition)*
Mr P. Toplis *(Primary Diets)*

Dr K.N. Boorman Prof P.J. Buttery Dr J.M. Dawson Dr P.C. Garnsworthy *(Secretary)* Prof G.E. Lamming Dr A.M. Salter Dr J. Wiseman *(Chairman)*	*University of Nottingham*

The conference was held at the University of Nottingham Sutton Bonington Campus, 5-7 January, 1999. The following persons registered for the meeting.

Adams, Dr C A	Kemin Europa NV, Industrie Zone Wolfstee, B-2200 Herentals, Belgium
Al-Shuweimi, Mr T	Arasco, P O Box 53845, Riyadh 11593, Saudi Arabia
Allder, Mr M	Eurotec Nutrition Ltd, Unit 24 Lancaster Way Business Park, Ely, Cambridge CB6 3NW, UK
Allen, Dr J	Frank Wright Ltd, Blenheim House, Blenheim Road, Ashbourne, Derbyshire DE6 1HA, UK
Anderson, Mr K R	Duffield Nutrition, Saxlingham Thorpe Mills, Ipswich Road, Norwich NR15 1TY, UK
Ashington, Mr B	Trouw Nutrition, Wincham, Northwich, Cheshire CW9 6DF, UK
Aspland, Mr P	Aspland and James Ltd, Medcalfe Way, Bridge Street, Chatteris, Cambs PE16 6QZ, UK
Ball, Mr A	Roche Products, Heanor Gate, Heanor, Derbyshire DE75 7SG, UK
Bartram, Dr C	Dalgety Feed Ltd, Aston Mill, Aston, Nantwich, Cheshire CW5 8DH, UK
Bassett, Mr R W	Park Tonks Ltd, 48 North Road, Great Abington, Cambs, UK
Beardsworth, Dr P M	Roche Products Ltd, Heanor Gate, Heanor, Derbyshire DE75 7SG, UK

Beaumont, Mr D	Pancosma (UK) Ltd, Crompton Road Industrial Estate, Ilkeston, Derbyshire, UK
Beer, Mr J H	Conifers, Pooley Bridge, Penrith, Cumbria CA11 OLL, UK
Bell, Miss J F	SAC Crichton Royal Farm, Mid Park, Bankend Road, Dumfries DG1 4SU, UK
Bennett, Mr R	Rhone-Poulenc Nutrition, 42 Avenue Aristide Briand, BP 100, 82154 Antony Cedex, France
Best, Mr P	Feed International, 18 Chapel Street, Petersfield, Hampshire GU32 3DZ, UK
Bishop, Ms R	Spillers Speciality Foods, Old Wolverhampton Road, Milton Keynes MK12 5PZ, UK
Blake, Dr J	Highfield, Little London, Andover, Hants SP11 6JE, UK
Bone, Mr P	Thompson and Joseph Ltd, 119 Plumstead Road, Norwich NR1 4JT, UK
Boorman, Dr KN	University of Nottingham, Sutton Bonington Campus, Loughborough, Leics LE12 5RD, UK
Brackenbury, Miss J	Nutec, Eastern Avenue, Lichfield, Staffs WS13 7SE, UK
Brooks, Prof P H	University of Plymouth, Seale Hayne Faculty, Newton Abbot, Devon TQ12 6NQ, UK
Brown, Mr G J P	Roche Products Ltd, Heanor Gate, Heanor, Derbyshire DE75 7SG, UK
Brown, Mr J M	Britphos Ltd, Rawdon House, Green Lane, Leeds LS19 7BY, UK
Burt, Dr A W A	Burt Research Ltd, 23 Stow Road, Kimbolton, Huntingdon, Cambs PE18 OHU, UK
Buss, Miss J	Farmers Weekly, Quadrant House, Sutton, Surrey SM2 5AS, UK
Buttery, Prof P J	University of Nottingham, Sutton Bonington Campus, Loughborough, Leics LE12 5RD, UK
Cameron, Miss J	University of Nottingham, Sutton Bonington Campus, Loughborough, Leics LE12 5RD, UK
Campiani, Dr I	F.lli Martini and C. S.p.A, Via Emilia 2614, Longiano (FO), Italy
Carrick, Mr I M	Axient Farm Business Solutions, Genus pls, Westmere Drive, Crewe, Cheshire CW1 6ZY, UK
Charles, Dr D R	D C R&D, 62 Main Street, Willoughby, Loughborough, Leics LE12 6SZ, UK
Cheek, Ms C	Feed Milling International, Armstrong House, 38 Market Square, Uxbridge, Middx UB8 1TG, UK
Clark, Dr M	MSF Ltd, County Mills, The Butts, Worcester WR1 3NU, UK
Clarke, Miss E	University of Nottingham, Sutton Bonington Campus, Loughborough, Leics LE12 5RD, UK
Clarke, Mr N	Britphos Ltd, Rawdon House, Green Lane, Leeds LS19 7BY, UK
Clay, Mr J	Alltech (UK) Ltd, Alltech House, Ryhall Road, Stamford PE9 1TZ, UK
Close, Dr W	129 Barkam Road, Wokingham, Berks RG41 2RD, UK
Coates, Mr G	Bugico S.A., 16 rue des Bugnons, CH-1217 Meyrin, Switzerland

Coenen, Mr H D N	ORFFA Belgium, Ovdemanstraat 13, B-1140 Londerzeel , Belgium
Cole, Dr D	Nottingham Nutrition International, 14 Potters Lane, East Leake, Loughborough LE12 6NG, UK
Cole, Dr M	SCA Nutrition, Maple Mill, Dalton, Thirsk YO7 3HE, UK
Collyer, Mr M	Kemin UK Ltd, Becor House, Green Lane, Lincoln LN6 7DL, UK
Connolly, Mr T	Hanford, Bourne Park, Piddlehinton, Dorchester, Dorset DT2 7TU, UK
Cook, Miss J D	Dalgety Feed Ltd, Springfield House, Springfield Road, Grantham NG31 7BG, UK
Cooke, Dr B C	1 Jenkins Orchard, Wick St Lawrence, Weston-Super-Mare, N Somerset BS22 7YP, UK
Coope, Mr R	Pen Mill Feeds Ltd, Babylon View, Pen Trading Estate, Yeovil, Somerset BA21 5HR, UK
Cooper, Dr A	University of Plymouth, Seal Hayne Faculty, Newton Abbot, Devon TQ12 6NQ, UK
Corless, Mr J	92 Caragh Court, Naas, Co Kildare, Republic of Ireland
Cox, Mr N	SCA Nutrition Ltd, Maple Mill, Dalton, Thirsk YO7 3HE, UK
Creasey, Mrs A	BASF, Blenheim House, Blenheim Road, Ashbourne, Derbyshire DE6 1HA, UK
Curtis, Mr L	BASF, Blenheim House, Blenheim Road, Ashbourne, Derbyshire DE6 1HA, UK
D'Onofrio, Mr R I	c/o ORFFA Nederland Feed B.V., Burgsstraat 12, 4283 GG Giessen, The Netherlands
Dawson, Dr J M	University of Nottingham, Sutton Bonington Campus, Loughborough, Leics LE12 5RD, UK
Dawson, Mr W	Britphos Ltd, Rawdon House, Green Lane, Leeds LS19 7BY, UK
De Brabander, Mr D	Agriculture Research Centre - Ghent, Dept of Animal Nutrition, Scheldeweg 68, B-9090 Melle-Gontrode, Belgium
Devriendt, Mr M	Vitamex N V, Booiebos 5, B-9031 Drongen, Belgium
Dickens, Mr A	BOCM Pauls, P O Box 39, 47 Key Street, Ipswich IP4 1BX, UK
Dixon, Mr D	Brown and Gillmer Ltd, Florence Lodge, 199 Strand Road, Merrion, Dublin 4, Ireland
Doran, Mr B	Trouw Nutrition UK, Wincham, Northwich, Cheshire CW9 6DF, UK
Doreau, Dr M	Unite de Recherches sur les Herbivores, (URH), INRA Theix, F 63122 Saint Genes Champanelle, France
Downey, Mr N	SCA Nutrition, Maple Mill, Dalton, Thirsk, North Yorkshire YO7 3HE, UK
Drakley, Miss C	University of Nottingham, Suton Bonington Campus, Loughborough, Leics LE12 5RD, UK
Emmens, Miss M	Hendrix UDT, Veerstraat 38, 5831 JN Boxmeer, The Netherlands
Ewing, Mrs A	Dalgety Feed Ltd, Aston Mill, Aston, Nr Nantwich, Cheshire CW5 8DH, UK

Ewing, Dr W	Nutec, Eastern Avenue, Lichfield, Staffs WS13 7SE, UK
Farley, Mr R	Trouw Nutrition, Wincham, Northwich, Cheshire CW9 6DF, UK
Farrell, Dr D	University of Queensland, School of Land and Food, St Lucia
Ferguson, Dr N S	University of Natal, Faculty of Agriculture, Private Bag X01, Scottsville, Pietermaritzburg, 3209, South Africa
Feuerstein, Dr D	BASF Ag, D-67056 Ludwigshafen, Germany
Filmer, Mr D	David Filmer Ltd, Wascelyn, Brent Street, Brent Knoll, Somerset, UK
Fisher, Dr C	Ross Breeders Ltd, Newbridge, Midlothian EH28 8SZ, UK
Fitt, Dr T	Roche Products Ltd, Heanor Gate, Heanor, Derbyshire DE75 7SG, UK
Ford, Dr M	Hydro Agri Europe, Hydro Agri Deutschland GmbH, Hanninghof 35/D-48249 Dulmen, Germany
Foulds, Mr S	Fayrefield Nutrition, Englesea House, Barthomley Road, Crewe, Cheshire CW1 1UF, UK
Fulford, Mr G W	Fishers Feeds Ltd, Cranswick, Driffield, East Yorkshire YO25 9PF, UK
Fullarton, Mr P J	Forum Products Ltd, Faraday House, Hookstone Oval, Harrogate, North Yorkshire HG2 8QE, UK
Gair, Ms D	Forum Products Ltd, 41-51 Brighton Road, Redhill RH1 6YS, UK
Garland, Mr P	BOCM Pauls Ltd, P O Box 39, 47 Key Street, Ipswich IP4 1BX, UK
Garnsworthy, Dr P C	University of Nottingham, Sutton Bonington Campus, Loughborough, Leics LE12 5RD, UK
Garwes, Dr D	MAFF, 649 St Christopher House, Southwark Street, London SE1 OUD, UK
Gatnau, Dr R	APC Europe, c/Tarragonna 161, 12tho, 08014 Barcelona, Spain
Gibson, Mr J E	Parnutt Foods Ltd, Hadley Road, Woodbridge Industrial Estate, Sleaford, Lincs NG34 7EG, UK
Gillespie, Miss F T	U M Group, Stretton House, Derby Road, Stretton, Burton on Trent, Staffs DE13 ODW, UK
Golds, Mr R A	University of Nottingham, Sutton Bonington Campus, Loughborough, Leics LE12 5RD, UK
Golds, Mrs S P	University of Nottingham, Sutton Bonington Campus, Loughborough, Leics LE12 5RD, UK
Gooderham, Mr B	Pye Milk Products, Lansil Way, Lancaster LA1 3QY, UK
Gordon, Prof F	Agricultural Research Institute of Northern Ireland, Large Park, Hillsborough, Co. Down, BT26 6DR, UK
Gotterbarm, Dr G	Hydro Agri Europe, Hydro Nutrition, Hydro Agri Deutschland GmbH, Hanninghof 35/D-48249 Dulmen, Germany
Gould, Mrs M	Volac International Ltd, Volac House, Orwell, Royston, Herts SG8 5QX, UK
Graham, Mr M	SVG Intermol Ltd, Shell Road, Royal Edward Dock, Avonmouth, Bristol BS11 9BW, UK
Gray, Mr W	Kemira Chemicals (UK) Ltd, Orm House, 2 Hookstone Park, Harrogate, North Yorkshire HG2 7DB, UK

Hambrecht, Miss E	Nutreco Swine Research Centre, P O Box 240, 5830 AE Boxmeer, The Netherlands
Haresign, Prof W	Welsh Institute of Rural Studies, University of Wales, Aberystwyth SY23 3AL, UK
Harland, Dr J I	Harland Hall, Bridge Cottage, Castle Eaton, Swindon SN6 6JZ, UK
Harrison, Mrs J	Sciantec Analytical Services Ltd, Main Site, Dalton, Thirsk YO7 3JA, UK
Haythornthwaite, Mr A	Farm Sense, Wild Goose House, Goe Lane, Freckleton, Lancs PR4 1XA, UK
Hazzledine, Mr M	Dalgety Feed Ltd, Springfield House, Springfield Road, Grantham NG31 7BG, UK
Hegeman, Mr F	AVEBE Business Unit Starch & Feed, Nijverheidsstraat 8, P O Box 15, AA Veendam, The Netherlands
Hemke, Mr G	P O Box 200, 5460 BC Veghel, The Netherlands
Higginbotham, Dr J D	United Molasses, Stretton House, Derby Road, Stretton, Burton on Trent, Staffs DE13 ODW, UK
Hissa, Mr P	Suomen Rehu Ltd, P O Box 75, 00501 Helsinki, Finland
Hoetink, Mr H	APC Europe, c/Tarragonna 161, 12tho, 08014 Barcelona, Spain
Hoffman, Dr R	University of Connecticut, Dept of Animal Science, 3636 Horsebarn Road Ext U-40, Storrs, CT 06269-4040, USA
Holder, Mr P	SVG Intermol, Shell Road, Royal Edward Dock, Avonmouth, Bristol BS11 9BW, UK
Holl, Dr E	Dr Eckel GmbH, Bausenbergweg 11, D-56661, Niederzissen, Germany
Holma, Mrs M	Raisio Feed Ltd, P O Box 101, FIN 21201 Raisio, Finland
Houseman, Dr R	M.C.S, 8 Bedern Bank, Ripon HG4 1PE, UK
Howie, Mr A	Nutrition Trading (International) Ltd, Orchard House, Manor Drive, Morton Bagot, Studley, Warks B80 7ED, UK
Jackson, Mr J	Nutec, Eastern Avenue, Lichfield, Staffs WS13 7SE, UK
Jagger, Dr S	Dalgety Feed Ltd, Springfield House, Springfield Road, Grantham NG31 7BG, UK
Jardine, Mr G	Guttridge Milling, 1 Mount Terrace, York YO2 4AR, UK
Johnson-Ord, Mrs S	Kemira Chemicals (UK) Ltd, Orm House, 2 Hookstone Park, Harrogate, North Yorkshire HG2 7DB, UK
Jones, Miss G R	Tithebarn Limited, P O Box 20, Weld Road, Southport PR8 2LY, UK
Jones, Mr H	Heygate & Sons Ltd, Bugbrooke Mills, Northampton NN7 3QH, UK
Kennedy, Mr D	International Additives, Old Gorsey Lane, Wallasey, Merseyside, UK
Kenyon, Mr S	Alltech (UK) Ltd, Alltech House, Ryhall Road, Stamford PE9 1TZ, UK
Keys, Mr J	32 Holbrook Road, Stratford upon Avon, Warwickshire CV37 9DZ, UK
King, Mr D	Rhone Poulenc Animal Nutrition, c/o Rhodia Ltd, Oak House, Reeds Crescent, Watford WD1 1QH, UK
Klein-Holkenborg, Mr A B	Gist Brocades BV, P O Box 1, 2600 MA Delft, The Netherlands
Knight, Dr R	Trouw Nutrition, Wincham, Northwich, Cheshire CW9 6DF, UK

Lake, Mr P	64 Over Norton Road, Chipping Norton, Oxfordshire OX7 5NR, UK
Lamming, Prof G E	University of Nottingham, Sutton Bonington Campus, Loughborough, Leics LE12 5RD, UK
Leek, Mr A B G	University College Dublin, Lyons Estate, Newcastle, Co Dublin, Ireland
Lima, Mr S	Felleskjopet Rogaland-Agder, P O Box 208, N4001 Stavanger, Norway
Lister, Mr M	Computer Applications Ltd, Rivington House, Drumhead Road, Chorley, Lancs PR6 7BX, UK
Lock, Mr A	University of Nottingham, Sutton Bonington Campus, Loughborough, Leics LE12 5RD, UK
Lowe, Dr J	Gilbertson and Page, P O Box 321, Welwyn Garden City, Herts AL7 1LF, UK
Lowe, Mr R	Frank Wright Ltd, Blenheim House, Blenheim Road, Ashbourne, Derbyshire, DE6 1HA, UK
MacDonald, Mr P C	David Moore (Flavours) Ltd, Ryhall Road, Stamford, Lincs, UK
Mafo, Mr A	Hi Peak Feeds, Proctors (Bakewell) Ltd, 12 Ashbourne Road, Derby, UK
Malandra, Dr F	Sildamin-Agribaanos, 27010 Sessa, Pavia, Italy
Marsden, Dr M	ABN Ltd, ABN House, P O Box 250, Oundle Road, Woodston, Peterborough, UK
Marsman, Dr G	Borculo Domo Ingredients, P O Box 46, 7270 AA Borculo, The Netherlands
Martyn, Mr S	International Additives, Old Gorsey Lane, Wallasey, Merseyside, UK
McCarthy, Mr P	Volac Feed Ltd, Killeshandra, Co Cavan, Ireland
McGrane, Mr M	206 Ryevale Lawns, Leiylip, Co Kildare, Ireland
McIlmoyle, Dr W A	Nutrition Consultants, 20 Young Street, Lisburn BT27 5EB, Northern Ireland
Mealey, Mr I	Format International Ltd, Format House, Poole Road, Woking, Surrey GU21 1DY, UK
Metcalf, Dr J A	Borregaard UK Ltd, Clayton Road, Risley, Warrington WA3 6QQ, UK
Metz, Dr S H M	Provimi Research and Tech. Centre, Lenneke Marelaan 2, B-1932 Sint-Stevens-Woluwe, Belgium
Meunier-Salaun, Dr M C	INRA, Station Recherches Porcines, 35590 Saint Gilles, France
Miller, Dr H M	University of Leeds, School of Biology, Leeds LS2 9JT, UK
Millward, Mr J J	Royal Pharmaceutical Society of Great Britain, Animal Medicines Division, 1 Lambeth High Street, London SE1 7JN, UK
Mitchell, Mr P P	Crown Chicken Ltd, Green Farm, Edge Green, Kenninghall, Norwich, Norfolk NR16 2DR, UK
Moller, Dr P E H	Hamlet Protein A/S, P O Box 130, DK-8700 Horsens, Denmark
Moody, Mr D	Diamond V Mills, 30 Church Street, Helmdon, Brackley, Northants NN13 5QJ, UK
Mosenthin, Prof R	Hohenheim University, Institute of Animal Nutrition (450), D-70593 Stuttgart, Germany

Mounsey, Mr A D	HGM Publications, HGM House, Nether End, Baslow, Bakewell DE45 1SR, UK
Mounsey, Mr S P	Feed Compunder, HGM House, Nether End, Baslow, Bakewell, Derbyshire DE45 1SR, UK
Mudd, Dr A	Roche Products Ltd, Heanor Gate, Heanor, Derbyshire DE75 7SG, UK
Murray, Mr F	Dairy Crest Ingredients, Dairy Crest House, Portsmouth Road, Surbiton, Surrey KT6 5QL, UK
Nelson, Ms J	UKASTA, 3 Whitehall Court, London SW1A 2EQ, UK
Newbold, Dr J R	BOCM Pauls Ltd, 47 Key Street, Ipswich IP4 1BX, UK
Newcombe, Mrs J	University of Nottingham, Sutton Bonington Campus, Loughborough, Leics LE12 5RD, UK
Nordang, Dr L	Felleskjopet, Forutvikling, N-7005 Trondheim, Norway
O'Grady, Dr J	IAWS Group plc, 151 Thomas Street, Dublin 8, Ireland
Offer, Dr J	SAC Veterinary Science Division, Dairy Health Unit, Auchincruive, Ayr, UK
Offer, Dr N	c/o SAC Crichton Royal Farm, Mid Park, Bankend Road, Dumfries DG1 4SU, UK
Overbeek, Dr G J	Borculo Domo Ingredients, P O Box 46, 7270 AA Borculo, The Netherlands
Overend, Dr M	Nutec, Eastern Avenue, Lichfield, Staffs WS13 7SE, UK
Owers, Dr M	BOCM Pauls, P O Box 39, 47 Key Street, Ipswich IP4 1BX, UK
Packington, Mr A	Roche Products Ltd, Heanor Gate, Heanor, Derbys DE75 7SG, UK
Pallister, Dr S M	Orffa UK Ltd, Park Street, Congleton, Cheshire, UK
Papasolomontos, Dr S	Kego S.A., 1st KM Artaki-Psachna Road, N Artaki, Greece
Parilo, Mrs A L	ORFFA Nederland Feed B.V, Burgsstraat 12, 4283 GG Giessen, The Netherlands
Partridge, Dr G	Finnfeeds, P O Box 777, Marlborough SN8 1XN, UK
Partridge, Mr M	Pen Mill Feeds Ltd, Babylon View, Pen Trading Estate, Yeovil, Somerset BA21 5HR, UK
Pass, Mr R T	United Distillery and Vintners, 33 Ellersly Road, Edinburgh EH12 6JW, UK
Perrott, Mr G	Trident Feeds, P O Box 11, Oundle Road, Peterborough, UK
Phillips, Mr G	Silo Guard Europe, Greenway Farm, Charlton Kings, Cheltenham, UK
Pickford, Mr J R	Bocking Hall, Bocking Church Street, Braintree, Essex CM7 5JY, UK
Pike, Dr I H	IFOMA Limited, 2 College Yard, St Albans AL3 4PA, UK
Piva, Prof G	Faculty of Agriculture, Via E Parmense, 84, 29100 Piacenza, Italy
Plowman, Mr G B	G W Plowman & Son Ltd, Selby House, High Street, Spalding, Lincs PE11 1TW, UK
Pluske, Dr J	Massey University, Monogastric Research Centre, Private Bag 11-222 Palmerston North, New Zealand
Powell, Mr P	Agil Ltd, Hercules 2, Calleva Park, Aldermaston RG19 3LG, UK

Pritchard, Mr S	Premier Nutrition Products Ltd, The Levels, Rugeley, Staffs WS15 1RD, UK
Probert, Miss L	University of Nottingham, Sutton Bonington Campus, Loughborough, Leics LE12 5RD, UK
Putnam, Mr M	61 Hempstead Lane, Potten End, Berkhamsted, Herts HP4 2RZ, UK
Reeve, Dr A	ICI Nutrition, Alexander House, Crown Gate, Runcorn WA7 2UP, UK
Retter, Dr W	Heygate & Sons Ltd, Bugbrooke Mills, Northampton NN7 3QH, UK
Revett, Mr S	Aspland & James Ltd, Medcalfe Way, Bridge Street, Chatteris, Cambs PE16 6QZ, UK
Richards, Mr K	Computer Applications Ltd, Rivington House, Drumhead Road, Chorley, Lancs PR6 7BX, UK
Richards, Dr S	Nutec, Eastern Avenue, Lichfield, Staffs WS13 7SE, UK
Robinson, Mr S	Nottingham University Press, Manor Farm, Thrumpton, Nottingham NG11 OAX, UK
Rogers, Mr M	Volac International Ltd, Volac House, Orwell, Royston, Herts SG8 5QX, UK
Rosen, Dr G D	66 Bathgate Road, London, SW19 5PH, UK
Routh, Mr M	Kemira Chemicals (UK) Ltd, Orm House, 2 Hookstone Park, Harrogate HG2 7DB, UK
Russell, Ms S	The Cottage, 9 New Road, Heage, Derbyshire, UK
Rymer, Dr C	ADAS Feed Evaluation & Nutrition Sciences, Drayton, Alcester Road, Strattford CV37 9RQ, UK
Salter, Dr A M	University of Nottingham, Sutton Bonington Campus, Loughborough, Leics LE12 5RD, UK
Shepperson, Dr N	Nutec, Eastern Avenue, Lichfield, Staffs WS13 7SE, UK
Shorrock, Dr C	FSL Bells, Hartham, Corsham, Wilts, UK
Short, Dr F	University of Nottingham, Sutton Bonington Campus, Loughborough, Leics LE12 5RD, UK
Smart, Mr A	International Additives, Old Gorsey Lane, Wallasey, Merseyside, UK
Stainsby, Mr A K	B.A.T.A. Ltd, Norton Road, Malton, North Yorkshire YO17 9RU, UK
Steinbock, Mr Mr	Forum Products Ltd, 41-57 Brighton Road, Redhill RH1 6YS, UK
Swarbrick, Mr J	Borculo Domo Ingredients, River Lane Saltney, Chester CH4 8RQ, UK
Sylvester, Mr D	Roche Products Ltd, Heanor Gate, Heanor, Derbyshire DE75 7SG, UK
Taylor, Dr A J	Roche Products Ltd, Heanor Gate, Heanor, Derbyshire DE75 7SG, UK
Taylor, Dr S J	Nutec Ltd, Greenhills Centre, Tallaght, Dublin, Ireland
ten Doeschate, Dr R	Dalgety Feed Ltd, Springfield House, Springfield Road, Grantham NG31 7BG, UK
Theophilou, Mr N	Silver and Baryte, 21A Amerikis Street, 10672 Athens, Greece
Thomas, Prof C	SAC, Crichton Royal Farm, Mid Park, Bankend Road, Dumfries DG1 4SZ, UK

Thompson, Mr D	Right Feeds Ltd, Castlegarde, Cappamore, Co Limerick, Ireland
Thompson, Miss J	University of Nottingham, Sutton Bonington Campus, Loughborough, Leics LE12 5RD, UK
Thompson, Mr M T	Sheldon Jones Agriculture, Royal Portbury Dock, Bristol BS20 9KS, UK
Tiekstra, Mr H	AVEBE, Business Unit Starch & Feed, P O Box 15, 9640 AA Veendam, Nijverheldsstraat 8, 9648 JA Wildervank, The Netherlands
Toplis, Mr P	Primary Diets Ltd, Melmerby Industrial Estate, Melmerby, Ripon, North Yorks HG4 5HP, UK
Touchburn, Prof S	McGill University, MacDonald Campus, Ste-Anne-de-Bellevue, Quebec H9X 3V9, Canada
Tsakiris, Mr J	Silver and Baryte Ores, 21A Amerikis Street, 10672 Athens, Greece
Twigge, Mr J	Trouw Nutrition, Wincham, Northwich, Cheshire, UK
Van Cauwenberghe, Mrs S	Eurolysine, 153 Rue de Courcelles, 75817 Paris Cedex 17, France
Van der Ploeg, Mr H	Stationsweg 4, 3603 EE Maarssen, The Netherlands
Van Hoecke, Mr P	Cerestar, Havenstraat 84, 1800 Vilvoorde, Belgium
Van Straalen, Dr W	Institute for Animal Nutrition, "De Schothorst", P O Box 533, 8200 AM Lelystad, The Netherlands
Varley, Dr M	Nutec, Eastern Avenue, Lichfield, Staffs WS13 7SE, UK
Vik, Mr K R	Storhollen AS, Box 44, N-5270 Vaksdal, Norway
Wakeman, Miss W	Roche Products Ltd, Heanor Gate, Heanor, Derbyshire DE75 7SG, UK
Waldron, Dr L	Finnfeeds, P O Box 777, Marlborough SN8 1XN, UK
Wallace, Mr J	Nutrition Trading (International) Ltd, Orchard House, Manor Drive, Morton Bagot, Studley, Warwickshire B80 7ED, UK
Waters, Dr J	Nutritech International Ltd, P O Box 62-121, Auckland, New Zealand
Webb, Prof R	University of Nottingham, Sutton Bonington Campus, Loughborough, Leics LE12 5RD, UK
Webster, Mrs M	Format International Ltd, Format House, Poole Road, Woking, Surrey GU21 1DY, UK
Whalley, Mrs L	Trouw (UK) Ltd, Wincham, Northwich, Cheshire, UK
Wilcock, Dr P	J Bibby Agriculture, ABN House, P O Box 250, Oundle Road, Woodston, Peterborough, UK
Willemsen, Mr M	Ross Breeders Ltd, Newbridge, Midlothian EH28 8SZ, UK
Williams, Mr P	Akzo Nobel, 23 Grosvenor Road, St Albans, Herts AL1 3AW, UK
Wiseman, Dr J	University of Nottingham, Sutton Bonington Campus, Loughborough, Leics LE12 5RD, UK
Yeo, Dr G W	Premier Nutrition Products Ltd, The Levels, Rugeley, Staffs WS15 1RD, UK
Zwart, Mr S	Tessenderlo Chemie Rotterdam B.V, Maassluissedijk 103, 3133 KA Vlaardingen, The Netherlands

INDEX